Isolation and Control of Fru
Vegetable Rot Fungi

Isolation and Control of Fruit and Vegetable Rot Fungi

Guest Editors

Nengguo Tao
Xiaoli Tan

MDPI

Basel • Beijing • Wuhan • Barcelona • Belgrade • Novi Sad • Cluj • Manchester

Guest Editors

Nengguo Tao
College of Chemical Engineering
Xiangtan University
Xiangtan
China

Xiaoli Tan
School of Chemical Engineering
Xiangtan University
Xiangtan
China

Editorial Office
MDPI AG
Grosspeteranlage 5
4052 Basel, Switzerland

This is a reprint of the Special Issue, published open access by the journal *Journal of Fungi* (ISSN 2309-608X), freely accessible at: https://www.mdpi.com/journal/jof/special_issues/8XZP7U8FAE.

For citation purposes, cite each article independently as indicated on the article page online and as indicated below:

Lastname, A.A.; Lastname, B.B. Article Title. *Journal Name* **Year**, *Volume Number*, Page Range.

ISBN 978-3-7258-6708-0 (Hbk)
ISBN 978-3-7258-6709-7 (PDF)
https://doi.org/10.3390/books978-3-7258-6709-7

Contents

Preface

Fruits and vegetables are vital for human health and the global economy, yet their high susceptibility to fungal pathogens after harvest leads to severe losses and waste. Confronting this issue requires a dual approach: the precise identification of the causative agents and the development of effective, environmentally sound control methods. This Reprint presents a curated selection of studies that embody this approach. The contributions within first underscore the importance of pathogen isolation and characterization as the foundation of targeted intervention. They further advance our understanding of fungal pathogenicity and host–pathogen interactions at both biochemical and molecular levels. Significantly, this volume highlights the shift towards sustainable alternatives, featuring innovative work on biocontrol agents, plant-derived antimicrobials, and other green technologies that disrupt pathogen viability while safeguarding produce quality and environmental health. We, the Guest Editors, are pleased to present this compilation, which we hope will serve as a useful resource and inspire further research into effective strategies for preserving our valuable fruit and vegetable resources.

Nengguo Tao and Xiaoli Tan
Guest Editors

Journal of
Fungi

Editorial

Isolation and Control of Fruit and Vegetable Rot Fungi

Xiaoli Tan * and Nengguo Tao *

School of Chemical Engineering, Xiangtan University, Xiangtan 411105, China
* Correspondence: tanxiaoli@xtu.edu.cn (X.T.); nengguotao@126.com (N.T.)

Fruits and vegetables play an important role in people's dietary health and economic development. However, fresh fruits and vegetables, due to their richness in nutrients, high water content, and active metabolism, are susceptible to pathogens at multiple stages including the planting, transportation, storage, and marketing processes, resulting in severe postharvest product decay and huge economic losses. In addition, there is a wide variety of pathogenic fungi, and the postharvest pathogenic species of fruits and vegetables vary depending on several factors, such as the type of fruit or vegetable, the variety, the geographic region, and the time of storage [1,2]. Therefore, isolation of the pathogens from specific postharvest products is essential for subsequent investigation of their pathogenic mechanisms and the development of strain-specific preservation measures. Wang et al. isolated *Diaporthe passiflorae* from rotting passion fruit brought from Fujian (China) and found that it could cause yellowing and increased cell membrane permeability at the infested site of postharvest passion fruit, leading to severe fruit decay [3]. Lv et al. isolated nine fungi causing postharvest disease in fresh *Codonopsis pilosula* and found that *Mucor* was first observed on day 7, followed by root rot caused by *Fusarium* on day 14. Blue mold disease caused by *Penicillum expansum* was detected as the most serious postharvest disease on day 28. Pink rot disease caused by *Trichothecium roseum* was observed on day 56 [1].

The development of highly targeted control strategies can be aided by research into the pathogenic processes of pathogens and the role of key genes. *Penicillium expansum* is a major apple rot pathogenic fungus, and the PacC-pH signaling pathway strongly influences the development, pathogenicity, and malate production of *P. expansum* [4]. Zhuo et al. discovered that three Ena family genes (*PeEnaA*, *PeEnaB*, and *PeEnaC*) are significant downstream targets of *PePacC* [5]. Among them, the PeEna protein family is essential for *P. expansum*'s malate production. Ergosterol is an important component of the cell membrane of filamentous fungi; Han et al. discovered that *P. expansum*'s ergosterol synthesis and spore formation were mediated by all three of the *erg4* genes (*erg4A*, *erg4B*, and *erg4C*) [6]. Furthermore, *erg4B* and *erg4C* were revealed to have a role in spore morphogenesis, cell wall integrity, and the oxidative stress response. To repel infection, the host launches its own defense mechanisms in response to the pathogen's attack [7,8]. By using proteome sequencing, Xu et al. discovered that *P. expansum* infection boosted the activity and expression of defense enzymes and triggered defensive systems such as apple phenylpropanes, flavonoids, and hormones. When the pathogen attacks the host, the host also activates its own defense mechanisms to resist pathogen infestation [9]. These findings point to the possibility of controlling fruit diseases by enhancing host resistance by fortifying the fruit resistance system, while also preventing the growth and development of pathogens by undermining the function of the pathogens' essential genes.

Presently, the main control practice for postharvest fruit and vegetable disease is using synthetic chemical fungicides. But the issues surrounding the usage of these chemicals—health, the environment, and the rise of resistant strains—are coming to light more and more. Recent study has shown that biocontrol control agents (BCA), plant essential oils (EOs), and metal ion compounds with safe, effective, and eco-friendly qualities are becoming more and more popular [10–12]. The genus *Phytophthora* is destructive to crops, and there are relatively few strategies available to prevent it. Santos et al. have shown that *Trichoderma aggressivum* f. *europaeum*, *T. longibrachiatum*, *Paecilomyces variotii*, and *T. saturnisporum* had highly effective antagonistic activity against *Phytophthora capsici* and *P. parasitica*, which considerably reduced the severity of Phytophthora blight in pepper [12]. Similarly, *Meyerozyma caribbica* can act as a bio-antagonist to impede *Alternaria alternata* growth and reduce the occurrence of jujube black spot rot [13]. BCAs primarily function by either directly or indirectly inhibiting the growth of pathogens; hence, encouraging BCAs' growth will increase the effectiveness of their control. According to Fan et al. the cell-free supernatant of *Bacillus subtilis* demonstrated remarkable efficacy in stifling the proliferation of *Botryosphaeria dothidea* and reducing its pathogenicity on kiwifruit. Additional research verified that the pathogen is under oxidative stress, which mitigates the severity of kiwifruit soft rot [14]. The production of biofilms has been demonstrated in studies to aid microorganisms in absorbing nutrients more quickly and increasing their competitiveness. By adding $CaCl_2$ (5.14 g/L) to the medium, Zheng et al. observed a significant increase in the biocontrol strain *B. mojavensis* D50's ability to form biofilm, colonize roots, and exhibit antifungal activity, as well as a reduction in the incidence of tomato gray mold [15]. Chen et al. demonstrated that 0.1% β-glucan improved the biofilm-forming ability of marine yeast *Scheffersomyeces spartinae* W9, enhanced its tolerance to various stress, and improved the biocontrol ability of W9 against strawberry gray mold [16]. Additionally, phenylalanine treatment increased the secretion of *M. caribbica* phenylethanol, which in turn encouraged the formation of biofilms, decreased *A. alternata* colonization on jujube fruit, and enhanced the biocontrol efficiency of the fruit against jujube black spot rot [13]. EOs have also been widely used in the control of postharvest diseases of fruits and vegetables. Chen et al. found that perillaldehyde fumigation reduces postharvest rot of sweet potatoes by stimulating the production of excessive ROS as well as inducing severe oxidative damage in *F. solani* [16]. According to Ouyang et al., trans-2-Hexenal can be utilized to suppress *Geotrichum citri-aurantii* and lower the occurrence of citrus sour rot. The primary explanation for trans-2-Hexenal's antifungal activity could be cell membrane damage brought on by decreased lipid and ergosterol levels [10]. Metal ion compounds such as ferric chloride can be used to control citrus anthracnose by triggering autophagy in *Colletotrichum gloeosporioides* to inhibit its spore germination and reduce pathogenicity [17]. Finding and developing more green control measures for fruit and vegetable diseases and elucidating their mechanisms of action will help to reduce postharvest losses and extend the shelf life of fruits and vegetables.

In conclusion, "Isolation and Control of Fruit and Vegetable Rot Fungi" comprises the latest research findings on the isolation and characterization of fruit and vegetable decay fungi, as well as control strategies. The studies cover a wide range of aspects from biofilm formation, mechanisms of action of antifungal agents, biological characterization of specific pathogens, to the development of biological control strategies. These studies not only provide insights for understanding pathogen–host interactions, but also provide a scientific basis for developing new, environmentally friendly approaches to disease management. Through these research results, it can be expected that the reliance on chemical fungicides will be reduced in the future and sustainable agriculture will be promoted.

Funding: This research received no external funding.

Conflicts of Interest: The author declares no conflict of interest.

References

1. Lv, B.; Yang, X.; Xue, H.; Nan, M.; Zhang, Y.; Liu, Z.; Bi, Y.; Shang, S. Isolation of Main Pathogens Causing Postharvest Disease in Fresh Codonopsis pilosula during Different Storage Stages and Ozone Control against Disease and Mycotoxin Accumulation. *J. Fungi* **2023**, *9*, 146. [CrossRef] [PubMed]
2. Wang, L.; Hu, J.; Li, D.; Reymick, O.O.; Tan, X.; Tao, N. Isolation and Control of *Botrytis cinerea* in Postharvest Green Pepper Fruit. *Sci. Hortic.* **2022**, *302*, 111159. [CrossRef]

3. Wang, H.; Chen, H.; Lin, Y.; Li, M.; Liu, Q.; Lin, Y.; Jiang, X.; Chen, Y. Insights into the Isolation, Identification, and Biological Characterization Analysis of and Novel Control Strategies for Diaporthe passiflorae in Postharvest Passion Fruit. *J. Fungi* **2023**, *9*, 1034. [CrossRef] [PubMed]
4. Chen, Y.; Li, B.; Xu, X.; Zhang, Z.; Tian, S. The pH-responsive PacC Transcription Factor Plays Pivotal Roles in Virulence and Patulin Biosynthesis in *Penicillium expansum*. *Environ. Microbiol.* **2018**, *20*, 4063–4078. [CrossRef] [PubMed]
5. Zhuo, R.; Chen, Y.; Xing, M.; Zhang, Z.; Tian, S.; Li, B. Ena Proteins Respond to PacC-Mediated pH Signaling Pathway and Play a Crucial Role in Patulin Biosynthesis. *J. Fungi* **2023**, *9*, 806. [CrossRef] [PubMed]
6. Han, Z.; Zong, Y.; Zhang, X.; Gong, D.; Wang, B.; Prusky, D.; Sionov, E.; Xue, H.; Bi, Y. Erg4 Is Involved in Ergosterol Biosynthesis, Conidiation and Stress Response in *Penicillium expansum*. *J. Fungi* **2023**, *9*, 568. [CrossRef] [PubMed]
7. Kumar, S.; Abedin, M.M.; Singh, A.K.; Das, S. Role of Phenolic Compounds in Plant-Defensive Mechanisms. *Plant Phenolics Sustain. Agric.* **2020**, *1*, 517–532.
8. Kaur, S.; Samota, M.K.; Choudhary, M.; Choudhary, M.; Pandey, A.K.; Sharma, A.; Thakur, J. How Do Plants Defend Themselves Against Pathogens-Biochemical Mechanisms and Genetic Interventions. *Physiol. Mol. Biol. Plants* **2022**, *28*, 485–504. [CrossRef]
9. Xu, M.; Wang, K.; Li, J.; Tan, Z.; Godana, E.A.; Zhang, H. Proteomic Analysis of Apple Response to *Penicillium expansum* Infection Based on Label-Free and Parallel Reaction Monitoring Techniques. *J. Fungi* **2022**, *8*, 1273. [CrossRef]
10. Ouyang, Q.; Shi, S.; Liu, Y.; Yang, Y.; Zhang, Y.; Yuan, X.; Tao, N.; Li, L. Inhibitory Mechanisms of trans-2-Hexenal on the Growth of *Geotrichum citri-aurantii*. *J. Fungi* **2023**, *9*, 930. [CrossRef] [PubMed]
11. Pan, C.; Yang, K.; Erhunmwunsee, F.; Wang, B.; Yang, D.; Lu, G.; Liu, M.; Li, Y.; Tian, J. Antifungal Activity of Perillaldehyde on *Fusarium solani* and Its Control Effect on Postharvest Decay of Sweet Potatoes. *J. Fungi* **2023**, *9*, 257. [CrossRef] [PubMed]
12. Santos, M.; Dianez, F.; Sanchez-Montesinos, B.; Huertas, V.; Moreno-Gavira, A.; Esteban Garcia, B.; Garrido-Cardenas, J.A.; Gea, F.J. Biocontrol of Diseases Caused by *Phytophthora capsici* and *P. parasitica* in Pepper Plants. *J. Fungi* **2023**, *9*, 360. [CrossRef] [PubMed]
13. Deng, Q.; Lei, X.; Zhang, H.; Deng, L.; Yi, L.; Zeng, K. Phenylalanine Promotes Biofilm Formation of *Meyerozyma caribbica* to Improve Biocontrol Efficacy against Jujube Black Spot Rot. *J. Fungi* **2022**, *8*, 1313. [CrossRef] [PubMed]
14. Fan, Y.; Liu, K.; Lu, R.; Gao, J.; Song, W.; Zhu, H.; Tang, X.; Liu, Y.; Miao, M. Cell-Free Supernatant of *Bacillus subtilis* Reduces Kiwifruit Rot Caused by *Botryosphaeria dothidea* through Inducing Oxidative Stress in the Pathogen. *J. Fungi* **2023**, *9*, 127. [CrossRef] [PubMed]
15. Zheng, L.; Gu, X.; Sun, L.; Dong, M.; Gao, A.; Han, Z.; Pan, H.; Zhang, H. Adding Metal Ions to the *Bacillus mojavensis* D50 Promotes Biofilm Formation and Improves Ability of Biocontrol. *J. Fungi* **2023**, *9*, 526. [CrossRef] [PubMed]
16. Chen, X.; Wei, Y.; Zou, X.; Zhao, Z.; Jiang, S.; Chen, Y.; Xu, F.; Shao, X. β-Glucan Enhances the Biocontrol Efficacy of Marine Yeast *Scheffersomyeces spartinae* W9 against *Botrytis cinerea* in Strawberries. *J. Fungi* **2023**, *9*, 474. [CrossRef] [PubMed]
17. Wang, Y.; Wu, X.; Lu, Y.; Fu, H.; Liu, S.; Zhao, J.; Long, C. Ferric Chloride Controls Citrus Anthracnose by Inducing the Autophagy Activity of *Colletotrichum gloeosporioides*. *J. Fungi* **2023**, *9*, 230. [CrossRef] [PubMed]

Journal of
Fungi

MDPI

Article

Proteomic Analysis of Apple Response to *Penicillium expansum* Infection Based on Label-Free and Parallel Reaction Monitoring Techniques

Meng Xu [1], Kaili Wang [1], Jun Li [2], Zhuqing Tan [1], Esa Abiso Godana [1] and Hongyin Zhang [1,*]

[1] School of Food and Biological Engineering, Jiangsu University, Zhenjiang 212013, China
[2] Analysis & Testing Center, Jiangsu University, Zhenjiang 212013, China
* Correspondence: zhanghongyin126@126.com; Tel.: +86-511-88790211; Fax: +86-511-88780201

Abstract: Blue mold, caused by *Penicillium expansum*, is the most destructive fungal disease of apples and causes great losses during the post-harvest storage of the fruit. Although some apple cultivars are resistant to *P. expansum*, there has been little information on the molecular mechanism of resistance. In this study, differential proteomic analysis was performed on apple samples infected and uninfected with *P. expansum*. Parallel reaction monitoring (PRM) technology was used to target and verify the expression of candidate proteins. The label-free technique identified 343 differentially expressed proteins, which were mainly associated with defense responses, metal ion binding, stress responses, and oxidative phosphorylation. The differential expression of enzymes related to reactive oxygen species (ROS) synthesis and scavenging, the activation of defense-related metabolic pathways, and the further production of pathogenesis-related proteins (PR proteins) during *P. expansum* infection in apples, and direct resistance to pathogen invasion were determined. This study reveals the mechanisms of apple response at the proteomic level with 9 h of *P. expansum* infection.

Keywords: apple; label-free; *Penicillium expansum*; defense response

1. Introduction

Apple is a widely consumed fruit, rich in vitamins, dietary fiber, and many other nutrients [1]. During post-harvest storage, apples are susceptible to a variety of pathogens, including blue mold, grey mold, and black spot disease. Blue mold caused by *P. expansum* is the most widespread and damaging disease, causing serious economic losses to the apple industry. Patulin (PAT) produced by *P. expansum* also poses a great risk to human health [2]. Chemical fungicides are often used for disease control, but due to environmental concerns, safer and more convenient control methods are increasingly sought. Therefore, an in-depth study of the mechanisms of action between *P. expansum* and apple could help find more suitable methods for disease control.

When plants are exposed to pathogens, a complex response is generated to mount an immune response. The plant cell wall is the first barrier against pathogens, and when pathogens break through it, they are recognized by pattern receptors on the plant surface, which activate pathogen-associated molecular pattern (PAMP)-triggered immunity (PTI) and trigger many downstream responses. Plant pattern recognition receptors (PRRs) are located on the cell surface, including plasma membrane-bound receptor-like kinases (RLKs) and receptor-like proteins (RLPs). PTI has a broad-spectrum defense role and is not sufficient to fully resist pathogen invasion. The pathogen releases effectors to overcome PTI, and the plant resistance (R) protein senses pathogen-associated avirulence basic (Avr) proteins to trigger the expression of the physiological defense response in the plant, called effector-triggered immunity (ETI). To defend against pathogenic bacteria, both PTI and ETI induce the production of a range of antimicrobial peptides, pathogenesis-related (PR) proteins, and other physiological defense substances [3].

Some good developments has been made in extending the research on the interaction of the molecular mechanisms between *P. expanusm* and apple. The transcription factor CreA (a global regulator of carbon catabolism) was assessed to be toxic and associated with PAT synthesis both in vitro and in vivo; knocking out *CreA* resulted in *P. expansum* being virtually nontoxic and not producing PAT, and these mutants could not be successfully colonized on apple [4]. *PacC* plays an important role in fungal adaptation to environmental pH. Chen et al. [5] found that *PePacC* can act as an effector for a variety of target proteins and plays an important role in extending the virulent synthesis of *P. expansum* [5]. Prusky et al. [6] found that *PePG1* expression is strongly correlated with environmental pH on apples and citrus, and the environmental acidification plays an important role in increasing the pathogenicity of pathogenic fungi [6]. Levin et al. [7] revealed 18 possible genes encoding the LysM protein for the first time and studied the regulatory role of LysM in the spread of *P. expansum* infection by gene knockout and other techniques [7].

Proteomic studies have shown their importance in studying plant–pathogen interactions. Unlike genomic tools, proteomics helps to understand the identity, abundance, turnover, post-translational modifications (PTMs), and interactions of different proteins in a mixture. Yang et al. [8] compared wheat inoculated and uninoculated with *Puccinia striiformis* f. sp. *tritici* (*Pst*) by quantitative proteomics and showed that 530 proteins were differentially expressed and several proteins were involved in the response to *Pst* infection [8]; Ravi et al. [9] compared the proteomes of *Xanthomonas oryzae* pv. *oryzae* (*Xoo*)-sensitive and -resistant rice plants and analyzed the differential proteins and screened 23 defense candidate proteins to improve resistance to *Xoo* by overexpressing arginase 1(*OsArg1*) in sensitive plants [9]. Lu et al. [10] compared leaf blight (LB)-resistant and LB-sensitive plants of orchids by proteomics and identified proteins in the reactive oxygen species metabolic pathway and found that Cu/Zn superoxide dismutase (BsSOD1) protein abundance and its gene expression were higher in LB-resistant plants than in LB-sensitive plants [10]. Proteomics can tell us what is happening in a particular state of a cell, a tissue, or an organism and what the protein–protein interactions are. Changes that occur within specific tissues and cells can be better predicted, revealing their mechanisms of action [11].

In this study, we used high-throughput proteomics to analyze the changes in apple protein levels before and after *P. expansum* infection and bioinformatics tools to analyze the differential proteins and investigate the defense mechanisms of apples during *P. expansum* infection. This study provides some theoretical references to the molecular mechanisms of apple defense against *P. expansum* infection.

2. Materials and Methods

2.1. Pathogen

The *P. expansum* strain was isolated by our research team (maintained at the China Center for Type Culture Collection and is numbered as CCTCC AF 2022039). The activated strain was incubated on PDA medium at 25 °C for seven days before the experiment. Then the spores were collected and adjusted to a concentration of 1×10^8 spores/mL in 0.9% saline.

2.2. Fruit

Apple (*Malus domestica*) fruits cv red Fuji at commercial maturity with uniform shape and size and no pests, scars, or mechanical damage (fruit hardness 7 N/cm^2, 13% soluble solid material) were selected as the test object. The selected fruits were soaked in 0.2% sodium hypochlorite solution for 2 min and then carefully rinsed with tap water for 5 min to wash off the surface dirt, as well as the sodium hypochlorite solution, and then air-dried.

2.3. Determination of Sampling Time Points and Preparation of Apple Samples

After the apples were air dried, three wounds of size 5 mm × 4 mm were punched evenly along the equatorial line of the apples with a hole puncher. After drying for 20 min, the prepared *P. expansum* spore suspension was injected into the wounds with a pipette,

30 μL per wound, and the control group was injected with the same volume of saline. The apples were placed evenly in clean frames, one frame for every 12 apples, wrapped in cling film, and placed in an incubator at 25 °C and RH 95%. Then, at 1 h, 3 h, 6 h, 9 h, 12 h, and 24 h after inoculation with molds, samples were taken at the wound site with a sterile scalpel, and the germination of mold spores was observed under a microscope after staining and preparation with lactate phenol cotton blue staining solution, so as to determine the best sampling time point. Then sampling was carried out at the best sampling time point. After the decaying part of the wound was cut with a sterile scalpel, 2–3 mm flesh was taken as a sample and quickly placed into liquid nitrogen. The samples were then stored in a refrigerator at −80 °C for further experiments.

2.4. Proteome Sample Preparation

The extraction of apple proteins was according to Tan's experimental method. Briefly, 5 g of apple tissue was ground in liquid nitrogen, washed three times with 15 mL of cold acetone, dried, and added to 15 mL of extraction buffer (0.1 M Tris pH 8.8, 10 mM ethylenediaminetetraacetic acid (EDTA), 0.4% β-mercaptoethanol, 0.9 M sucrose) and 15 mL tris-saturated phenol (pH 8.8), mixed well on ice and centrifuged at 12,000× *g* for 20 min. Then, the phenol phase was collected in a new tube, repeated once, and the two phenol phases were combined and centrifuged at 10,000× *g* for 15 min to remove the supernatant. The supernatant was incubated with 25 mL of a methanol solution containing 0.1 M at −20 °C overnight, then centrifuged and the supernatant was discarded. The precipitate was washed with acetone three times and air-dried, stored at −80 °C in the refrigerator [12]. The protein precipitate was dissolved with 8 M urea, and the protein supernatant was transferred to a new EP tube; the concentration of the protein was detected by the BCA protein quantification kit, and the integrity of the protein bands was measured by SDS-PAGE. Protein samples were desalted with C18 small columns, and the treated samples were lyophilized with a freeze dryer, stored in a −80 °C refrigerator, and redissolved with 0.1% formic acid water before mass spectrometry analysis.

2.5. Proteomic Mass Spectrometry

On-board operation was performed using a Thermo Q-Exactive mass spectrometer with the following parameters: column type: 75 μm × 15 cm, 2 μm particle size, C18, 100 Å, Acclaim PepMap. with trap enrichment column. Mobile phase: A: aqueous solution (0.1% formic acid) B: 100% ACN (0.1% formic acid); detection mode: full MS/dd-MS2; ion source voltage: 2.0 KV; scan range: 350–1800 m/z; ion source temperature 320 °C; s-lens: 50 °C resolution: 70,000; secondary mass spectrometry: starting fixed m/z 110. resolution 70,000; secondary mass spectrometry: starting fixed m/z 110, resolution 70,000; secondary rupture mode: HCD; AGC setting: primary 3×10^5, secondary 1×10^5; dynamic exclusion time: 60 ms.

2.6. PRM Verification of DEPs

According to the proteomic results, four proteins were randomly selected to conduct targeted qualitative analysis of the target proteins by LC–PRM/MS. After the peptides were separated by high-performance liquid chromatography, a Q-Exactive HF mass spectrometer (Thermo Fisher Scientific, 81 Wyman Street, Waltham, MA, USA) was used for PRM mass spectrometry analysis. Finally, Skyline 3.5.0 was used to analyze the data of the PRM original files.

2.7. Data Analysis

Data from this trial were searched for databases using Maxquant 1.6.17.0 and quantified for LFQ, including the quantification of proteins, peptides, and PSM. The apple protein database was compared with UP000290289 (*Malus domestica*) from UniProt (http://www.uniprot.org/, accessed on 26 November 2021). GO functional annotation was searched on UniProtprot (http://www.uniprot.org, accessed on 24 February 2022); KEGG function was annotated with KOBAS (http://kobas.cbi.pku.edu.cn/, accessed on 5 March 2022). The prediction of the subcellular localization of key proteins was performed using WoLF PSORT (https://wolfpsort.hgc.jp/, accessed on 27 November 2022).

3. Results

3.1. Result Analysis

3.1.1. Selection of Key Time Points for Sampling

Samples were taken, and wounds were made at different time points during *P. expansum* infection. It was found by microscope observation that 1 h after inoculation with the bacterial suspension, mold spores began to change from an oval shape to a shape with a tip. Then, 3 h after inoculation with the bacterial suspension, spore deformation was obvious, and budding tubes emerged. Then, 6 h after inoculation with the bacterial suspension, the budding tubes of conidia directly invaded the apple tissue, and the tails of some conidia expanded and formed appressorium. After 9 h inoculation with the bacterial suspension, the infecting nails grew on the appressorium and infected the apple tissue. The bud tube extended into the apple tissue and continued to infect the apple tissue. Then, 12 h after inoculation, primary hyphae continued to extend into apple tissues. After inoculation with the bacterial suspension for 24 h, new spores were formed on the secondary mycelia. At 9 h after inoculation, *P. expansum* successfully infected apples, so the sampling time was determined to be 9 h (Figures 1 and 2).

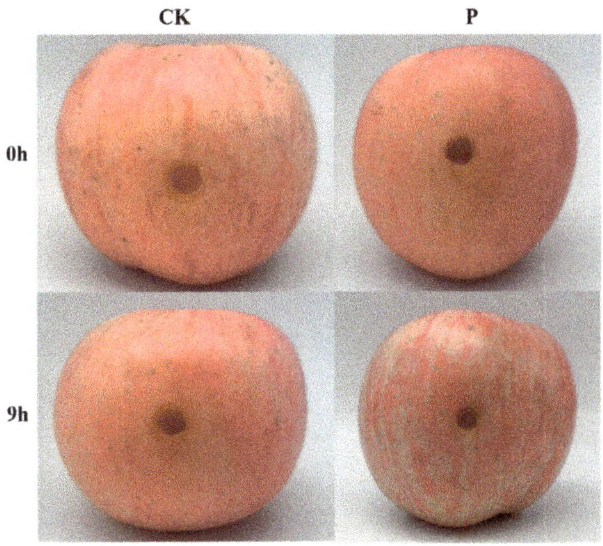

Figure 1. The phenotype of apple inoculated at 0 h and 9 h in control and experimental groups.

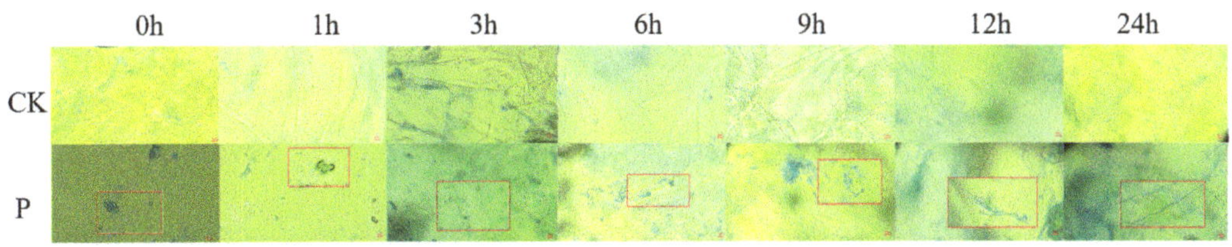

Figure 2. Spore germination status of *P. expansum* at different time points of inoculation.

3.1.2. Screening for Differentially Expressed Proteins

A total of 4164 proteins were identified in the *P. expansum* infected and non-infected apple samples. In the quantitative results, the screening for significantly expressed differential proteins was based on FC > 1.5 or FC < 0.67 and $p < 0.05$. The results showed that a total of 343 differential proteins were screened. Among them, 131 proteins were downregulated and 212 proteins were upregulated. Functional annotation and enrichment analysis were performed for all differentially expressed proteins. Figure 3 shows volcano plot of the differentially expressed proteins, where the upregulated proteins are in red and the downregulated proteins are in blue.

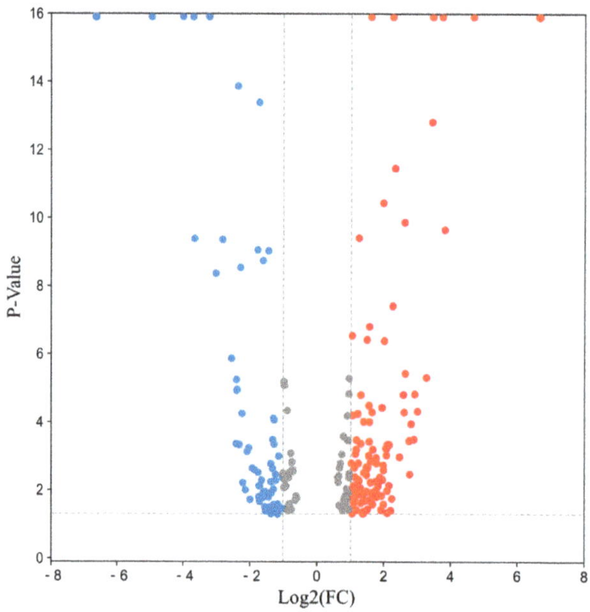

Figure 3. Volcano plot of total differential protein expression. Red represents significantly upregulated differential proteins, and blue represents significantly downregulated differential proteins.

3.1.3. Subcellular Localization of Differentially Expressed Proteins

Studying the mechanisms and patterns of protein localization in cells and predicting the subcellular localization of proteins are important for understanding protein structure, properties, functions, and protein interactions. In this study, it can be seen from the prediction results of subcellular localization that the differentially expressed proteins are mainly located in nucleus, cycloplasm, and chloroplast (Figure 4). The upregulated proteins and downregulated proteins have similar subcellular localization. These proteins may interact in these organelles and perform important biological functions.

3.1.4. GO Enrichment Analysis of Differentially Expressed Proteins

The DEPs were annotated into the GO database, and 319 of the 343 differentially expressed proteins were annotated into the GO database for cellular component, molecular function, and biological process classification. Among the cellular components, cytosol, cytoplasm, an integral component of the membrane, and ribosome were the most enriched, indicating that most biological processes in apple defense occur in the cytoplasm and membrane. The subclasses enriched in molecular functions are ATP binding, metal ion binding, oxidoreductase activity, GTP binding, and hydrolysis activity. This suggests that ATP synthesis, metal ions, and some oxidoreductases involved in oxidation and reduction reactions play important roles in the defense response of apple. The subclasses that were more enriched in biological processes were translation, protein transport, tricarboxylic acid cycle, defense response, and protein folding. The GO enrichment of the differential proteins is shown in Figure 5, where only the top 15 pathways were enriched for differential proteins at three levels of classification.

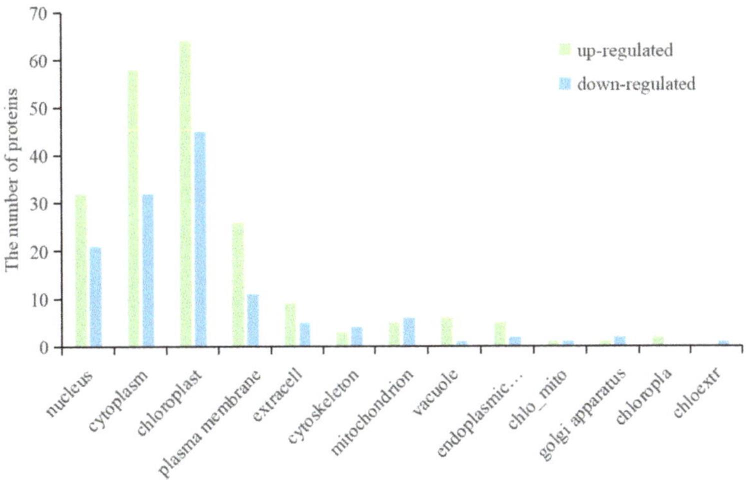

Figure 4. Subcellular localization of DEPs.

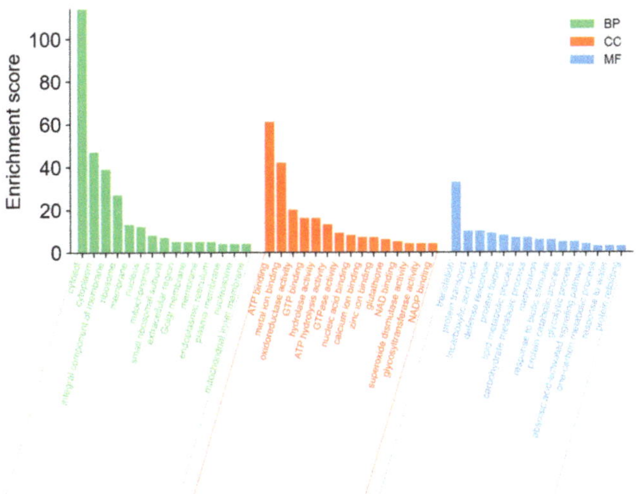

Figure 5. GO enrichment analysis of total DEPs. Divided into three subcategories: biological process, cellular component, and molecular function.

3.1.5. KEGG Enrichment Analysis of Differentially Expressed Proteins

The results of the annotation of the DEPs in the KEGG database showed that the differentially expressed proteins could be annotated into a total of 19 secondary subclasses of six primary classifications (Figure 6), with 11 secondary subclasses common to both the up- and downregulated proteins, of which glycan biosynthesis and metabolism, signal transduction, nucleotide metabolism, replication and repair, and signal transduction were unique to the downregulated proteins. Glycan biosynthesis and metabolism, the metabolism of terpenoids and polyketides, transcription, and signal transduction were unique to the downregulated proteins. Polyketides, transcription, and translation were unique to the upregulated proteins. These specific metabolic pathways may play important functions in the response of apples to extended *P. expansum* infection. KEGG pathways with $p < 0.05$ were considered significantly enriched pathways, and a total of 32 pathways belonging to 11 secondary subclasses were screened. The most enriched KEGG pathway was metabolic pathways, with 61 proteins annotated to this pathway.

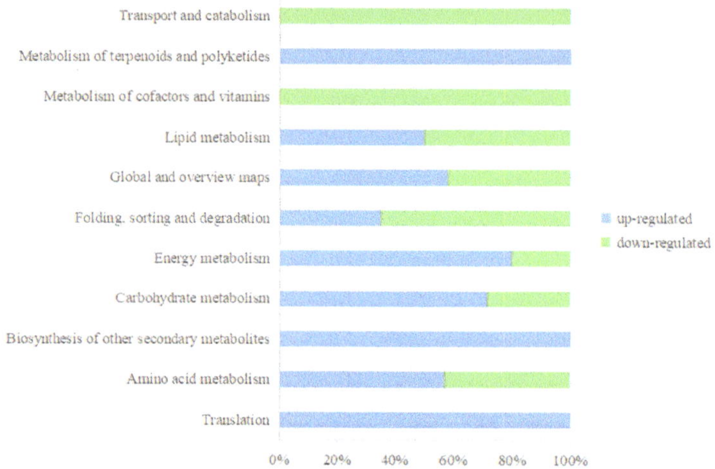

Figure 6. Secondary classification of KEGG functional annotations of DEPs. Blue represents upregulated proteins, and green represents downregulated proteins.

When apples sense pathogenic invasion, multiple metabolic pathways are activated in vivo, such as the ubiquitin–proteasome pathway mediating protein degradation; the production of secondary metabolites, such as lignans and flavonoids, to enhance resistance through the phenylpropane biosynthesis pathway and the flavonoid biosynthesis pathway; the activation of the mitogen-activated protein kinase (MAPK) signaling pathway in response to defense plant signals; and the activation of reactive oxygen species production and the promotion of downstream immune protein expression through the plant–pathogen-interactions pathway. Table 1 lists some of the metabolic pathways associated with the apple defense response to *P. expansum*.

Table 1. Important metabolic pathways related to apple defense and the number of DEPs.

Pathways	The Number of Upregulated DEPs	The Number of Downregulated DEPs
Plant-pathogen interaction	3	2
Pyruvate metabolism	4	2
Oxidative phosphorylation	4	0
Alpha-linolenic acid metabolism	1	1
Phenylpropanoid biosynthesis	5	0
Flavonoid biosynthesis	1	0
Glutathione metabolism	1	2
MAPK signaling pathway—plant	0	1
Ubiquitin-mediated proteolysis	0	1

3.1.6. Validation of Selected Candidates by PRM

According to the results of the proteomics of apple infected by *P. expansum*, four proteins were randomly selected and analyzed by PRM. The results showed that the expression of the four identified proteins (A0A498JTJ2, A0A498JUX9, A0A498KA74, and A0A498KKS6) was consistent with the trend of proteomic data, indicating that the proteomic results were reliable (Figure 7).

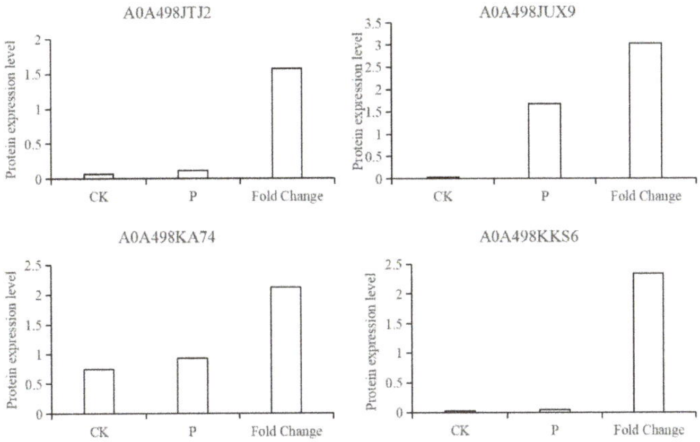

Figure 7. PRM verification of candidate proteins. CK and P represent the protein expression in PRM verification, and fold change is from the proteomic data.

3.1.7. Expression of Pathogenesis-Related Proteins

Pathogenesis-related proteins (PRs) are low-molecular-weight proteins that are selectively soluble at low pH and are mainly found in the vesicles and extracellular spaces. PR proteins are classified into 17 families, PR-1 to PR-17, based on their molecular weight, biochemical function, and properties [13]. Among the differential proteins of apple, a total of nine proteins have been identified as PR proteins (Figure 8), of which four proteins belonging to the Bet-v-1 family and three proteins belonging to the Mal d family with ribonuclease activity belong to PR-10, one β-1,3-endoglucanase belongs to PR-2, and one peroxidase (POD) belongs to PR-9. PR-10 is a highly conserved protein that is induced by biotic and abiotic stresses. PR-10 proteins all have a common fold that acts as a binding site for different small molecules, mainly cytokinins and sterols. Furthermore, it has been shown that different protein ligands affect the stability of Bet-v-1. This dual characteristic provides a plausible explanation for the differential expression of bet-v-1 in apples. PR9 has POD activity and participates in the synthesis of lignin, enhancing the structure of the cell wall and the scavenging of reactive oxygen species in the fruit.

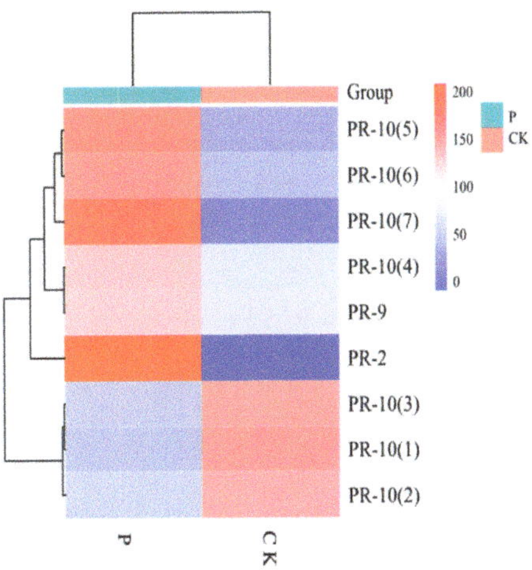

Figure 8. Clustering of pathogenesis-related proteins expression. The depth of color indicates how much different proteins are expressed in the sample.

4. Discussion

Plants usually mount an immune response in the face of pathogen infection through two defense systems. One is the plant's immune system, and the other is a pathogen-induced defense system. The cross-response between the two systems triggers the activation of multiple signaling pathways in the plant, changes in hormone levels, and the production of disease-resistant proteins to respond to the infection.

In the present study, the differential expression of many proteins was induced under *P. expansum* infection, and most of these DEPs were associated with plant defense mechanisms against pathogenic invasion. Proteins associated with biological processes, such as the MAPK cascade, plant–pathogen interactions, the stimulation of metabolite biosynthesis, and amino acid metabolism were significantly upregulated in expression. They are involved in various processes such as reactive oxygen species (ROS) burst and scavenging, hypersensitive response (HR), redox reactions, and defense-related protein induction.

The KEGG metabolic pathway shows that DEPs are involved in several metabolic pathways related to plant defense, including MAPK (plant) and plant–pathogen interactions (Figure 9). The pathway diagram shows that pathogen stimulation causes an increase in the expression of calmodulin and enhanced disease susceptibility 1 (EDS1) and heat shock protein 90 (HSP90), which in turn trigger a hypersensitive response in apple, and the accumulation of EDS1, an important signaling molecule in the regulation of plant immunity, is essential for the subsequent amplification of disease resistance signals during pathogen infection [14]. HSP90 is widely distributed in plant cells and is a molecular chaperone protein); in addition, MPK4 expression, was downregulated at 9 h of *P. expansum* infection. It has been reported that MPK4 plays a negative regulatory role in plant immunity and can be activated by PAMP [15,16]; Tereza et al. [17] showed that the Gnomon molecular chaperone protein plays an important role in plant stress resistance [17]. The expression of HSP90 was significantly increased in the face of pathogenic infection and may perform an important role in the defense of apple. In addition, calcium-dependent MPK4s play a negative regulatory role in SA accumulation and defense responses but a positive regulatory role in growth and development regulation, and their function is conserved among plant species [18]. CDPK is the largest family of calcium-regulated protein kinases in plants. It rapidly senses changes in transient Ca^{2+} signals in plants, recognizes and phosphorylates specific substrates, and then triggers various physiological responses through signal cascade transduction, thereby regulating plant growth and development and response to a variety of stresses. ZmCDPK1 negatively regulates the stress response to cold in maize, and the cotransfection of maize leaf protoplasts with ZmCDPK1 inhibited the expression of the cold-inducible marker gene *Zmerf*3 [19]. CDPK and respiratory burst oxidase homolog (Rboh) are involved in biotic/abiotic stress processes and cell death in plants, and the expression of Rboh plays a key role in the accumulation of ROS, which in turn increases the level of plant defense against pathogens [20]. CDPK expression was significantly downregulated during defense, leading to the hypothesis that CDPK plays a negative regulatory role in the apple defense response to *P. expansum* infection.

The production of ROS is an important way for plants to resist pathogens in their presence. Studies have shown that ROS can be produced in several parts of plant cells, including the cell wall, plasma membrane, peroxisome, and endoplasmic reticulum [21]. In response to external environmental stimuli, a variety of enzymes in the cell wall can produce ROS, and reduced coenzyme II (NADPH) oxidase in the cytoplasmic membrane is the main pathway for hydrogen peroxide (H_2O_2) production. The large accumulation of ROS, on the one hand, improve the resistance of plants to pathogens and, on the other hand, can damage the cellular tissue structure of the plant [22]. Thus, in response to the changing external environment, plants have evolved ROS scavenging mechanisms to maintain homeostasis in vivo, mainly through enzymatic and non-enzymatic mechanisms. During *P. expansum* infection, three superoxide dismutases (Cu/Zn-SOD) were identified as differentially downregulated among all differential proteins, together with glutathione peroxidase (GPX). SOD and GPX are enzymes that play key roles in common ROS scavenging mechanisms, and their expression is usually elevated under common biotic, as well as abiotic, stresses. However, SOD and GPX are key enzymes in the common ROS scavenging mechanism. Chang et al. (2009) used the total nonsense transgenic strains AS-*cpGPX* and *Arabidopsis* (At) *GPX7* mutant material located in the chloroplast of *Arabidopsis thaliana* and found that At*GPX1* was responsible for the reduction of H_2O_2 under photo-oxidative stress and that Cu/Zn-SOD and Mn-SOD activities were reduced in the AS-*cpGPX* strain under strong light stress, indicating that chloroplast GPX and SOD jointly regulate the balance of O^{2-} and H_2O_2 in chloroplasts [23]. We, therefore, hypothesize that SOD and GPX expression is suppressed in the pre-infection period, causing a burst of ROS in response to the infection of *P. expansum*. In the meantime, POD, glutathione transferase (GST), key enzymes in the lignin synthesis pathway, and the accumulation of lignin and callosein the cell wall strengthen the mechanical strength of the cell wall [24]. GST can be induced to be expressed under various biotic or abiotic stresses to enhance plant resistance to adversity and is an important enzyme in the antioxidant defense system as a channel for pathogen dispersal between cells, thus limiting further pathogen invasion into the protoplasm and reproduction in tissues [25]. In the studies of the defense mechanisms between wheat and powdery mildew, the expression of multiple ROS scavenging genes was observed, and GST was shown to contribute to resistance to powdery mildew [26]. The overexpression of GST genes in crops, such as rice and tomato, can improve plant resistance to a variety of biotic stresses [27,28]. These protein changes associated with oxidative stress suggest that the homeostatic levels of intracellular ROS are well maintained during the defense of apples against *P. expansum*.

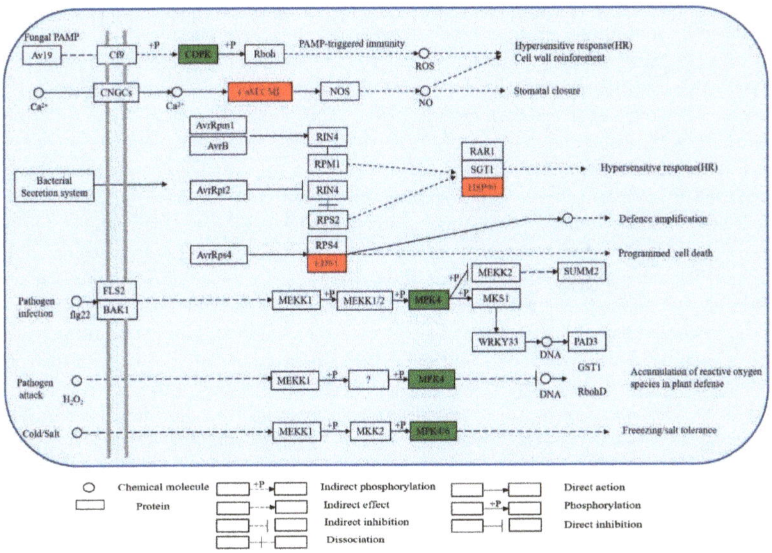

Figure 9. Part of the metabolic pathways involved in defense responses in apples. Red indicates upregulated DEPs, and green indicates downregulated DEPs.

Induced by pathogenic bacteria, the response is usually accompanied by changes in the content of plant hormones. Ethylene (ET), abscisic acid (ABA), jasmonic acid (JA), salicylic acid (SA), and growth hormones are common hormone species that play a key function in plant defense mechanisms. The JA-, SA-, and ET-mediated signaling pathways are among the metabolic pathways that are associated with disease resistance signals [29,30]. SA is an endogenous phenolic substance that usually accumulates in large quantities in plants after infection by pathogenic fungi, leading to the development of systemic acquired resistance (SAR), which in turn activates the expression of a series of related PR proteins (PR-1, PR-2, and PR-5) [31]. JA and ET usually accumulate in large quantities when pathogens invade, and ET acts as a defense against a wide range of pathogens and acts synergistically with JA in the defense process [32]. In a transcriptomic study of cucumber wilt, inoculation with *Fusarium oxysporum* f. sp. *Cucumerinum* (Foc) the expression of ethylene-related genes was significantly increased in cucumbers inoculated with Foc, and the treatment of cucumbers inoculated with exogenous ET improved resistance to Foc [33]. In the present study, infection by *P. expansum* activated the accumulation of the tryptophan metabolic pathway, the phenylalanine metabolic pathway, the α-linolenic acid metabolic pathway, and other key enzymes for hormone synthesis such as methyltransferase, phenylalanine deaminase, and lipoxygenase in apple cells, inducing hormone synthesis and improving apple resistance.

When a plant is susceptible to a disease, a series of changes in the plant's defense enzyme system occur that favor plant defense. In this study, the expression of lipoxygenase (LOX), phenylalanine ammonia-lyase (PAL), polyphenol oxidase (PPO), and NADPH-cytochrome P450 reductase (CPR) enzymes were altered to varying degrees (Table 2). PAL is one of the key enzymes in the phenylpropanoid metabolic pathway in plants. Both pathogen infection and pathogenic toxin treatment-induced enhanced PAL activity, and the enhanced enzyme activity was positively correlated with disease resistance. In maize and sugarcane mosaic virus studies (SCMV), the downregulation of ZmPAL expression leads to increased SCMV infection and viral accumulation, as revealed by metabolomic studies, leading to the accumulation of lignin, as well as other metabolites [34]. The exposure of gene-silenced peppers to low temperatures by virus-induced gene silencing revealed a reduction in PAL activity and tolerance to low temperatures, suggesting that PAL may have an important role in the resistance of peppers to low-temperature stress [35]. The LOX-mediated oxylipin synthesis pathway is known as the lipoxygenase pathway, and there are several branching pathways downstream of LOX, such as the allene oxide synthase (AOS) pathway and the hydroperoxide lyase (HPL) pathway. LOX plays an important role when plants are subjected to a variety of biotic and abiotic stresses. ZmLOX10 is an important pest resistance gene in maize that directly resists pests by inducing the production of substances such as JA [36]. Analysis using the *Arabidopsis* 9-LOX deletion mutant LOX1/LOX5 and the oxylipin-insensitive mutant nonresponding to oxylipins (noxy)2-2 and related mutants of brassinosteroids (BRs) showed that oxylipin from 9-LOX induces BR synthesis and signal transduction and activates cell wall responses, such as callus deposition, to limit pathogen invasion and prevent pathogen infection [37]. In a study of walnut resistance to *Xanthomonas arboricola* pv. *juglandis* (*Xaj*), PPO was significantly induced at the site of infection in most walnut varieties, and transgenic strains showed higher PPO activity and lower levels of disease [38]. Niranjan et al. [39] analyzed the role of the isolated PPO gene in downy mildew pearl millet intercropping and showed that the gene accumulated significantly more and faster in downy-mildew-resistant pearl millet seedlings inoculated with *Sclerospora graminicola* PPO mRNA compared to the sensitive control [39]. Wilt and blast are important fungal diseases that are harmful to the health of soybeans and are positively correlated with POD and PPO in the plant, according to experiments [40]. Cytochrome P450 is an important stress-related enzyme in plants. P450s play an important role in plant defense through their involvement in the biosynthesis of plant antitoxins, hormone metabolism, and the biosynthesis of a number of other secondary metabolites [41]. In cytochrome-P450-mediated metabolism of endogenous and exogenous compounds, NADPH–cytochrome P450 reductase (CPR) plays an important role. Most P450 catalytic activity depends on NADPH–cytochrome P450 reductase for electron supply [42]. Thus, in apple, the upregulation of NADPH–CPR expression promoted cytochrome-P450-mediated resistance. Chalcone isomerase (CHI) is a key enzyme in the biosynthetic pathway of flavonoids and flavonoid substances, which have been shown to play an important metabolic role in plant stress resistance. The silencing of the land cotton CHI gene (*GhCHI*) using VIGS technology resulted in the loss of resistance to yellow wilt in cotton, confirming the role of the *GhCHI* gene in cotton resistance to yellow wilt [43]. In a study of the soybean CHI gene, it was found to be expressed in roots, stem, and leaves, and the overexpression of *GmCHI1A* in hairy roots revealed a shortening of spot length and a reduction in free spore germination, indicating that it is regulating the response of soybean to soybean blast [44]. In proteomic studies of apples, the upregulated expression of CHI proteins was likewise identified, which may play a key role in apple defense processes.

Table 2. Expression levels of key enzymes in apple response to *P. expansum* infection.

Proteins	Description	FC	*p*-Value
A0A498K3F2	Superoxide dismutase [Cu-Zn]	0.609	0.00244296
A0A498K8S2	Superoxide dismutase [Cu-Zn]	0.612	0.00281705
A0A498INN8	Superoxide dismutase	0.645	0.01964518
A0A498JNW6	Glutathione peroxidase	0.292	8.84×10^{-10}
S4UL39	Lipoxygenas	2.941	1.54×10^{-7}
A0A498HL71	Methyltransferase	2.158	0.00218763
A0A498JZC5	Phenylalanine ammonia-lyase	2.949	9.91×10^{-5}
A0A498K3I5	Chalcone isomerase	100	1.25×10^{-16}
A0A498HJB2	NADPH–cytochrome P450 reductase	100	1.25×10^{-16}
A0A498IC26	NADPH–cytochrome P450 reductase	2.202	0.00085584
Q93XM8	Polyphenol oxidase 2	100	1.25×10^{-16}
A0A540LHP9	Peroxidase	1.855	6.51×10^{-5}
A0A498HW70	Glutathione transferase	1.664	0.00087862

A further transmission of disease resistance signals triggers the accumulation of multiple PR proteins in apples, which further resist invasion by pathogenic fungi by increasing cell wall resistance and synthesizing plant antitoxins. Zhang et al. [45] determined that MdPR10-1 and MdPR10-2 promote leaf spot resistance of *Streptomyces* by inhibiting fungal growth, as determined by in vitro experiments [45]. Wang et al. [46] verified the interaction of VmEP1 with MdPR10 through a yeast two-hybrid assay and confirmed the positive regulatory role of MdPR10 in enhancing resistance in apples [46]. The increase in POD activity promoted the accumulation of lignin content and improved fruit disease resistance. PR-2 expression was induced by pathogenic fungi. β-1,3 glucanase can hydrolyze the β-1,3 glycosidic bond, which is an important component of pathogenic fungi, so it is thought that β-1,3 glucan hydrolase can cause pathogen cell lysis and death by hydrolyzing the pathogen cell wall. In vitro functional assays on three grape-derived EGases showed that EGase3 has strong anti-*P. viticola* activity [47]. In *Pectobacterium carotovorum* subsp. *carotovorum* (Pcc) induced higher expression of PR-2 in resistant varieties of tomato than insensitive strains, involved in the defense response of tomato to Pcc [48].

5. Conclusions

The cell wall was the first barrier to resist the invasion of *P. expansum*, and the mechanical strength of the apple cell wall was enhanced to resist the invasion of pathogenic fungi through the activation of lignin-synthesis-related pathways. The production of Ca^{2+} channels and the burst of ROS are important signals of the apple immune response, which can improve the defense ability through the production and clearance mechanism of ROS. The activation of PTI and ETI leads to the synthesis of many downstream defense pathways, along with the production of related plant hormones and the expression of defense enzymes. The expression of related PR proteins also directly improves the resistance to pathogens, all of which participated in the response of apples to *P. expansum* infection (Figure 10).

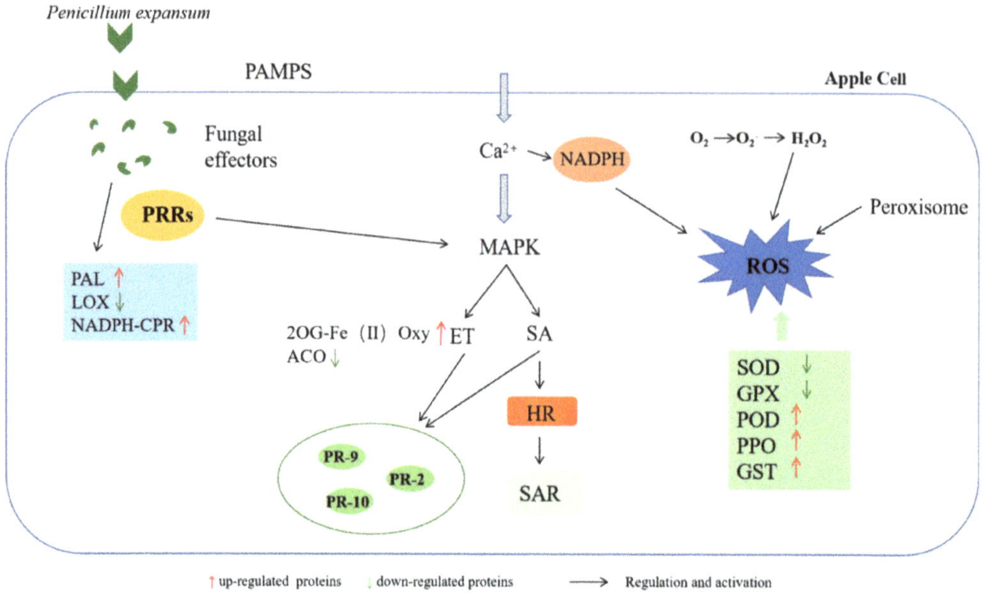

Figure 10. Map of defense mechanisms in response to *P. expansum* infection in apples.

Author Contributions: M.X.: investigation, data curation, formal analysis, writing—original draft. K.W.: investigation, supervision, data curation, methodology. J.L.: data curation, methodology. Z.T.: investigation, data curation. E.A.G.: writing—reviewing and editing. H.Z.: conceptualization, supervision, project administration, writing—reviewing and editing. All authors have read and agreed to the published version of the manuscript.

Funding: This research was supported by the National Natural Science Foundation of China (32072276, 32102030).

Informed Consent Statement: Not applicable.

Data Availability Statement: The datesets generated during and/or analysed during the current study are available from the corresponding author on reasonable request.

Conflicts of Interest: The authors declare that they have no known competing financial interest or personal relationships that could have appeared to influence the work reported in this paper.

References

1. Li, H.K.; Wan, Y.Z.; Wang, M.; Huo, X.X. History, status, and prospects of the apple industry in China. *J. Am. Pomol. Soc.* **2015**, *69*, 174–185.
2. Patriarca, A. Fungi and mycotoxin problems in the apple industry. *Curr. Opin. Food Sci.* **2019**, *29*, 42–47. [CrossRef]
3. Gebrie, S.A. Biotrophic fungi infection and plant defense mechanism. *J. Plant Pathol. Microbiol.* **2016**, *7*, 2. [CrossRef]
4. Tannous, J.; Kumar, D.; Sela, N.; Sionov, E.; Prusky, D.; Keller, N.P. Fungal attack and host defence pathways unveiled in near-avirulent interactions of *Penicillium expansum* creA mutants on apples. *Mol. Plant Pathol.* **2018**, *19*, 2635–2650. [CrossRef] [PubMed]
5. Chen, Y.; Li, B.Q.; Xu, X.D.; Zhang, Z.Q.; Tian, S.P. The pH-responsive PacC transcription factor plays pivotal roles in virulence and patulin biosynthesis in *Penicillium expansum*. *Environ. Microbiol.* **2018**, *20*, 4063–4078. [CrossRef] [PubMed]
6. Prusky, D.; McEvoy, J.L.; Saftner, R.; Conway, W.S.; Jones, R. Relationship between host acidification and virulence of *Penicillium* spp. on apple and citrus fruit. *Phytopathology* **2004**, *94*, 44–51. [CrossRef] [PubMed]
7. Levin, E.; Ballester, A.R.; Raphael, G.; Feigenberg, O.; Liu, Y.S.; Norelli, J.; Droby, S. Identification and characterization of LysM effectors in *Penicillium expansum*. *PLoS ONE* **2017**, *12*, e0186023. [CrossRef] [PubMed]
8. Yang, Y.H.; Yu, Y.; Bi, C.W.; Kang, Z.S. Quantitative proteomics reveals the defense response of wheat against *Puccinia striiformis* f. sp. *tritici*. *Sci. Rep.* **2016**, *6*, 1–16. [CrossRef]

9. Ravi, G.; Woo, M.C.; Seungmin, S.; Hyun, L.G.; Woo, J.J.; Wook, K.S.; Tae, K.S. Comparative proteome profiling of susceptible and resistant rice cultivars identified an arginase involved in rice defense against *Xanthomonas oryzae* pv. *oryzae*. *Plant Physiol. Biochem.* **2021**, *171*, 105–114. [CrossRef]
10. Lu, B.W.; An, F.X.; Cao, L.J.; Yang, Y.J.; Liu, P.M.; Wang, X.; Liu, J. Proteomic profiling uncovered the cytosolic superoxide dismutase BsSOD1 associated with plant defense in the herbal orchid *Bletilla striata*. *Funct. Plant Biol.* **2020**, *47*, 937–944. [CrossRef]
11. Grandellis, C.; Vranych, C.V.; Piazza, A.; Garavaglia, B.S.; Gottig, N.; Ottado, J. An overview of proteomics tools for understanding plant defense against pathogens. *Curr. Issues Mol. Biol.* **2016**, *19*, 129–135. [CrossRef] [PubMed]
12. Tan, X.Y.; Misran, A.; Daim, L.D.J.; Lau, B.Y.C. Optimization of protein extraction for proteomic analyses of fresh and frozen "Musang King" durian pulps. *Food Chem.* **2021**, *343*, 128471. [CrossRef] [PubMed]
13. Gharbi, Y.; Barkallah, M.; Bouazizi, E.; Hibar, K.; Gdoura, R.; Triki, M.A. Lignification, phenol accumulation, induction of PR proteins, and antioxidant-related enzymes are key factors in the resistance of *Olea europaea* to verticillium wilt of olive. *Acta Physiol. Plant.* **2017**, *39*, 1–15. [CrossRef]
14. Dongus, J.A.; Parker, J.E. EDS1 signaling: At the nexus of intracellular and surface receptor immunity. *Curr. Opin. Plant Biol.* **2021**, *62*, 102039. [CrossRef] [PubMed]
15. Lin, C.W.; Lott, A.A.; Zhu, W.; Dufresne, C.P.; Chen, S.X. Mitogen-activated protein kinase 4-regulated metabolic networks. *Int. J. Mol. Sci.* **2022**, *23*, 880. [CrossRef]
16. Zhang, T.; Schneider, J.D.; Lin, C.W.; Geng, S.S.; Ma, T.Y.; Lawrence, S.R.; Chen, S.X. MPK4 phosphorylation dynamics and interacting proteins in plant immunity. *J. Proteome Res.* **2019**, *18*, 826–840. [CrossRef]
17. Tereza, T.; Despina, S.; Anna, K.; Tereza, V.; Jozef, Š. Multifaceted roles of heat shock protein 90 molecular chaperones in plant development. *J. Exp. Bot.* **2020**, *71*, 3966–3985. [CrossRef]
18. Liu, J.Z.; Horstman, H.D.; Braun, E.; Graham, M.A.; Zhang, C.Q.; Navarre, D.; Whitham, S.A. Soybean homologs of MPK4 negatively regulate defense responses and positively regulate growth and development. *Plant Physiol.* **2011**, *157*, 1363–1378. [CrossRef]
19. Philipp, W.; Britta, E.; Tina, R. ZmCPK1, a calcium-independent kinase member of the *Zea mays* CDPK gene family, functions as a negative regulator in cold stress signalling. *Plant Cell Environ.* **2015**, *38*, 544–558. [CrossRef]
20. Yu, H.X.; Xiao, A.F.; Dong, R.; Fan, Y.Q.; Zhang, X.P.; Liu, C.; Zhang, Z.M. Suppression of innate immunity mediated by the CDPK-Rboh complex is required for rhizobial colonization in medicago truncatula nodules. *New Phytol.* **2018**, *220*, 425–434. [CrossRef]
21. Tamara, V.; Jorge, V.; Satish, K.; Mats, H.; Carmen, C. Emerging complexity in reactive oxygen species production and signaling during the response of plants to pathogens. *Plant Physiol.* **2010**, *154*, 444–448. [CrossRef]
22. Cejudo, F.J.; Sandalio, L.M.; Van Breusegem, F. Understanding plant responses to stress conditions: Redox-based strategies. *J. Exp. Bot.* **2021**, *72*, 5785–5788. [CrossRef]
23. Chang, C.C.C.; Slesak, I.; Jorda, L.; Sotnikov, A.; Melzer, M.; Miszalski, Z.; Karpinski, S. *Arabidopsis* chloroplastic glutathione peroxidases play a role in cross talk between photooxidative stress and immune responses. *Plant Physiol.* **2009**, *150*, 670–683. [CrossRef] [PubMed]
24. Sarah, K.N. A review on plant peroxidases. *Nova Biol. Reper.* **2019**, *5*, 428–437. [CrossRef]
25. Xu, J.; Tian, Y.S.; Xing, X.J.; Xu, Z.S.; Zhu, B.; Fu, X.Y.; Yao, Q.H. Enhancement of phenol stress tolerance in transgenic *Arabidopsis* plants overexpressing glutathione S-transferase. *Plant Growth Regul.* **2017**, *82*, 37–45. [CrossRef]
26. Wang, J.M.; Liu, H.Y.; Xu, H.M.; Li, M.; Kang, Z.S. Analysis of differential transcriptional profiling in wheat infected by *Blumeria graminis* f. sp. *Tritici* using GeneChip. *Mol. Biol. Rep.* **2012**, *39*, 381–387. [CrossRef] [PubMed]
27. Dipali, S.; Giti, V.; Singh, C.A.; Veena, P.; Debasis, C. Rice (*Oryza sativa* L.) tau class glutathione S-transferase (OsGSTU30) overexpression in *Arabidopsis thaliana* modulates a regulatory network leading to heavy metal and drought stress tolerance. *Metallomics* **2019**, *11*, 375–389. [CrossRef]
28. Yang, Q.; Liu, Y.J.; Zeng, Q.Y. Overexpression of three orthologous glutathione s-transferases from populus increased salt and drought resistance in *Arabidopsis*. *Biochem. Syst. Ecol.* **2019**, *83*, 57–61. [CrossRef]
29. Svoboda, T.; Thon, M.R.; Strauss, J. The role of plant hormones in the interaction of *Colletotrichum* species with their host plants. *Int. J. Mol. Sci.* **2021**, *22*, 12454. [CrossRef]
30. Zhu, F.Y.; Wang, Z.W.; Fang, Y.; Tong, J.H.; Xiang, J.; Yang, K.K.; Wang, R.Z. Study on the role of phytohormones in resistance to watermelon *Fusarium* wilt. *Plants* **2022**, *11*, 156. [CrossRef]
31. Song, K.; Chen, B.; Cui, Y.; Zhou, L.; Chan, K.G.; Zhang, H.Y.; He, Y.W. The plant defense signal salicylic acid activates the RpfB-dependent quorum sensing signal turnover via altering the culture and cytoplasmic pH in the phytopathogen *Xanthomonas campestris*. *Mbio* **2022**, *13*, e03644-21. [CrossRef] [PubMed]
32. Czekus, Z.; Kukri, A.; Hamow, K.A.; Szalai, G.; Tari, I.; Ordog, A.; Poor, P. Activation of local and systemic defence responses by flg22 is dependent on daytime and ethylene in intact tomato plants. *Int. J. Mol. Sci.* **2021**, *22*, 8354. [CrossRef] [PubMed]
33. Dong, J.P.; Wang, Y.A.; Xian, Q.Q.; Chen, X.H.; Xu, J. Transcriptome analysis reveals ethylene-mediated defense responses to *Fusarium oxysporum* f. sp. *Cucumerinum* infection in *Cucumis sativus* L. *BMC Plant Biol.* **2020**, *20*, 1–10. [CrossRef] [PubMed]
34. Yuan, W.; Jiang, T.; Du, K.T.; Chen, H.; Cao, Y.Y.; Xie, J.P.; Zhou, T. Maize phenylalanine ammonia-lyases contribute to resistance to sugarcane mosaic virus infection, most likely through positive regulation of salicylic acid accumulation. *Mol. Plant Pathol.* **2019**, *20*, 1365–1378. [CrossRef] [PubMed]

35. Cheng, G.X.; Sun, J.T.; Shang, J.P.; Gong, Z.H. Virus-induced gene silencing for phenylalanine ammonia-lyase affects pepper adaption to low temperature. *Biol. Plant.* **2019**, *63*, 601–609. [CrossRef]
36. Christensen, S.A.; Nemchenko, A.; Borrego, E.; Murray, I.; Sobhy, I.S.; Bosak, L.; Kolomiets, M.V. The maize lipoxygenase, ZmLOX10, mediates green leaf volatile, jasmonate, and herbivore-induced plant volatile production for defense against insect attack. *Plant J.* **2013**, *74*, 59–73. [CrossRef] [PubMed]
37. Ruth, M.; Yovanny, I.; Tamara, V.; Satish, K.; Tomás, C.; Mats, H.; Carmen, C. 9-lipoxygenase-derived oxylipins activate brassinosteroid signaling to promote cell wall-based defense and limit pathogen infection. *Plant Physiol.* **2015**, *169*, 2324–2334. [CrossRef]
38. Khodadadi, F.; Tohidfar, M.; Vahdati, K.; Dandekar, A.M.; Leslie, C.A. Functional analysis of walnut polyphenol oxidase gene (*JrPPO1*) in transgenic tobacco plants and PPO induction in response to walnut bacterial blight. *Plant Pathol.* **2020**, *69*, 756–764. [CrossRef]
39. Niranjan-Raj, S.; Nagaraju, L.S.; Chandra, N.S. Molecular cloning and characterization of pearl millet polyphenol oxidase and its role in defense against downy mildew. *J. Plant Prot. Res.* **2019**, *59*, 423–427. [CrossRef]
40. Yuldashov, U.X.; Matniyazova, H.X.; Azimov, A.A.; Sherimbetov, A.G.; Khamdullaev, S.A.; Rasulova, O.O.; Shavkiev, J.S. Dependence of peroxidase (PO) and polyphenol oxidase (PPO) enzymes activity on plant productivity under the influence of phytopathogen micromycetes in soybean plant (*Glycine max* (L.) Merr.). *Plant Cell Biotechnol. Mol. Biol.* **2021**, *22*, 293–303. Available online: https://www.ikprress.org/index.php/PCBMB/article/view/6383 (accessed on 4 May 2021).
41. Renault, H.; Bassard, J.E.; Hamberger, B.; Werck-Reichhart, D. Cytochrome P450-mediated metabolic engineering: Current progress and future challenges. *Curr. Opin. Plant Biol.* **2014**, *19*, 27–34. [CrossRef]
42. Jensen, K.; Møller, B.L. Plant NADPH-cytochrome P450 oxidoreductases. *Phytochemistry* **2009**, *71*, 132–141. [CrossRef] [PubMed]
43. Song, C.; Xia, S.; Wang, X.; Zhang, J.; Qin, H.; Zhang, Y.; Bie, S. Cloning and expression analysis of chalcone synthase and chalcone isomerase encoding genes in *Gossypium hirsutum*. *Agric. Biotechnol.* **2018**, *7*, 15–21+26. [CrossRef]
44. Zhou, Y.; Huang, J.L.; Zhang, X.L.; Zhu, L.M.; Wang, X.F.; Guo, N.; Xing, H. Overexpression of chalcone isomerase (CHI) increases resistance against *Phytophthora sojae* in soybean. *J. Plant Biol.* **2018**, *61*, 309–319. [CrossRef]
45. Zhang, Q.L.; Xu, C.R.; Wei, H.Y.; Fan, W.Q.; Li, T.Z. Two pathogenesis-related proteins interact with leucine-rich repeat proteins to promote *Alternaria* leaf spot resistance in apple. *Hortic. Res.* **2021**, *8*, 219. [CrossRef] [PubMed]
46. Wang, W.D.; Nie, J.J.; Lv, L.Q.; Gong, W.; Wang, S.L.; Yang, M.M.; Huang, L.L. A valsa mali effector protein 1 targets apple (*Malus domestica*) pathogenesis-related 10 protein to promote virulence. *Front. Plant Sci.* **2021**, *12*, 741342. [CrossRef]
47. Pere, M.; Gautier, A.; Marie-Christine, P.; Camille, R.; Christophe, R.; Didier, M.; Jean-François, C. Identification of a *Vitis vinifera* endo-β-1,3-glucanase with antimicrobial activity against *Plasmopara viticola*. *Mol. Plant Pathol.* **2017**, *18*, 708–719. [CrossRef]
48. Farahani, A.S.; Taghavi, S.M.; Afsharifar, A.; Niazi, A. Changes in expression of pathogenesis-related gene 1, pathogenesis-related gene 2, phenylalanine ammonia-lyase, and catalase in tomato in response to *Pectobacterium carotovorum* subsp. *carotovorum*. *J. Plant Pathol.* **2016**, *98*, 525–530. [CrossRef]

Journal of
Fungi

Article

Phenylalanine Promotes Biofilm Formation of *Meyerozyma caribbica* to Improve Biocontrol Efficacy against Jujube Black Spot Rot

Qian Deng [1], Xingmeng Lei [1], Hongyan Zhang [1], Lili Deng [1,2], Lanhua Yi [1,2] and Kaifang Zeng [1,2,3,*]

1 College of Food Science, Southwest University, Chongqing 400715, China
2 Food Storage and Logistics Research Center, Southwest University, Chongqing 400715, China
3 Chongqing Key Laboratory of Speciality Food Co-Built by Sichuan and Chongqing, Chongqing 400715, China
* Correspondence: zengkaifang@hotmail.com

Abstract: During storage and transportation after harvest, the jujube fruit is susceptible to black spot rot, which is caused by *Alternaria alternata*. The present study aimed to evaluate the effectiveness of the yeast *Meyerozyma caribbica* in controlling *A. alternata* in postharvest jujube fruits, and to explore the biofilm formation mechanism. The results showed that *M. caribbica* treatment significantly reduced the *A. alternata* decay in jujube fruits. *M. caribbica* could rapidly colonize jujube fruit wounds, adhering tightly to hyphae of *A. alternata*, and accompanied by the production of extracellular secretions. In *in vitro* experiments, we identified that *M. caribbica* adhered to polystyrene plates, indicating a strong biofilm-forming ability. Furthermore, we demonstrated that *M. caribbica* can secrete phenylethanol, a quorum sensing molecule which can affect biofilm development. Phenylalanine (a precursor substance for phenylethanol synthesis) enhanced the secretion of phenylethanol and promoted the formation of *M. caribbica* biofilms. Meanwhile, phenylalanine enhanced the biological control performance of *M. caribbica* against jujube black spot rot. Our study provided new insights that enhance the biological control performance of antagonistic yeast.

Keywords: *Meyerozyma caribbica*; jujube; black spot rot; biofilm formation; phenylalanine

1. Introduction

The jujube fruit is a nutrient-rich fruit and a functional food that is widely appreciated by consumers for its health benefits [1]. However, fresh jujube fruit is highly susceptible to pathogenic fungi, and black spot rot caused by *Alternaria alternata* can seriously damage the commercial value of fresh jujube [2]. Chemical fungicide-induced pathogen resistance, fungicide residues, and toxicological problems have increased the urgent need for safer alternative strategies [3]. Due to their ability to manage postharvest infections without producing toxins, antagonistic yeasts are recognized as commercially viable biocontrol agents. The yeasts *Rhodosporidium paludigenum*, *Metschnikowia pulcherrima*, and *Cryptococcus laurentii* were effective in inhibiting *A. alternata* infection in postharvest jujube [4–6]. Antagonistic yeasts resist pathogen infection through mechanisms such as competition for nutrients and space, parasitism, secretion of antimicrobial substances, and induction of fruit resistance. Among these, the primary antagonistic mechanism of yeast is recognized as competition with pathogens for the limited nutrients and physical space in the host [7]. Antagonistic yeasts successfully compete for limited nutrient factors in the host environment and, thus, colonize rapidly. In addition, yeasts can also limit the growth of pathogens by adhering to fruit wounds and occupying limited space. Adhesion usually occurs during the formation of yeast biofilms, a specific mechanism by which yeast competes for space [8].

Biofilm formation is an effective mechanism of action for the management of postharvest diseases using antagonistic yeast [9]. For example, *C. laurentii* competed for nutrients and space with *Colletotrichum gloeosporioides* via adhesion and biofilm formation [10]. The biofilm of *Pichia kudriavzevii* exhibited better biological control than that of yeast in controlling gray mold and anthracnose in pear fruit caused by *B. cinerea* and *C. gloeosporioides* [11]. Biofilms are dense structural networks formed by small molecules (e.g., proteins, nucleic acids, polysaccharides) secreted by microorganisms during growth [12]. The formation of biofilms depends on quorum sensing (QS) between microorganisms. Microorganisms achieve cell-to-cell communication by secreting substances called "quorum sensing molecules (QSMs)" [13,14]. When QSM concentrations in the microbial community reach a certain threshold, they promote the expression of specific genes and are involved in the regulation of population behavior [15].

QSMs such as tryptophol, farnesol, and phenylethanol are commonly reported signaling molecules that perform critical roles in fungal morphological transformation and biofilm growth [16]. Among these QSMs, phenylethanol was able to affect the filamentous growth and adhesion of *Kloeckera apiculata* in citrus fruit [17]. In addition, phenylethanol was able to induce the biofilm formation of *Debaryomyces nepalensis*, which contributed to its adhesion on the jujube fruit and, as a result, protected the fruit from *A. alternata* infection [18]. Phenylethanol promoted the expression of *FLO11*, which was responsible for encoding the flocculation protein, through a tpk2p-dependent mechanism. This conferred adhesion properties to yeast and contributed to invasive cell growth, and the production of pseudohyphae and biofilm formation [19,20]. However, using phenylethanol is limited in some ways. Phenylethanol produced by chemical synthesis is limited in its application due to safety and environmental issues, while natural phenylethanol isolated from plants is limited due to its high price [21]. The synthesis of phenylethanol by microorganisms is a safe and nontoxic method.

Phenylalanine (Phe), as a precursor substance for phenylethanol, is often used in microbial fermentation to produce natural phenylethanol, which is safer than chemical synthesis [22]. Phe is relatively inexpensive and widely regarded as safe. In addition, these amino acids have an important role in the yeast culture process, and some functional amino acids are able to enhance the survival of yeast cells in extreme environments [23]. For instance, proline is used as a stress protector to enhance the resistance of industrial yeast [24], and it also promotes biofilm formation in *M. citriensis* [25]. This may be related to proline raising the content of pulcherrimin, which is a potential signaling molecule that influences the biofilm development of *M. citriensis*.

The yeast *M. caribbica* is considered to be a potential probiotic strain [26], which is widely used in food fermentation, such as functional beverages, coffee, or fruit wines [27,28]. In addition, previous studies have demonstrated that *M. caribbica* is an effective class of biocontrol agent. It has exhibited an excellent biocontrol performance in controlling the development of postharvest diseases in fruits, including mango, kiwifruit, and passion fruit [29,30]. However, the effectiveness of *M. caribbica* in biologically controlling *A. alternata* on jujube fruits is not clear, nor are the effects of *M. caribbica* biofilm formation for disease control. Although several studies have suggested the contribution of phenylethanol to yeast biofilm development, its role in *M. caribbica* biofilm formation is still unclear. It is worth investigating how Phe affects the development of *M. caribbica* biofilms and its bioprotective efficacy on jujube. Thus, the present study aimed to investigate (1) the biocontrol performance of *M. caribbica* in the control of *A. alternata* on jujube fruits, (2) the capacity of *M. caribbica* to form biofilms, (3) the biofilm formation mechanism in *M. caribbica* and its possible QSM, and (4) the influence of Phe on biofilm formation and the biocontrol efficacy of *M. caribbica* for jujube black spot rot.

2. Materials and Methods

2.1. Yeast, Pathogen, and Jujube Fruit

The yeast *M. caribbica* (KC422423.1) was preserved in our laboratory and was stored in 30% glycerol at −80 °C. It was identified by the DNA sequence of the internally transcribed spacer region. The sequences were BLASTed against the NCBI database (http://www.ncbi.nlm.nih.gov (accessed on 22 October 2021)). *M. caribbica* was grown on a NYDA medium

(8 g beef paste, 5 g yeast paste, 10 g dextrose, 10 g agar, and 1000 mL distilled water; Aoboxing, Beijing, China) for 48 h before use, and it was then inoculated into the NYDB broth (NYDA without agar) to be cultured in a shaker incubator (200 rpm, 28 °C) for 16 h. Fresh *M. caribbica* cells were harvested after centrifugation (5000× *g*, 5 min) and washed twice with sterile distilled water (SDW). The *M. caribbica* cell suspension was adjusted to the desired test concentration using SDW (1×10^8 cells/mL).

The fungal pathogen *A. alternata* was obtained by isolation and identification from decayed jujube fruit. *A. alternata* was cultured on the PDA medium (100 g fresh potato, 10 g glucose, 10 g agar, and 500 mL distilled water; Chronchem, Qionglai, Sichuan, China) at 25 °C. After 7 days of incubation, an *A. alternata* spore suspension was collected with SDW and then adjusted to the proper concentration (1×10^6 spores/mL).

Jujube (Zizyphus jujuba cv. Dongzao) fruits were uniform in shape and color, free from both infestation and mechanical injury, and purchased from the market (Beibei, Chongqing, China). Before the experiments, the fruit was cleaned with 2% sodium hypochlorite for 2 min, rinsed in SDW, and air-dried until use.

2.2. Effect of M. caribbica on the Biological Control of A. alternata on Jujube Fruit

A sterile punch was used to make a 3 × 3 mm wound even on both equatorial sides of the jujube. An injection of 20 µL of *M. caribbica* cell suspension was applied to the injury of each jujube fruit, and sterile water was used as a control. After 4 h, an additional suspension of 10 µL of *A. alternata* spores was injected into each wound. Jujube fruits were air-dried before being placed in plastic baskets that maintained a stable temperature and humidity (25 °C, 90%). The detection of disease incidence (DI) and lesion diameter (LD), which define the disease progress in jujube fruit, were recorded every two days. Measurements of DI and LD were performed in accordance with the method of Liu et al. [31]. Each treatment was repeated three times, with ten fruits each time, and the entire assay was replicated twice.

2.3. Dynamics of M. caribbica Colonization on Jujube Wounds

Wounds were made in the fruit equator as described in Section 2.2. A 20 µL cell suspension of *M. caribbica* was incubated on the injury of each jujube. All jujube fruits were stored at 25 °C after inoculation. Jujube sample tissue was extracted from wounds on days 0 (2 h after inoculation), 1, 2, 3, 4, 5, 6, 7, 8, and 9 for population monitoring. The tissues (10 mm diameter) from the fruit wounds were macerated in 10 mL of phosphate-buffered saline (PBS, pH 7.2) and diluted 10-fold according to the sequence. The appropriate diluted yeast solution was evenly distributed on the NYDA medium. All plates were cultivated under 28 °C incubation conditions. After 48 h, the $\log_{10}$ CFU/wound was used to reflect the overall colonization of *M. caribbica* populations. Three replicates were included for each experiment.

2.4. The Adhesion of M. caribbica to A. alternata

2.4.1. Microscopy Observation of M. caribbica and A. alternata

Microscopy of *M. caribbica* adhesion to *A. alternata* was performed based on the method of Liu et al. [9]. A 10 µL suspension of *A. alternata* (1×10^6 spores/mL) was used to inoculate slides containing small pieces of the PDA medium (2 cm × 2 cm) and then incubated at 25 °C until a large number of visible mycelia were formed (approximately 24 h). The hyphal surface of *A. alternata* was infected with an equal amount of *M. caribbica* cell suspension (1×10^6 cells/mL). The mixed cultures were cultivated for 24 and 48 h and then rinsed with SDW. The state of the mixed cultures was observed using a light microscope (Olympus, Tokyo, Japan).

2.4.2. SEM Observation of M. caribbica and A. alternata in the Jujube Wound

The interactions of *M. caribbica* and *A. alternata* in jujube wounds were observed as described by Chen et al. [32]. The wound preparation, microbial inoculation, and subsequent storage of jujube fruit were conducted following the steps in Section 2.2. Two days after storage, the tissue (5 mm × 3 mm × 3 mm) was cut from the jujube wounds to prepare electron microscopy samples by using a sterile scalpel. Sample tissues were fixed overnight at 4 °C with electron microscopy fixative (containing 2.5% glutaraldehyde). Subsequently, sample tissues were impregnated with gradient ethanol to remove water, followed by soaking in gradient tert-butanol to displace the ethanol. The sample tissues were dried at 60 °C for 2.5 h using a vacuum desiccator (DZF-6051, China). Tissue surfaces were gold-plated and observed under a scanning electron microscope (SEM, Phenom Pro, Phenom World, Eindhoven, The Netherlands).

2.5. Detection of Biofilms Formed by M. caribbica

Biofilm formation by *M. caribbica* was evaluated using the assay mentioned by Parafati et al. [33]. Fresh *M. caribbica* cell suspension was added to YNB (Hopebio, Shandong, China) containing glucose at 100 mmol/L, with an adjusted concentration of 1×10^7 cells/mL. YNB without yeast addition was used as a control. One hundred microliters of *M. caribbica* cell suspension was incubated in polystyrene plates at 28 °C for 3, 8, 24, 48, and 72 h at 75 rpm. After incubation for the corresponding time periods, the polyethylene plates were removed, and the incubation solution was washed with PBS (pH 7.2). Then, an equal amount of 0.4% crystalline violet solution was added for 45 min of staining. Following staining, the unadsorbed stain was washed away with PBS, and 200 μL of 95% ethanol was added for decolorization. The processing of decolorization was completed after 45 min, and the decolorized solution absorbance value measured at 590 nm was used to represent the biofilm formation ability. Each measurement was replicated three times, and the experiment was conducted in triplicate.

2.6. Effect of Phenylethanol on Biofilm Formation of M. caribbica

2.6.1. Biofilm Formation of *M. caribbica* in CM Medium

The conditioned medium (CM) was prepared according to Albuquerque et al. [34]. *M. caribbica* was incubated in NYDB in a shaker incubator for 5 days (200 rpm, 28 °C). After centrifugation of the *M. caribbica* fermentation broth, the collected supernatant liquor was filtered through a 0.22 μm microporous membrane to obtain the CM medium. The cultured *M. caribbica* cells were suspended in YNB containing different levels of CM (final concentrations of 0%, 5%, 25%, 50%, 75%, and 100%) and adjusted to 1×10^7 cells/mL as the final concentration of *M. caribbica* cells. The biofilm measurements were performed following the steps in Section 2.5.

2.6.2. Evaluation of Phenylethanol Production from *M. caribbica* by HPLC

The phenylethanol concentration of CM was measured by HPLC according to the method of Lei et al. [18] with slight modifications. The CM medium was obtained following the treatment outlined in Section 2.6.1 and tested immediately. The experimental setup was as follows: a C_{18} column and a 260 nm UV detector at 30 °C. The detection of the CM medium was performed at a rate of 0.7 mL/min with a 20 μL injection volume. The liquids used were 0.6% acetic acid solution as mobile phase A and methanol as mobile phase B.

2.6.3. Effect of Phenylethanol on Biofilm Formation by *M. caribbica*

The cultured *M. caribbica* cells were suspended in a YNB medium containing various phenylethanol concentrations (final concentrations of 0, 1, 2, and 4 mmol/L) to adjust the *M. caribbica* cell concentration to 1×10^7 cells/mL. Biofilm formation assays were performed as outlined in Section 2.5.

2.7. Influence of Phe on the Biocontrol Efficiency of M. caribbica

2.7.1. Effect of Phe on Phenylethanol Production of *M. caribbica*

The cultured *M. caribbica* cells were suspended in NYDB with various concentrations of Phe (final concentrations of 0, 1, and 8 mmol/L) to achieve a *M. caribbica* cell concentration of 1×10^7 cells/mL and were cultured on a shaker (28 °C, 200 rpm) for 5 days. The content of phenylethanol was tested following the treatment outlined in Section 2.6.2.

2.7.2. Effect of Phe on Biofilm Formation of *M. caribbica*

Freshly cultured yeast cells were added to the YNB medium containing different concentrations of Phe (final concentrations of 0, 1, and 8 mmol/L) and reached a concentration of 1×10^7 *M. caribbica* cells/mL. The biofilm assay was performed as outlined in Section 2.5.

The state of *M. caribbica* in jujube wounds was observed by SEM. Fresh *M. caribbica* cells were inoculated in NYDB medium containing various concentrations of Phe (final concentrations of 0, 1, and 8 mmol/L). After 16 h of incubation, *M. caribbica* cells from the different treatment groups were collected. Jujube fruit and sample tissues were prepared according to Section 2.4.2. The *M. caribbica* cell suspension (20 μL; 1×10^8 cells/mL) from each Phe treatment group was injected into the jujube wounds, with SDW inoculated as a control. The jujube fruits were stored for 2 days and then sampled for observation.

2.7.3. Effect of Phe on the Biocontrol Assay of *M. caribbica*

The *M. caribbica* cell suspension obtained from each Phe treatment group was used to inoculate jujube fruit wounds prepared as described in Section 2.7.2. Jujubes inoculated with SDW served as a control. After 4 h, an additional suspension of 10 μL *A. alternata* spores was injected into each wound. After the jujube fruits were air-dried, they were stored in plastic baskets that maintained a stable temperature and humidity (25 °C, 90%). The DI and LD were recorded as described in Section 2.2. Each treatment was performed three times, with 10 fruits each time, and the experiment was replicated twice.

2.8. Statistical Analysis

SPSS 26.0 (SPSS Inc., Chicago, IL, USA) was used to conduct the data analysis for this study. The independent samples' *t*-test and Duncan's multiple comparison method were used for the ANOVA, and the statistical significance was denoted by $p < 0.05$.

3. Results

3.1. Efficiency of M. caribbica against Jujube Black Spot Rot

In vivo experiments illustrate that the *M. caribbica* treatment dramatically prevented the rate of *A. alternata* infection in the jujube fruit (Figure 1). After storage at 25 °C for 10 days, the DI and LD of the jujube fruit were decreased by 48.0% and 59.7%, respectively, compared with the control. The rate of LD increase in *M. caribbica*-treated jujube fruit was much slower than that in the control for the entire storage period. This result demonstrated that *M. caribbica* had good biological control of jujube black spot rot.

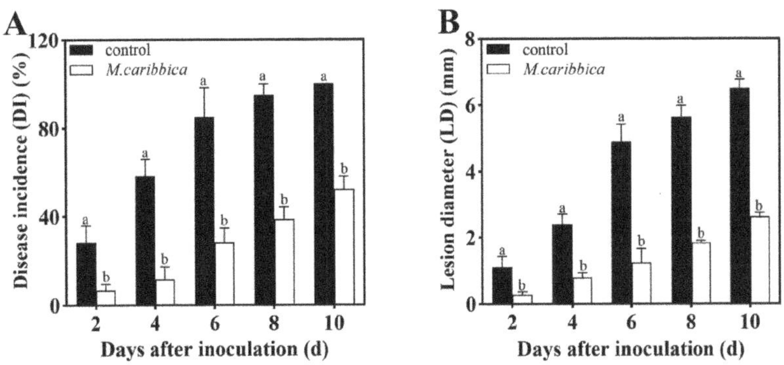

Figure 1. The biocontrol effect of *M. caribbica* inhibits *A. alternata* growth in jujube fruit. The (**A**) DI and (**B**) LD of jujube fruit were measured after inoculation at 25 °C. Standard errors of the means are indicated using vertical bars, and significant differences are denoted with different letters ($p < 0.05$).

3.2. Colonization of Jujube Wounds by M. caribbica

M. caribbica colonized jujube fruit wounds and grew rapidly under 25 °C storage conditions. This demonstrated the colonization changes of *M. caribbica* in jujube wounds over the entire storage period (Figure 2). The population of *M. caribbica* increased dramatically in the first 24 h, with the number increasing from an initial 5.99 to 7.50 $\log_{10}$ CFU/wound and then approaching a steady state. The viable count of *M. caribbica* was 7.46 $\log_{10}$ CFU/wound, which was a significant increase compared with the preliminary viable count after a 9-day storage period. These results indicated that *M. caribbica* had excellent colonization ability.

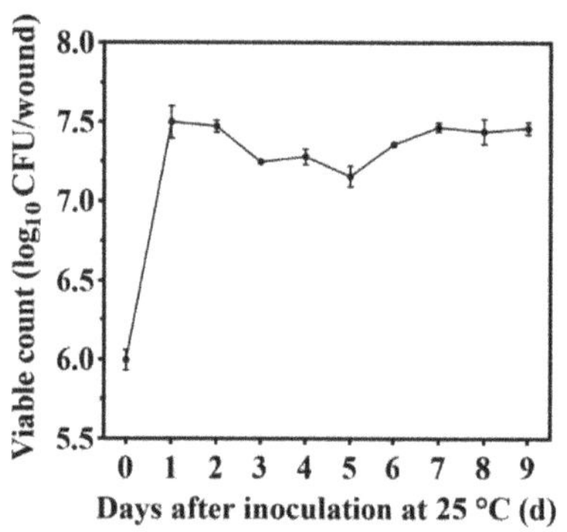

Figure 2. Colonization status of *M. caribbica* on jujube wounds. Vertical bars denote the standard error of the mean.

3.3. Adherence of M. caribbica to A. alternata

3.3.1. In Vitro Interaction Study between M. caribbica and A. alternata

The mixed cultures of *M. caribbica* and *A. alternata* on slides were observed by microscopy. After rinsing the cocultures with SDW for 30 s, *M. caribbica* was still able to adhere tightly to the *A. alternata* hyphae, indicating an adhesive effect (Figure S1). However, no breakage or deformation of *A. alternata* hyphae was observed, indicating that *M. caribbica* had no parasitic effect on *A. alternata*.

3.3.2. SEM Observation of M. caribbica and A. alternata on Jujube Wounds

A. alternata alone multiplied on the wounds of jujube fruit with a large amount of hyphal production, as shown in Figure 3A. In contrast, inoculation with *M. caribbica* reduced the hyphal production of *A. alternata*. In addition, *M. caribbica* was observed to adhere closely to the hyphae of *A. alternata*, effectively reducing the physical space available for *A. alternata* growth (Figure 3B). The rapid colonization ability of *M. caribbica* at the wounds exerted strong spatial competition pressure on *A. alternata*, further inhibiting its infestation of the jujube fruit.

3.4. Biofilm-Forming Ability of M. caribbica

The OD_{590} value was used to report the biofilm formation ability of *M. caribbica*. *M. caribbica* exhibited high adherence to polystyrene plates and remained stably adherent to the plates after repeated washing. As shown in Figure 4, after 3 h of culture, *M. caribbica* could adhere to the polystyrene plates with an OD_{590} value of 0.67. It reached the maximum value (OD_{590} was 0.92) after 8 h of culture and then began to decrease. After 48 and 72 h of culture, the OD_{590} showed no significant change compared with 3 h but still maintained a high biofilm formation ability.

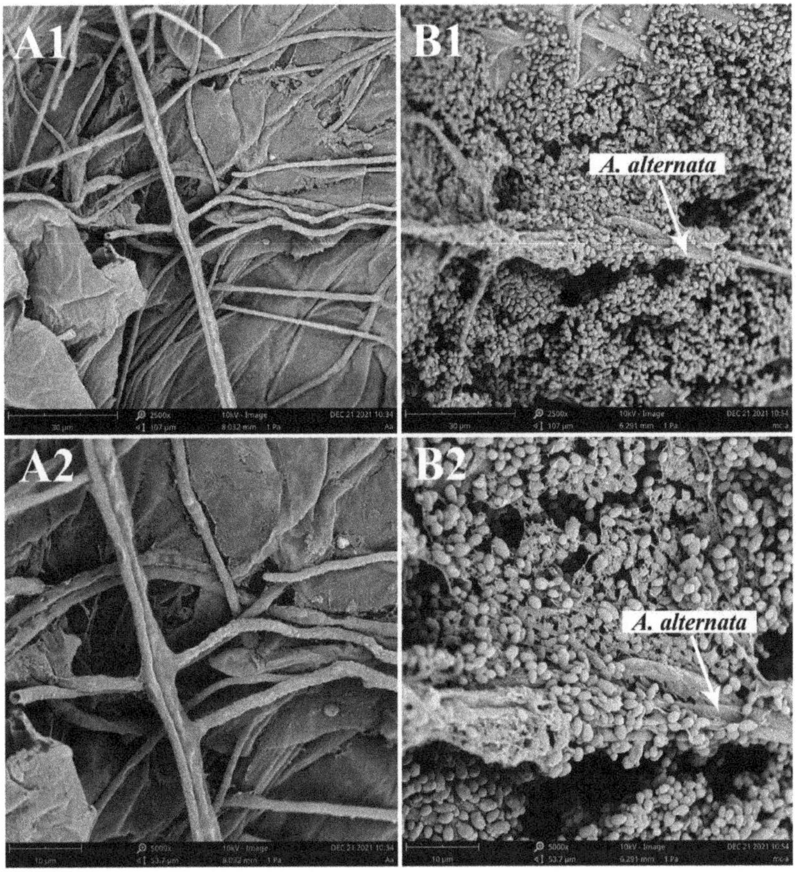

Figure 3. SEM of jujube wound tissues 48 h after inoculation. Wounds were incubated with (**A**) *A. alternata*, (**B**) *M. caribbica*, and *A. alternata*; (**A1**,**B1**): magnification of 2500×; (**A2**,**B2**): magnification of 5000×.

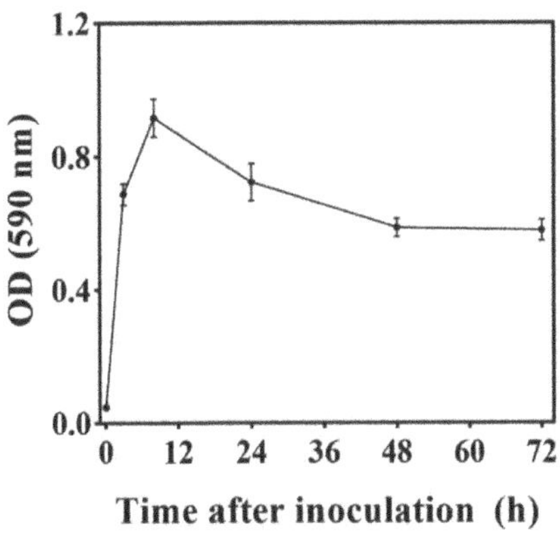

Figure 4. Biofilm formation ability on *M. caribbica* stained with crystal violet. The error bar reflects the standard deviation of the mean.

3.5. The Mechanism of Biofilm Formation in *M. caribbica*

3.5.1. Development of *M. caribbica* Biofilm by CM Medium

The efficacy of CM on *M. caribbica* cell growth and biofilm development is shown in Figure 5. The 25% CM medium significantly stimulated the growth of *M. caribbica* during 8–16 h of incubation (Figure 5A). According to the OD_{590} value, the 25% CM medium significantly promoted the formation of *M. caribbica* biofilms during the culture time (Figure 5B). This result suggests that the presence of a substance in the CM medium can act as a QSM, affecting the biofilm development of *M. caribbica*. In contrast, the 100% CM medium was unfavorable for the growth of *M. caribbica* biofilms. The possible reason for this is that the 100% CM medium has few of the nutrients required for yeast and cannot sustain its normal growth.

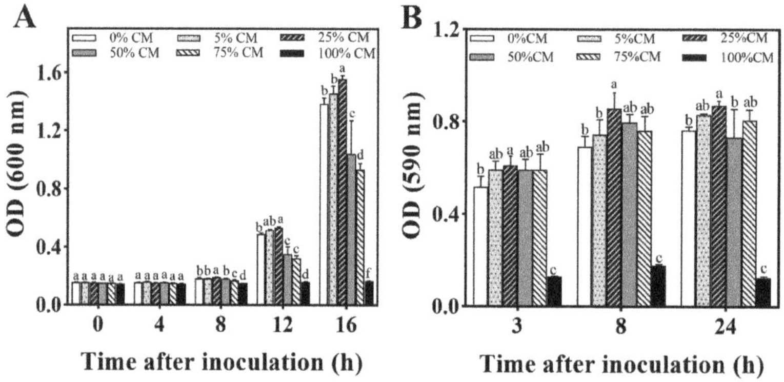

Figure 5. Effect of various concentrations of CM on (**A**) cell growth and (**B**) biofilm formation. Error bars indicate the standard deviation, and different letters indicate significant differences ($p < 0.05$).

3.5.2. Phenylethanol Determination in CM Medium by HPLC

The presence of phenylethanol in the CM medium was measured by HPLC, as shown in Figure S2. The CM medium sample was examined, and a retention peak time similar to that of the phenylethanol standard was observed (Figure S2A,B), while the corresponding peak did not appear in the NYDB sample without the yeast addition (Figure S2C). This result suggests that phenylethanol is one of the metabolites of *M. caribbica* that is secreted into the medium.

3.5.3. Biofilm Formation in *M. caribbica* Induced by Phenylethanol

As shown in Figure 6, different concentrations of phenylethanol were demonstrated to have an effect on the formation of *M. caribbica* biofilms. During the initial 3 h incubation time, 1, 2, and 4 mmol/L of phenylethanol significantly enhanced the biofilm of *M. caribbica*. At 8 h of incubation, 1 and 2 mmol/L phenylethanol significantly enhanced the biofilm formation of *M. caribbica*. At 24 h of incubation, 1 mmol/L phenylethanol significantly enhanced biofilm formation. Phenylethanol (1 mmol/L) was shown to promote biofilm formation throughout the culture time, and it significantly enhanced the growth of yeast during the incubation time of 8 to 16 h. This result suggested that phenylethanol is a QSM that influences the biofilm development of *M. caribbica*.

3.6. Influence of Phe on the Biocontrol Efficiency of *M. caribbica*

3.6.1. Efficacy of Phe on Phenylethanol Production in *M. caribbica*

The content of phenylethanol in the CM medium was influenced by the addition of Phe. The high concentration of Phe was beneficial for the production of phenylethanol. The content of phenylethanol in CM supplemented with 8 mmol/L Phe was significantly higher than that in CM supplemented with 1 mmol/L Phe and CM without Phe (Figure 7A).

The results indicate that the addition of Phe can increase the amount of phenylethanol produced by *M. caribbica*.

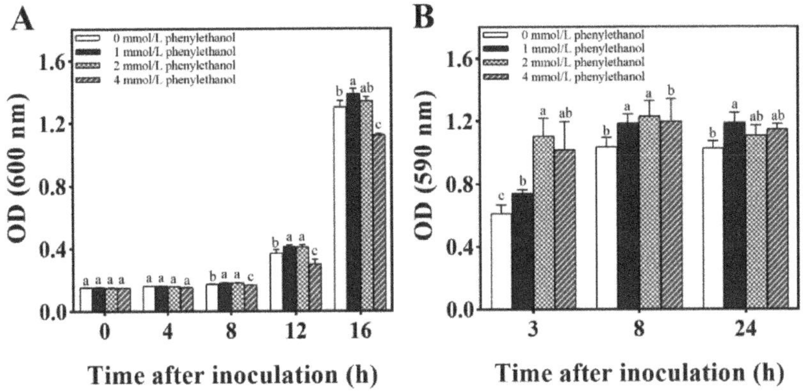

Figure 6. Effect of various concentrations of phenylethanol on (**A**) cell growth and (**B**) biofilm formation. Error bars indicate the standard deviation, and different letters indicate significant differences ($p < 0.05$).

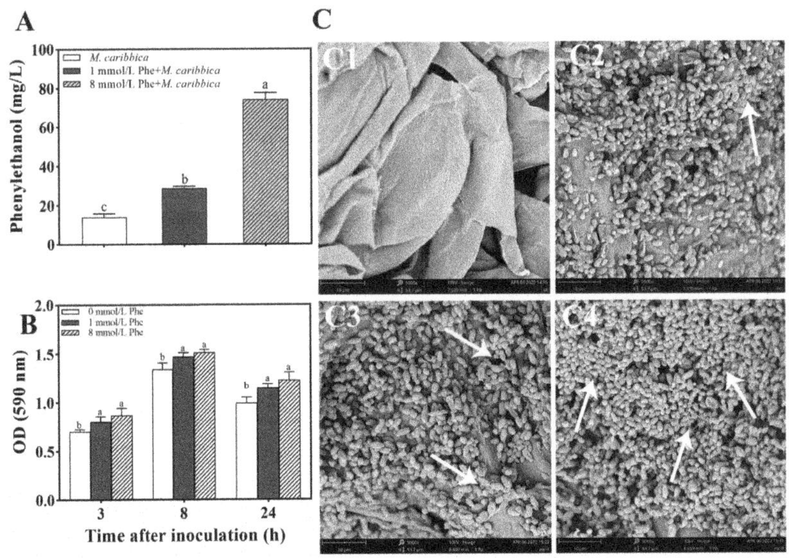

Figure 7. Effect of Phe on phenylethanol production and biofilm formation of *M. caribbica*. (**A**) Influence of different Phe concentrations on the phenylethanol content produced by *M. caribbica*. The vertical bar denotes the standard deviation of the mean. (**B**) Influence of different Phe concentrations on biofilm formation of *M. caribbica*. (**C**) SEM of jujube wound tissues 48 h after inoculation. Wounds were incubated with (**C1**) SDW, (**C2**) *M. caribbica*, (**C3**) *M. caribbica* obtained from NYDB medium with 1 mmol/L Phe, and (**C4**) *M. caribbica* obtained from NYDB medium with 8 mmol/L Phe. (**C1–C4**): magnification of 5000×. White arrows indicate the extracellular matrix secreted by *M. caribbica*. Error bars indicate the standard deviation, and different letters indicate significant differences ($p < 0.05$).

3.6.2. Induction of Biofilm Formation in *M. caribbica* by Phe

The effect of various concentrations of Phe on the formation of biofilms by *M. caribbica* is depicted in Figure 7B. At 3 h of culture, Phe demonstrated the ability to increase the OD_{590} value compared with the group with no Phe addition. The addition of 1 and 8 mmol/L Phe significantly promoted the biofilm formation of *M. caribbica* during 3–24 h of incubation. The above results indicate that Phe enhances the biofilm formation of *M. caribbica*.

The state of *M. caribbica* in the jujube wounds is shown in Figure 7C2–C4. As seen in the scanning electron micrographs, secretion of the yeast extracellular matrix was observed in all samples except the control. *M. caribbica* treated with Phe secreted more of the extracellular matrix and aggregated to a greater extent between yeast cells, while yeast untreated with Phe showed less extracellular matrix secretion. This suggests that Phe promotes the formation of *M. caribbica* biofilms and adhesion in jujube wounds.

3.6.3. Enhancement of Biocontrol Performance in *M. caribbica* by Phe

The efficacy of the Phe addition to the NYDB broth of *M. caribbica* in reducing jujube black spot rot is shown in Figure 8. Each *M. caribbica*-treated group significantly inhibited the growth of *A. alternata* in jujube fruit compared with the uninoculated yeast control. After a 10-day storage at 25 °C, *M. caribbica* cultured in NYDB broth with 8 mmol/L Phe showed the best efficacy in controlling black spot rot, as the DI and LD of the jujube fruits were significantly reduced by 12.67% and 21.75%, respectively, compared with those of the group treated with only *M. caribbica*. The results indicate that 8 mmol/L Phe significantly improves the biocontrol efficiency of *M. caribbica* against *A. alternata* in jujube fruit.

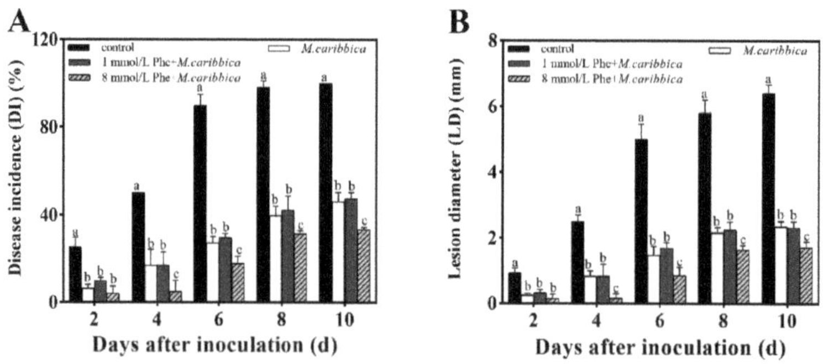

Figure 8. The effect of different concentrations of Phe on the biocontrol of jujube black spot rot by *M. caribbica*. The DI (**A**) and LD (**B**) were recorded for jujube black spot rot in fruit stored at 25 °C for 10 d. Error bars indicate the standard deviation, and different letters indicate significant differences ($p < 0.05$).

4. Discussion

Antagonistic yeast has proven to be a useful biocontrol tool for postharvest fungal diseases, is perceived as an appropriate alternative to chemical fungicides, and has an important role in biological control [35]. The capacity to target numerous pathogens in fruit is one of the criteria for being an ideal biocontrol agent [36]. The yeast *M. caribbica* has been reported to control postharvest anthracnose on mango and papaya fruit caused by *C. gloeosporioides* [37,38]. In our research, *M. caribbica* significantly reduced the fungal infection caused by the artificial inoculation of *A. alternata* in jujube fruits.

For our study, an important strategy used by *M. caribbica* against *A. alternata* was the competition for nutrients and available space. The excellent biocontrol performance of antagonistic yeast was associated with a high population density on the fruit [39]. *M. caribbica* rapidly colonized the jujube wounds after inoculation and maintained a stable population density throughout the entire duration of storage (Figure 2). The high population density of antagonistic yeast was conducive to biofilm coverage at the fruit wounds [40]. The mixed cultures of *M. caribbica* and *A. alternata* were observed using microscopy, and it was found that *M. caribbica* was not parasitic to *A. alternata* (Figure S1). The hyphae of *A. alternata* did not show deformation or breakage due to the attachment of *M. caribbica*, as observed by SEM. However, we found that *M. caribbica* firmly adhered to the surface of *A. alternata* hyphae (Figure S1). Around jujube wounds, the extracellular matrix encasing encapsulating yeast, as well as the secretion of aggregated yeast cells, were visible (Figure 3B). This indicated that *M. caribbica* competed for nutrient factors in the fruit wounds to rapidly multiply and form biofilms, occupying space in the wound sites to suppress *A. alternata* growth. Through in vitro experiments, *M. caribbica* was found to adhere tightly to polystyrene plates, demonstrating its strong biofilm formation ability (Figure 4).

Biofilm formation is regulated by QS, and QSMs have essential functions in biofilm development [41]. It was reported that QSMs should follow a density-dependent accumulation during microbial growth, and thus be able to induce a coordinated population response after reaching a specific concentration. Moreover, QSMs are able to reproduce the QS phenotype by the exogenous addition of substances [42]. In our study, it was demonstrated that QS was involved in regulating the formation of *M. caribbica* biofilms, and the 25% CM medium significantly improved the biofilm growth of *M. caribbica* (Figure 5B). This also suggests that it is possible for some metabolites in the CM medium to act as QSMs of *M. caribbica*. The major volatile metabolic organisms reported for *M. caribbica* include 1-butanol, 3-methyl, phenylethanol, and ethyl acetate [43,44]. Phenylethanol is a reportedly important communication molecule that regulates the population behavior and biofilm formation of *S. cerevisiae* and *D. hansenii* [45,46]. In our study, phenylethanol, which is the secondary metabolite of *M. caribbica*, was assayed in the CM medium by HPLC (Figure S2). Moreover, the formation of *M. caribbica* biofilms was enhanced by the exogenous addition of phenylethanol, which indicated that phenylethanol acted as a QSM of *M. caribbica*.

The synthesis of phenylethanol in microorganisms is based on the precursor substance Phe. Interestingly, we have found that some yeasts are able to convert Phe into phenylethanol: a method commonly used in the industrial production of natural phenylethanol [47]. *S. cerevisiae*, for example, can ferment Phe via the Ehrlich pathway to produce more phenylethanol [48]. In the present research, Phe stimulated the biofilm formation of *M. caribbica* and contributed to more extracellular matrix secretion by *M. caribbica*, which assisted in the adhesion to jujube fruit wounds (Figure 7C). HPLC revealed that Phe could stimulate *M. caribbica* to secrete more phenylethanol into the culture. Moreover, a positive correlation was found between Phe, *M. caribbica* biofilm formation, and phenylethanol content (Figure S3). This suggests that Phe can influence the biofilm formation of *M. caribbica* at the level of phenylethanol. It is interesting to note that some QSM-producing precursors, such as amino acids, also affect the morphological structure and biofilm formation of yeast. Indeed, 2-Methyl-1-butanol is one of the QSMs of the dimorphic fungus *Ophiostoma ulmi*, which can affect its morphological transformation, while isoleucine (a precursor of 2-methyl-1-butanol) has analogous effects on the morphological transformation of *O. ulmi* [49]. Nitrogen sources can alter the morphology of *Pichia fermentans*, and methionine, valine, and Phe can induce pseudohyphal morphology [50].

In the in vivo experiment, *M. caribbica* treated with 8 mmol/L Phe performed the best biocontrol effect against black spot rot, and the addition of Phe was negatively correlated with jujube disease incidence (Figure S3). A negative correlation was also found between the formation of *M. caribbica* biofilm and fruit diseases. This result suggested that Phe had a positive impact on the effectiveness of *M. caribbica* as a biocontrol agent, which greatly depends on its biofilm enhancement. Notably, in a study conducted by Klein and Kupper [51], 1% ammonium sulfate stimulated the production of *Aureobasidium pullulans* biofilms, thereby increasing the inhibitory efficiency against *G. citri-aurantii* and effectively controlling citrus decay. Additionally, Wang's study showed that the biocontrol potency of *M. citriensis* against citrus sour rot was enhanced by arginine treatment by promoting its adhesion to citrus fruit and inducing an increase in its oxidative stress tolerance [52]. These studies emphasized that the enhancement of antagonistic yeast biofilms is associated with improving their biocontrol efficacy.

5. Conclusions

In the current study, our results demonstrated that *M. caribbica* exhibited excellent efficacy in the control of jujube black spot rot. *M. caribbica* can rapidly colonize fruit wounds and has a potent biofilm-forming ability. The formation of *M. caribbica* biofilms was influenced by QSMs. Phenylethanol, as a QSM, regulated the biofilm growth of *M. caribbica*. In addition, Phe, as a precursor substance of phenylethanol, promoted the formation of *M. caribbica* biofilms by increasing the content of phenylethanol, thus enhancing the ability of *M. caribbica* to control black spot rot on the jujube. The present results are of great significance for enhancing the bioprotective efficacy of antagonistic yeast, and also provide a promising strategy for the control of postharvest disease in jujube fruits.

Supplementary Materials: The following supporting information can be downloaded at: https://www.mdpi.com/article/10.3390/jof8121313/s1, Figure S1: Interaction between *M. caribbica* and *A. alternata* for (A1,B1) 24 h/(A2,B2) 48 h after incubation. The *A. alternata* hyphae alone were used as a control. A,B: magnification of 1000×. Figure S2: HPLC analysis of phenylethanol in CM (A–C). (A) CM medium, (B) Standard: commercial phe-nylethanol, and (C) NYDB medium. Figure S3: Pearson's correlation analysis among Phe, phenylethanol content, biofilm, and disease incidence. Each square color scale represents the correlation coefficient. * indicates significant correlation ($p < 0.05$), ** indicates highly significant correlation ($p < 0.01$).

Author Contributions: Conceptualization, Q.D. and K.Z.; formal analysis, Q.D. and X.L.; resources, K.Z.; data curation, Q.D. and H.Z.; writing—original draft preparation, Q.D. and X.L.; writing—review and editing, K.Z., L.Y. and L.D.; supervision, K.Z.; funding acquisition, K.Z. All authors have read and agreed to the published version of the manuscript.

Funding: This research was supported by the National Key Research and Development Program of China (2019YFD1002300).

Institutional Review Board Statement: Not applicable.

Informed Consent Statement: Not applicable.

Data Availability Statement: All the data supporting the findings of this study are included in this article.

Conflicts of Interest: The authors declare no conflict of interest.

References

1. Li, W.; Yuan, S.; Li, Q.; Sang, W.; Cao, J.; Jiang, W. Methyl p-coumarate inhibits black spot rot on jujube fruit through membrane damage and oxidative stress against *Alternaria alternata*. *Postharvest Biol. Technol.* **2018**, *145*, 230–238. [CrossRef]
2. Yuan, S.; Yan, J.; Wang, M.; Ding, X.; Zhang, Y.; Li, W.; Cao, J.; Jiang, W. Transcriptomic and metabolic profiling reveals 'green ring' and 'red ring' on jujube fruit upon postharvest *Alternaria alternata* infection. *Plant Cell Physiol.* **2019**, *60*, 844–861. [CrossRef] [PubMed]
3. Dukare, A.S.; Paul, S.; Nambi, V.E.; Gupta, R.K.; Singh, R.; Sharma, K.; Vishwakarma, R.K. Exploitation of microbial antagonists for the control of postharvest diseases of fruits: A review. *Crit. Rev. Food Sci. Nutr.* **2019**, *59*, 1498–1513. [CrossRef] [PubMed]
4. Wang, Y.; Tang, F.; Xia, J.; Yu, T.; Wang, J.; Azhati, R.; Zheng, X.D. A combination of marine yeast and food additive enhances preventive effects on postharvest decay of jujubes (*Zizyphus jujuba*). *Food Chem.* **2011**, *125*, 835–840. [CrossRef]
5. Guo, D.; Zhu, L.; Hou, X. Combination of UV-C treatment and *Metschnikowia pulcherrimas* for controlling *Alternaria rot* in postharvest winter jujube fruit. *J. Food Sci.* **2015**, *80*, M137–M141. [CrossRef]
6. Qin, G.Z.; Tian, S.P. biocontrol of postharvest diseases of jujube fruit by *Cryptococcus laurentii* combined with a low dosage of fungicides under different storage conditions. *Plant Dis.* **2004**, *88*, 497–501. [CrossRef]
7. Wang, Z.; Sui, Y.; Li, J.; Tian, X.; Wang, Q. Biological control of postharvest fungal decays in citrus: A review. *Crit. Rev. Food Sci. Nutr.* **2022**, *62*, 861–870. [CrossRef]
8. Freimoser, F.M.; Rueda-Mejia, M.P.; Tilocca, B.; Migheli, Q. Biocontrol yeasts: Mechanisms and applications. *World J. Microbiol. Biotechnol.* **2019**, *35*, 154. [CrossRef]
9. Liu, Y.; Yao, S.; Deng, L.; Ming, J.; Zeng, K. Different mechanisms of action of isolated epiphytic yeasts against *Penicillium digitatum* and *Penicillium italicum* on citrus fruit. *Postharvest Biol. Technol.* **2019**, *152*, 100–110. [CrossRef]

10. Bautista-Rosales, P.U.; Calderon-Santoyo, M.; Servín-Villegas, R.; Ochoa-Álvarez, N.A.; Vázquez-Juárez, R.; Ragazzo-Sánchez, J.A. Biocontrol action mechanisms of *Cryptococcus laurentii* on *Colletotrichum gloeosporioides* of mango. *Crop. Prot.* **2014**, *65*, 194–201. [CrossRef]
11. Chi, M.; Li, G.; Liu, Y.; Liu, G.; Li, M.; Zhang, X.; Sun, Z.; Sui, Y.; Liu, J. Increase in antioxidant enzyme activity, stress tolerance and biocontrol efficacy of *Pichia kudriavzevii* with the transition from a yeast-like to biofilm morphology. *Biol. Control* **2015**, *90*, 113–119. [CrossRef]
12. Spadaro, D.; Droby, S. Development of biocontrol products for postharvest diseases of fruit: The importance of elucidating the mechanisms of action of yeast antagonists. *Trends Food Sci. Technol.* **2016**, *47*, 39–49. [CrossRef]
13. Rodrigues, C.F.; Černáková, L. Farnesol and Tyrosol: Secondary metabolites with a crucial *quorum-sensing* role in *Candida* biofilm development. *Genes* **2020**, *11*, 444. [CrossRef] [PubMed]
14. Mehmood, A.; Liu, G.; Wang, X.; Meng, G.; Wang, C.; Liu, Y. Fungal quorum-sensing molecules and inhibitors with potential antifungal activity: A review. *Molecules* **2019**, *24*, 950. [CrossRef]
15. Barriuso, J.; Hogan, D.A.; Keshavarz, T.; Martínez, M.J. Role of quorum sensing and chemical communication in fungal biotechnology and pathogenesis. *FEMS Microbiol. Rev.* **2018**, *42*, 627–638. [CrossRef]
16. Wongsuk, T.; Pumeesat, P.; Luplertlop, N. Fungal quorum sensing molecules: Role in fungal morphogenesis and pathogenicity. *J. Basic Microbiol.* **2016**, *56*, 440–447. [CrossRef]
17. Pu, L.; Jingfan, F.; Kai, C.; Chao-an, L.; Yunjiang, C. Phenylethanol promotes adhesion and biofilm formation of the antagonistic yeast *Kloeckera apiculata* for the control of blue mold on citrus. *FEMS Yeast Res.* **2014**, *14*, 536–546. [CrossRef]
18. Lei, X.; Deng, B.; Ruan, C.; Deng, L.; Zeng, K. Phenylethanol as a quorum sensing molecule to promote biofilm formation of the antagonistic yeast *Debaryomyces nepalensis* for the control of black spot rot on jujube. *Postharvest Biol. Technol.* **2022**, *185*, 111788. [CrossRef]
19. Yang, L.; Zheng, C.; Chen, Y.; Ying, H. *FLO* genes family and transcription factor mig1 regulate *Saccharomyces cerevisiae* biofilm formation during immobilized fermentation. *Front. Microbiol.* **2018**, *9*, 1860. [CrossRef]
20. Zhang, D.; Wang, F.; Yu, Y.; Ding, S.; Chen, T.; Sun, W.; Liang, C.; Yu, B.; Ying, H.; Liu, D.; et al. Effect of quorum-sensing molecule 2-phenylethanol and ARO genes on *Saccharomyces cerevisiae* biofilm. *Appl. Microbiol. Biotechnol.* **2021**, *105*, 3635–3648. [CrossRef]
21. Sekar, B.S.; Lukito, B.R.; Li, Z. Production of natural 2-phenylethanol from glucose or glycerol with coupled *Escherichia coli* strains expressing L -phenylalanine biosynthesis pathway and artificial biocascades. *ACS Sustain. Chem. Eng.* **2019**, *104*, 177. [CrossRef]
22. Mo, Q.; Chen, H.; Fan, C.; Zhang, D.; Liu, L.; Fu, B.; Yuan, J. Utilization of a styrene-derived pathway for 2-phenylethanol production in budding yeast. *Appl. Microbiol. Biotechnol.* **2021**, *105*, 2333–2340. [CrossRef] [PubMed]
23. Takagi, H. Metabolic regulatory mechanisms and physiological roles of functional amino acids and their applications in yeast. *Biosci. Biotech. Bioch.* **2019**, *83*, 1449–1462. [CrossRef]
24. Takagi, H. Proline as a stress protectant in yeast: Physiological functions, metabolic regulations, and biotechnological applications. *Appl. Microbiol. Biotechnol.* **2008**, *81*, 211–223. [CrossRef] [PubMed]
25. Liu, Y.; Yi, L.; Ruan, C.; Yao, S.; Deng, L.; Zeng, K. Proline increases pigment production to improve oxidative stress tolerance and biocontrol ability of *Metschnikowia citriensis*. *Front. Microbiol.* **2019**, *10*, 1273. [CrossRef] [PubMed]
26. Simões, L.A.; Cristina de Souza, A.; Ferreira, I.; Melo, D.S.; Lopes, L.A.A.; Magnani, M.; Schwan, R.F.; Dias, D.R. Probiotic properties of yeasts isolated from Brazilian fermented table olives. *J. Appl. Microbiol.* **2021**, *131*, 1983–1997. [CrossRef]
27. Amorim, J.C.; Piccoli, R.H.; Duarte, W.F. Probiotic potential of yeasts isolated from pineapple and their use in the elaboration of potentially functional fermented beverages. *Food Res. Int.* **2018**, *107*, 518–527. [CrossRef] [PubMed]
28. Martinez, S.J.; Bressani, A.P.P.; Simão, J.B.P.; Pylro, V.S.; Dias, D.R.; Schwan, R.F. Dominant microbial communities and biochemical profile of pulped natural fermented coffees growing in different altitudes. *Food Res. Int.* **2022**, *159*. [CrossRef] [PubMed]
29. Navarro-Herrera, Y.Y.; Ortíz-Moreno, M.L. Yeast strains with antagonist activity against *Colletotrichum gloeosporioides* (Penz.) Penz. & Sacc. and their phenotypic characterization. *Egypt. J. Biol. Pest Control* **2020**, *30*, 120. [CrossRef]
30. Qiu, J.-e.; Zhao, L.; Jiang, S.; Abiso Godana, E.; Zhang, X.; Zhang, H. Efficacy of *Meyerozyma caribbica* in the biocontrol of blue mold in kiwifruit and mechanisms involved. *Biol. Control* **2022**, *173*, 105000. [CrossRef]
31. Liu, Y.; Wang, W.; Zhou, Y.; Yao, S.; Deng, L.; Zeng, K. Isolation, identification and *in vitro* screening of Chongqing orangery yeasts for the biocontrol of *Penicillium digitatum* on citrus fruit. *Biol. Control* **2017**, *110*, 18–24. [CrossRef]
32. Chen, O.; Yi, L.; Deng, L.; Ruan, C.; Zeng, K. Screening antagonistic yeasts against citrus green mold and the possible biocontrol mechanisms of *Pichia galeiformis* (BAF03). *J. Sci. Food Agric.* **2020**, *100*, 3812–3821. [CrossRef] [PubMed]
33. Parafati, L.; Vitale, A.; Restuccia, C.; Cirvilleri, G. Biocontrol ability and action mechanism of food-isolated yeast strains against *Botrytis cinerea* causing post-harvest bunch rot of table grape. *Food Microbiol.* **2015**, *47*, 85–92. [CrossRef] [PubMed]
34. Albuquerque, P.; Nicola, A.M.; Nieves, E.; Paes, H.C.; Williamson, P.R.; Silva-Pereira, I.; Casadevall, A. Quorum sensing-mediated, cell density-dependent regulation of growth and virulence in *Cryptococcus neoformans*. *mBio* **2013**, *5-e00986*. [CrossRef] [PubMed]
35. Zhang, X.; Li, B.; Zhang, Z.; Chen, Y.; Tian, S. Antagonistic yeasts: A promising alternative to chemical fungicides for controlling postharvest decay of fruit. *J. Fungi* **2020**, *6*, 158. [CrossRef] [PubMed]
36. Nunes, C.A. Biological control of postharvest diseases of fruit. *Eur. J. Plant Pathol.* **2012**, *133*, 181–196. [CrossRef]
37. Aguirre-Güitrón, L.; Calderón-Santoyo, M.; Lagarón, J.M.; Prieto, C.; Ragazzo-Sánchez, J.A. Formulation of the biological control yeast *Meyerozyma caribbica* by electrospraying process: Effect on postharvest control of anthracnose in mango (*Mangifera indica* L.) and papaya (*Carica papaya* L.). *J. Sci. Food Agric.* **2022**, *102*, 696–706. [CrossRef]

38. Aguirre-Güitrón, L.; Calderón-Santoyo, M.; Bautista-Rosales, P.U.; Ragazzo-Sánchez, J.A. Application of powder formulation of *Meyerozyma caribbica* for postharvest control of *Colletotrichum gloeosporioides* in mango (*Mangifera indica* L.). *Lebensm.-Wiss. Technol.* **2019**, *113*, 108271. [CrossRef]
39. Li, Q.; Li, C.; Li, P.; Zhang, H.; Zhang, X.; Zheng, X.; Yang, Q.; Apaliya, M.T.; Boateng, N.A.S.; Sun, Y. The biocontrol effect of *Sporidiobolus pararoseus* Y16 against postharvest diseases in table grapes caused by *Aspergillus niger* and the possible mechanisms involved. *Biol. Control* **2017**, *113*, 18–25. [CrossRef]
40. Wang, S.; Ruan, C.; Yi, L.; Deng, L.; Yao, S.; Zeng, K. Biocontrol ability and action mechanism of *Metschnikowia citriensis* against *Geotrichum citri-aurantii* causing sour rot of postharvest citrus fruit. *Food Microbiol.* **2020**, *87*, 103375. [CrossRef]
41. Bandara, H.M.H.N.; Lam, O.L.T.; Jin, L.J.; Samaranayake, L. Microbial chemical signaling: A current perspective. *Crit. Rev. Microbiol.* **2012**, *38*, 217–249. [CrossRef] [PubMed]
42. Albuquerque, P.; Casadevall, A. Quorum sensing in fungi—A review. *Med. Mycol.* **2012**, *50*, 337–345. [CrossRef] [PubMed]
43. Iñiguez-Moreno, M.; Ragazzo-Sánchez, J.A.; Barros-Castillo, J.C.; Sandoval-Contreras, T.; Calderón-Santoyo, M. Sodium alginate coatings added with *Meyerozyma caribbica*: Postharvest biocontrol of *Colletotrichum gloeosporioides* in avocado (*Persea americana* Mill. cv. Hass). *Postharvest Biol. Technol.* **2020**, *163*, 111123. [CrossRef]
44. Choińska, R.; Piasecka-Jóźwiak, K.; Chabłowska, B.; Dumka, J.; Łukaszewicz, A. Biocontrol ability and volatile organic compounds production as a putative mode of action of yeast strains isolated from organic grapes and rye grains. *Anton. Leeuw. Int. J. G.* **2020**, *113*, 1135–1146. [CrossRef]
45. Gori, K.; Knudsen, P.B.; Nielsen, K.F.; Arneborg, N.; Jespersen, L. Alcohol-based quorum sensing plays a role in adhesion and sliding motility of the yeast *Debaryomyces hansenii*. *FEMS Yeast Res.* **2011**, *11*, 643–652. [CrossRef]
46. Avbelj, M.; Zupan, J.; Raspor, P. Quorum-sensing in yeast and its potential in wine making. *Appl. Microbiol. Biotechnol.* **2016**, *100*, 7841–7852. [CrossRef]
47. Mitri, S.; Koubaa, M.; Maroun, R.G.; Rossignol, T.; Nicaud, J.-M.; Louka, N. Bioproduction of 2-phenylethanol through yeast fermentation on synthetic media and on agro-industrial waste and by-products: A review. *Foods* **2022**, *1*, 109. [CrossRef]
48. Dai, J.; Xia, H.; Yang, C.; Chen, X. Sensing, uptake and catabolism of L-phenylalanine during 2-phenylethanol biosynthesis via the Ehrlich pathway in *Saccharomyces cerevisiae*. *Front. Microbiol.* **2021**, *12*, 601963. [CrossRef]
49. Berrocal, A.; Navarrete, J.; Oviedo, C.; Nickerson, K.W. Quorum sensing activity in *Ophiostoma ulmi*: Effects of fusel oils and branched chain amino acids on yeast-mycelial dimorphism. *J. Appl. Microbiol.* **2012**, *113*, 126–134. [CrossRef]
50. Sanna, M.L.; Zara, S.; Zara, G.; Migheli, Q.; Budroni, M.; Mannazzu, I. *Pichia fermentans* dimorphic changes depend on the nitrogen source. *Fungal Biol.* **2012**, *116*, 769–777. [CrossRef]
51. Klein, M.N.; Kupper, K.C. Biofilm production by *Aureobasidium pullulans* improves biocontrol against sour rot in citrus. *Food Microbiol.* **2018**, *69*, 1–10. [CrossRef] [PubMed]
52. Wang, S.; Zhang, H.; Qi, T.; Deng, L.; Yi, L.; Zeng, K. Influence of arginine on the biocontrol efficiency of *Metschnikowia citriensis* against *Geotrichum citri-aurantii* causing sour rot of postharvest citrus fruit. *Food Microbiol.* **2022**, *101*, 103888. [CrossRef] [PubMed]

Journal of
Fungi

MDPI

Article

Cell-Free Supernatant of *Bacillus subtilis* Reduces Kiwifruit Rot Caused by *Botryosphaeria dothidea* through Inducing Oxidative Stress in the Pathogen

Yezhen Fan [1,†], Kui Liu [1,2,†], Ruoxi Lu [1], Jieyu Gao [1], Wu Song [1], Hongyan Zhu [1], Xiaofeng Tang [1], Yongsheng Liu [3,4] and Min Miao [1,*]

1 School of Food and Biological Engineering, Hefei University of Technology, Hefei 230036, China
2 Institute of Botany, The Chinese Academy of Sciences, Beijing 230094, China
3 Ministry of Education Key Laboratory for Bio-Resource and Eco-Environment, State Key Laboratory of Hydraulics and Mountain River Engineering, College of Life Science, Sichuan University, Chengdu 610064, China
4 School of Horticulture, Anhui Agricultural University, Hefei 230036, China
* Correspondence: minmiao@hfut.edu.cn
† These authors contributed equally to this work.

Abstract: Biological control of postharvest diseases has been proven to be an effective alternative to chemical control. As an environmentally friendly biocontrol agent, *Bacillus subtilis* has been widely applied. This study explores its application in kiwifruit soft rot and reveals the corresponding mechanisms. Treatment with cell-free supernatant (CFS) of *Bacillus subtilis* BS-1 significantly inhibits the mycelial growth of the pathogen *Botryosphaeria dothidea* and attenuates the pathogenicity on kiwifruit in a concentration-dependent manner. In particular, mycelial growth diameter was only 21% of the control after 3 days of treatment with 5% CFS. CFS caused swelling and breakage of the hyphae of *B. dothidea* observed by scanning electron microscopy, resulting in the leakage of nucleic acid and soluble protein and the loss of ergosterol content. Further analysis demonstrated that CFS significantly induces the expression of *Nox* genes associated with reactive oxygen species (ROS) production by 1.9–2.7-fold, leading to a considerable accumulation of ROS in cells and causing mycelial cell death. Our findings demonstrate that the biocontrol effect of *B. subtilis* BS-1 CFS on *B. dothidea* is realized by inducing oxidative damage to the mycelia cell.

Keywords: biocontrol; *Botryosphaeria dothidea*; postharvest disease; oxidative damage; kiwifruit

1. Introduction

Kiwifruit is one of the most resourceful, nutritious, and economically healthy fruits. The freshly ripened kiwifruit is juicy, unique in flavor, rich in vitamins and minerals, and deeply accepted by the market [1–4]. However, as commercial kiwifruit cultivation and production continue to expand, various diseases have emerged during the harvest and storage of kiwifruit and have become increasingly prominent and severe. These serious diseases include soft rot, grey mold, and black spot disease [5–7]. Among them, the soft rot of kiwifruit is the dominant postharvest disease. Despite that there is no difference between the appearance of a diseased fruit and that of a healthy one, the inside of the diseased fruit turns dirty brown and decays, losing edible value. The occurrence of soft rot seriously affects the quality and yield of kiwifruit and is a common cause of major economic losses worldwide [8,9]. Numerous scholars have been working on isolating and identifying the causal agent of the soft rot of kiwifruit. Several major pathogens have been reported to be associated with postharvest rot of kiwifruit, including *Botryosphaeria dothidea*, *Diaporthe actinidiae*, *Phomopsis sp.*, and *Alternaria alternata* [10–13]. On balance, *Botryosphaeria dothidea* is considered one of the primary pathogens of kiwifruit soft rot worldwide [14]. *B. dothidea*, a necrotrophic pathogen responsible for diseases in a broad range of plant hosts, belongs to the class *Dothideomycetes* of the phylum *Ascomycota* and is distributed worldwide. It is a killer of woody plant species, causing branches' death, ulcers, gum flow, and postharvest decay of fruits. Kiwifruits infected by *B. dothidea* show a water-soaked lesion spot on the peeled flesh with a light brown center and a dark green ring edge [15,16].

Although certain sorts of chemicals are effective against these pathogens, mixtures of pyraclostrobin and boscalid could change the morphology of conidia and hyphae branches, resulting in suppressed growth of *B. dothidea*. However, these chemical fungicide residues are potentially hazardous to human health and the environment [12,17]. As public concern about the quality and safety of agricultural products rises, researchers are devoted to exploring efficient biological control methods to manage plant diseases. Microbial antagonists against plant pathogens provide an environmentally friendly approach and meet the urgent demand for controlling kiwifruit soft rot diseases [18,19]. *Bacillus spp.* is one of the bacteria broadly distributed in the soil rhizosphere and is widely regarded as a powerful biological control agent against a wide range of pathogenic fungi on a variety of fruits. Volatile organic compounds released by *Bacillus subtilis* strain CL2 inhibited the growth of four pathogenic fungi, contributing to the reduction of postharvest disease in wolfberry [20]. The culture suspension of *B. subtilis* L1-21 effectively inhibited the growth of *Botrytis cinerea* by up to 86.57%, resulting in effective control of postharvest tomato grey mold [21]. In addition, many countries have approved *B. subtilis* as a food supplement, and the toxicity of *B. subtilis* has been evaluated by acute and chronic doses in guinea pigs and rabbits, showing no harmfulness in animals under tested conditions [22]. Although *B. subtilis* is a broad-spectrum antagonistic bacterium suited for disease control in various fruits and vegetables, most works focused on the biological control effect. Moreover, the molecular regulatory mechanism of *B. subtilis* that destroyed pathogens and inhibited their growth is poorly understood. In addition, the inhibition mechanism of *B. subtilis* on kiwifruit soft rot, especially on the primary pathogen *B. dothidea*, has not been extensively studied.

In this study, the biocontrol application of *B. subtilis* was extended to prevent and control the soft rot of kiwifruit. We mainly aimed to examine the inhibitory effect of CFS on soft rot caused by *B. dothidea* in vitro and in vivo and further explore the damage mechanism of CFS on *B. dothidea*. This work will assist in providing a practical, sustainable, and safe biological approach to managing soft rot in kiwifruit.

2. Materials and Methods

2.1. Microbial Materials

The soft rot fungal pathogen *Botryosphaeria dothidea* was isolated and identified from the perspective of morphology, molecular biology, and Koch's rule and shares 100% identity with the ITS information of *B. dothidea* strain CMW8000 (AY236949.1). The *B. dothidea* was stored in a −80 °C refrigerator and cultured in potato dextrose agar medium (PDA) at 25 °C.

Bacillus subtilis strain BS-1 was kindly provided by Professor Liu Jia from Chongqing College of Arts and Sciences. Its morphologic properties, growth characteristics, and DNA analysis results met the criteria for identifying *B. subtilis*. The 16sRNA sequence of strain BS-1 was 99% identical to that of *B. subtilis* strain CNPMS22169 (MH358457.1).

2.2. Preparation of the Cell-Free Supernatant (CFS)

B. subtilis strain BS-1 was activated in an LB plate at 28 °C for 16–24 h, and individual clones were transferred to sterile test tubes containing 2 mL LB medium and cultured until the cell cultures OD_{600} reached 0.6 as a seed broth. Fermentation cultures were expanded at 1% of the seed broth inoculum for 72 h and centrifuged at 8000 rpm for 15 min. The supernatant was purified by passing through a 0.22 μm sterile extractor to give a cell-free supernatant (CFS). The CFS was diluted with distilled water to 1%, 2%, and 5% of the final working volume fraction.

2.3. Fruits

Mature kiwifruits (*Actinidia deliciosa* cv. Xuxiang) were purchased from supermarkets during the marketing season from September to October with the characteristics of similar size, without decay or physical damage, 8–10% soluble solids content (SSC), and 30–35 N hardness. Experimental kiwifruits were disinfected with 1% (*v/v*) sodium hypochlorite for 10 min, rinsed three times with water, and air-dried at room temperature [23].

2.4. Effectiveness of the CFS against Kiwifruit Soft Rot Pathogen In Vitro and In Vivo

Mycelial discs (5 mm diameter) from one-week-old cultured *B. dothidea* were placed in the center of PDA plates, amended with CFS at each working volume fraction (1%, 2%, and 5% *v/v*) or without CFS, and further incubated at 25 °C. The inhibition rate was calculated according to the equation [24], and the assay was performed 3 times with 5 replicates. The experimental kiwifruits were divided into five groups and soaked in different concentrations of CFS for 5 min. Punched holes about 2 mm in diameter and 3 mm in depth were created in the equatorial part of the treated kiwifruit with a sterile inoculation spike and inoculated with mycelial discs. The inoculation holes were wrapped in sealing film, and the kiwifruits were incubated under 22–25 °C and 80–90% humidity. The area of rotten spots was calculated with the cross method after 3 and 5 days of affection. The experiment was replicated three times with 12 fruits per treatment group.

2.5. Ultrastructural Morphology of Pathogens Treated with the CFS

Mycelial discs (5 mm diameter) from one-week-old cultures of *B. dothidea* were added to culture flasks containing 30 mL of PDB medium and incubated for 48 h at 28 °C on a shaker at 120 rpm. Then, CFS with final volume fractions of 0%, 1%, 2%, and 5% was added to the culture. Mycelia were taken at 0 h, 24 h, and 48 h, then washed three times with 0.05 M PBS (pH 7.2) buffer, fixed with 2.5% (*v/v*) glutaraldehyde (Solarbio, Beijing, China) for 24 h and then dehydrated with graded ethanol (30%, 50%, 70%, and 90%, *v/v*) and freeze-dried as described. The structural features of the mycelium were then observed using a scanning electron microscope (SEM) (Hitachi S-4800, JAPAN) [25].

2.6. Effect of the CFS on Cell Death of Pathogenic Mycelium

Fresh mycelium samples treated in step 2.4 were stained using the trypan blue staining kit (Sangon Biotech, Shanghai, China) and observed with a light microscope. Each group was treated with three samples and the experiment was repeated three times.

2.7. Determination of the Integrity of Cell Membranes

A few mycelia treated in step 2.4 were incubated with 5 mM propidium iodide (PI) for 20 min at 30 °C and washed twice with 0.05 M pH = 7.4 PBS buffer to remove the stain. Each processing sample was poked and dispersed, then observed with a laser confocal microscope (ZEISS LSM710, Germany). The green light was excited at 405 ± 10 nm, and the red light was absorbed at 633 ± 10 nm [26].

2.8. Determination of Ergosterol, Determination of the Leakage of Cellular Contents

Fresh mycelium samples treated in step 2.4 were taken and rinsed twice with 0.05 M PBS (pH 7.2) buffer to remove residual media. Each sample was placed in a drying oven at 80 °C, dried to a constant mass, ground to a powder with liquid nitrogen, weighed at 0.5 g of the dry powder in a test tube with anhydrous ethanol (1:10, m/V), mixed thoroughly for 10 min, and centrifuged at 3500 rpm for 10 min. The supernatant was filtered through a 0.22 μm microporous membrane, and the ergosterol content was determined at 292 nm [27]. Each group was treated with three samples, and the experiment was repeated three times.

Nucleic acid and protein exudation were determined by reference to the methods of Bradford et al. with appropriate modifications [28]. Mycelial discs (5 mm diameter) of one-week-old *B. dothidea* were added to shake flasks containing 30 mL of PDB medium and incubated at 25 °C at 120 rpm. After 48 h of incubation, the mycelium was combined and washed with sterile distilled water. Subsequently, 0.5 g of mycelium was suspended in CFS solutions at final volume fractions of 0%, 2%, and 5% and further incubated for 24 and 48 h at 25 °C in a shaker at 120 rpm. After filtering the mycelial samples, the resulting filtrate of the culture was assayed with NANODROP 2000 to determine the concentration of nucleic acids and soluble proteins. Each group was treated with three samples, and the experiment was repeated three times.

2.9. The CFS induces Malondialdehyde (MDA) Production and Reactive Oxygen Species (ROS) Accumulation

The MDA content was determined using the thiobarbituric acid (TBA) method [18]. 2′,7′-dichlorodihydrofluorescein diacetate (H2DCFDA; Invitrogen, Eugene, OR, USA) was used to detect ROS accumulation. Mycelium samples treated in step 2.4 were resuspended in 0.05 M PBS (pH 7.2) and co-incubated for 20 min with 10 μM H2DCFDA, then washed three times with PBS to remove the dye. The ROS accumulation was observed using a confocal laser microscope (LSM800; Zeiss, Oberkochen, Germany) [29]. Each group was treated with three samples, and the experiment was repeated three times. Hydrogen peroxide (H_2O_2) content and superoxide anion ($O_2^{\cdot -}$) production rate were individually determined using commercial kits (Nanjing Jiancheng Institute of Biological Engineering, Nanjing, China).

2.10. Enzyme Activity Assay and Real-Time Fluorescence Quantitative PCR Assay

SOD, POD, and CAT enzyme activities were determined using commercial kits (Jiancheng Institute of Biological Engineering, Nanjing, China). Mycelium samples were collected at 0 h and 6 h according to the method for culturing pathogenic mycelium in step 2.4 and stored at 4 °C and −80 °C until further processing. The total RNA was extracted from mycelium using an RNA Extraction Reagent (Vazyme, R401-01, China), and the cDNA was obtained using a cDNA Synthesis kit (Vazyme, R323-01, China). Real-time qRT-PCR was performed in 10 μL reactions using ChamQTM SYBR® qPCR Master Mix (Vazyme, China) in a fluorescence quantitative PCR instrument (Bio-Rad CFX96). The PCR reaction conditions were as follows: one cycle at 95 °C with initial denaturation for 30 s; then 40 cycles of 15 s at 95 °C and 30 s at 60 °C; and a final cycle of 95 °C at 25 s, 60 °C at 60 s, and 95 °C at 15 s. The gene expression level was normalized by the internal reference gene *Actin* (WWBZ02000009.1) with the $2^{-\Delta\Delta Ct}$ method [30]. The specific primers for the genes evaluated are listed in Table 1. Each treatment consisted of three biological replicates, and the experiments were performed twice.

Table 1. Primers Used for the RT-qPCR Analysis.

Gene Name	Accession Number	Primer Name	Primer Sequences (5′→3′)
Nox1	WWBZ02000033.1	*BdNox1-F1*	CGAGTCGATATGGTTCCACAG
		BdNox1-R1	GCCTGGGTATAGTGAATCTGG
Nox1f	WWBZ02000073.1	*BdNox1f-F1*	TGGCATCTACCTTTTCGAGC
		BdNox1f-R1	CCAGGCTTGTACTTCATCGAG
Nox3	WWBZ02000020.1	*BdNox3-F1*	CGTTTCCAAGGTTATTCAGCAC
		BdNox3-R1	TTGTGAGAGTGAAAGGGTGG
Sod	WWBZ02000033.1	*BdSod-F1*	GTCGGTGACAACTCTGGC
		BdSod-R1	GGTAATGGATCACGATGCTG
Pod	WWBZ02000001.1	*BdPod-F1*	TCTCGGCTACAAGATCGGAG
		BdPod-R1	TCACAGAATCCGGCAAGC
Cat	WWBZ02000022.1	*BdCat-F1*	TATTCCTCAGGCACGATCTTG
		BdCat-R1	GTCTATTGAGAAGGGTCGGTTC
Actin	WWBZ02000009.1	*BdActin-F1*	GGTTCAACTACCACCTCAAGAATG
		BdActin-R1	GCCGTGGGCGTCAGAAAT

2.11. Effect of N-Acetylcysteine (NAC) on the Recovery of Mycelial Growth

2.11.1. Plate Recovery Experiments

Mycelial discs (5 mm diameter) of one-week-old *B. dothidea* cultures were placed in the center of the PDA discs, with one group containing final volume fractions of 0%, 2%, and 5% CFS and 10 mM NAC. The control group had an identical final volume fraction of CFS and equal amounts of sterile water. Colony diameter was measured after 5 days of inoculation [31].

2.11.2. Mycelial ROS Accumulation and Mycelial Cell Damage/Integrity Recovery Test

Mycelial discs (5 mm diameter) from one-week-old cultures of *B. dothidea* were added to shake flasks containing 30 mL of PDB medium and incubated at 25 °C at 120 rpm. After 2 days of incubation, the cultures were divided into two groups. The treatment group contained different fractions of CFS (0%, 2%, and 5%) and 10 mM NAC, while the control group contained the exact final volume fraction of CFS alone. The two group cultures were incubated at 28 °C and 120 rpm for 24 h. The incubation of mycelium was followed by ROS accumulation staining and the trypan blue staining described previously.

2.12. Statistical Analysis

Statistical analyses were performed using SPSS 23.0, plotted using GraphPad Prism8, analyzed using a one-way analysis of variance (ANOVA), and shown as standard errors of the mean ± least squares mean (SEM). A *p*-value < 0.05 was accepted as statistically significant according to Duncan's multiple polar difference test. Significant differences are indicated by letters, with different letters indicating significant differences.

3. Result

3.1. The CFS of Bacillus Subtilis BS-1 Inhibits Vegetative Growth and Pathogenicity of Botryosphaeria Dothidea

Mycelial discs (5 mm in diameter) of uniformly sized one-week-old *B. dothidea* cultures were inoculated on PDA plates containing different concentrations of BS-1 CFS to investigate the vegetative growth of *B. dothidea* affected by CFS. After 3 days of incubation, *B. dothidea* growth was suppressed by CFS in a concentration-dependent manner, with mycelial growth at about 66%, 26%, and 21% of the control's colony diameter under exposure to 1%, 2%, and 5% of CFS, respectively (Figure 1A,B). The growth inhibition properties of *B. dothidea* on the fifth day, similarly to that on the culture on day 3, were affected by the dosage mode of CFS. CFS-treated fruits with the inoculation of mycelial discs were stored at 25 °C to explore CFS's effects on the virulence of *B. dothidea*. Fruit with CFS treatment developed a substantially and significantly less rotten area than control fruit in a dose-dependent way (Figure 1C,D). Consequently, the optimal suppression was obtained with the 5% CFS treatment; the lesion area was reduced by more than half compared to the control group (Figure 1C,D). *B. subtilis* CFS vitality assays demonstrated dose-response inhibitory effects against *B. dothidea* in vitro and in vivo, showing a promising biocontrol result.

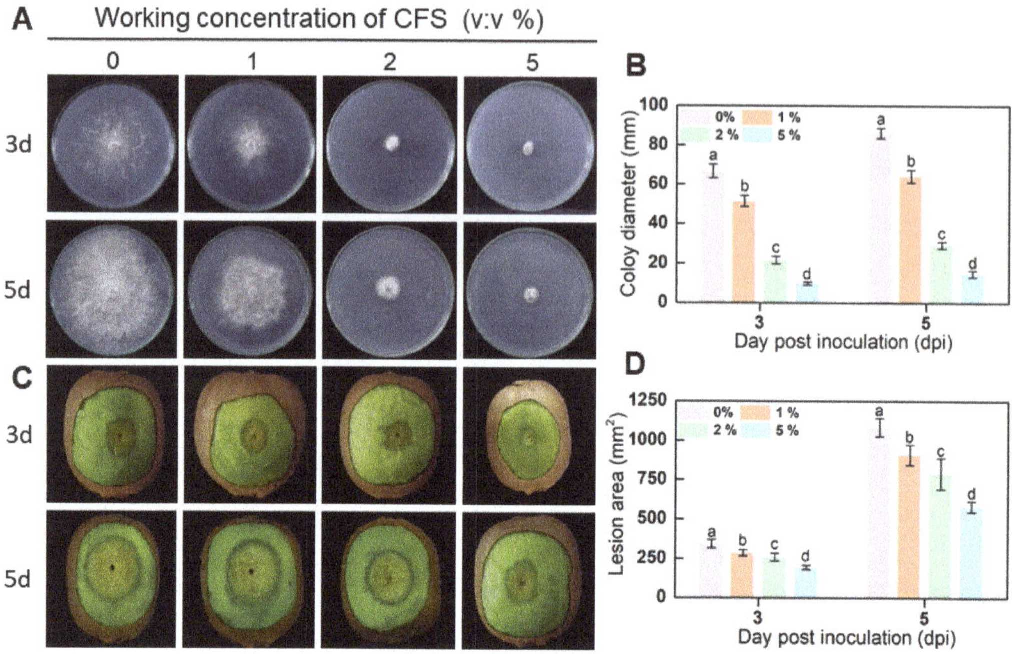

Figure 1. The CFS inhibited vegetative growth of *B. dothidea* on the PDA (potato dextrose agar) plate (**A**,**B**) and reduced the pathogenicity of *B. dothidea* in kiwifruit (**C**,**D**). (**A**) Images were recorded after 3 and 5 days of incubation. (**B**) Statistical analysis for colony diameters. (**C**) Photographs of diseased fruits were taken 3 and 5 days after treatment. (**D**) Lesion area on harvested kiwifruit was measured using Image J. Histograms with different letters indicate significant differences according to Duncan's multiple range tests at $p < 0.05$.

3.2. The CFS Damages and Ruptures the Hyphae of *B. dothidea*

The normal hyphal morphology and the integrity of hyphal structure are necessary for the growth and pathogenicity of *B. dothidea*. The morphological effects of CFS on growing hyphae of *B. dothidea* were examined by microscopy. In the absence of CFS, hyphae exhibited ordinary status over a 24 h incubation period. Compared with the control group, CFS strongly induced striking changes in the hyphal morphology, depending on the CFS concentration. The changes observed included the increasing formation of branches and swelling of hyphal cells (Figure 2A). Based on light microscopic observation, 2% CFS and 5% CFS had a pronounced effect on hyphae. The effects of 1% CFS on pathogenicity inhibition and hyphae morphology were not significantly different compared with the control (Figure 2A); the main focus was explored through the two concentrations of 2% and 5% CFS in further analysis. Then, the scanning electron microscopic analysis further demonstrated that the hyphae in the CFS treatment group were rough and shriveled, in contrast to the plump and smooth hyphae in the control group (Figure 2B). The scanning electron microscopic observation showed that the cell inclusions almost disappeared, but part of collapsed and shriveled hyphae was left (Figure 2B). These results indicated that the CFS primarily affected the normal formation and structure of growing hyphae, and the swelling hyphae cells were accompanied by excessive lateral branching and hyphae fracture that ultimately led to the destruction of the whole hyphae.

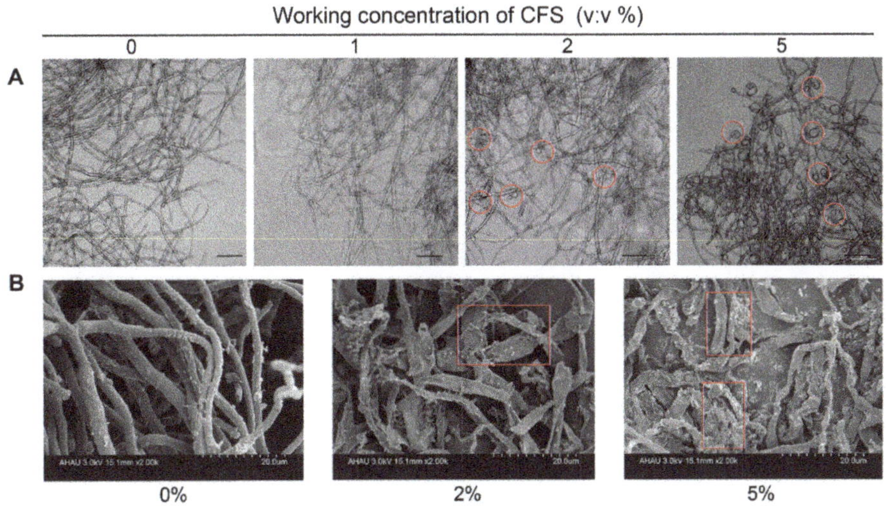

Figure 2. Effects of CFS on the morphological structure of *B. dothidea*. (**A**) Effects of CFS on the morphological structure of *B. dothidea* were tested in a PDB medium containing the indicated concentration of CFS. The hyphae were collected after 24 h of treatment, and images were recorded using a microscope. Black bars represent 50 μm. Hyphae swollen into spheres are marked by red circles. (**B**) Scanning electron micrographs of *B. dothidea* treated with CFS at different working concentrations. The shriveled hyphae are marked by red rectangles.

3.3. The CFS Destroys the Hyphal Cytomembrane and Cause Hyphae death of B. dothidea *In Vitro*

Shriveled hyphae were observed in the CFS treatment group, suggesting that CFS may damage the cell membrane of hyphae and cause hyphal death. To test this hypothesis, propidium iodide (PI) and trypan blue were employed to detect the membrane integrity of *B. dothidea* cells since the dead cells with incomplete cell membranes could be stained in red or blue by propidium iodide or trypan blue staining, respectively. As shown in Figure 3A, hyphae in the treatment group showed stronger red fluorescence, and 2% of CFS was sufficient to cause severe damage to the cell membranes of hyphae. These results were repeated on the same batch of samples following trypan blue staining. Similar results were obtained in this assay, and the CFS-treated hyphae showed a darker blue than the control (Figure 3B), further proving that CFS retains the ability to destroy cell membranes.

Cell membrane damage is often accompanied by changes in membrane substances, including the malondialdehyde (MDA), a product of membrane lipid peroxidation, and ergosterol, an essential component of fungal cell membranes. The content of MDA and ergosterol in hyphae revealed major changes between CFS-free and CFS-treated samples. The significantly increased MDA content was detected upon CFS presence at volume fractions of 2% and 5% (Figure 3C). Accordingly, ergosterol decreased significantly after CFS treatment and showed a dose- and time-dependent effect (Figure 3D). The peroxidation-damaged cell membrane may lead to the exudation of cell contents, mainly protein and nucleic acid. Therefore, the concentrations of the above two substances in the hyphal culture medium with or without CFS were further detected. Figure 3E,F show that the protein and nucleic acid content increased conspicuously depending on CFS concentration. These results indicate that the CFS destroyed the integrity of the hyphal membrane of *B. dothidea* and caused cell death by promoting the realization of hyphae membrane peroxidation damage.

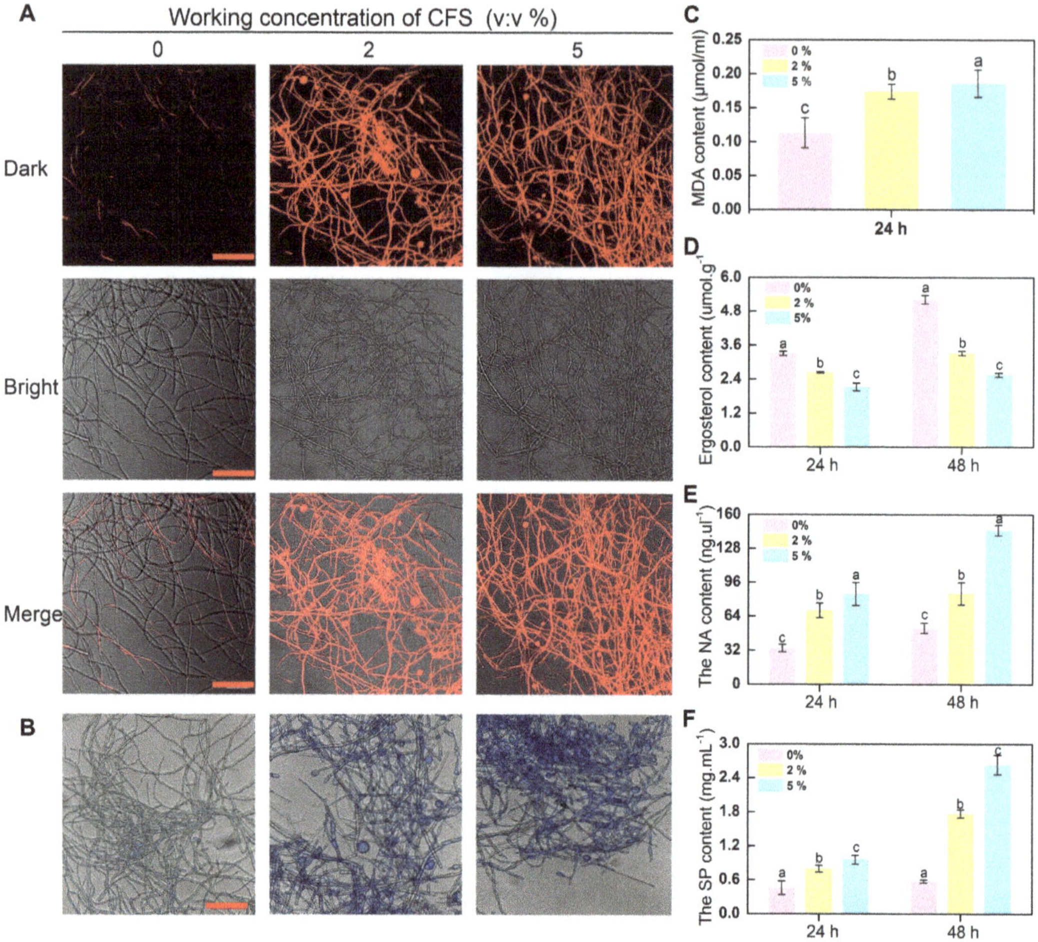

Figure 3. Injury efficacy of CFS on the hyphal cytomembrane of *B. dothidea*. (**A**) Detecting the integrity of the membrane of hyphal cells stained by propidium iodide (PI). The fluorescence signal was detected using a laser scanning confocal microscope at 405 nm excitation and 633 nm emission. (**B**) Detecting the integrity of membrane and activity of hyphal cells at 24 h after the application of CFS stained by trypan blue. The images were recorded using a microscope. The red bars represent 50 µm. (**C**,**D**) The content of MDA and ergosterol in hyphae after treatment with CFS for 24 h and 48 h. (**E**,**F**) Nucleic acid (NA) and Soluble protein (SP) contents were measured in a hyphal culture medium after different amounts of CFS t reatment at indicated period. Vertical bars represent standard deviations of the means. Columns followed by different letters are statistically different by the Duncan's multiple ranges ($p < 0.05$).

3.4. The CFS Promotes the Explosion of Reactive Oxygen Species in Hyphal Cells of B. dothidea

In the staining image of CFS-treated hyphae of *B. dothidea*, an incomplete cell membrane was observed. Primary cellular contents of hyphae were also found in the culture medium with CFS. These results lead us to the possible mechanism that CFS is active against fungal cell membrane integrity, probably due to provoking reactive oxygen species (ROS) burst. The cell-permeable fluorescence probe H2DCFDA was applied to detect intracellular ROS in hyphae. After 24 h of treatment with 2% CFS, most of the hyphae gave

ROS green fluorescence (Figure 4A). The same fluorescence signal accumulation was also observed in samples treated with a higher concentration of CFS, and such fluorescence increased in extent with an increasing CFS amount. The proliferation of ROS in hyphae triggered by CFS ultimately results in peroxidation of the cell membrane and collapse of the whole hyphae. ROS are chemically active free radical molecules, which could be as simple as molecules of superoxide or more complex molecules, such as hydrogen peroxide. The detection of superoxide and hydrogen peroxide are significant indicators of reactive oxygen species accumulation. As shown in Figure 4B, the mycelium treated with CFS had a significantly higher hydrogen peroxide content and superoxide anion production rate than the control mycelium. Moreover, the production rate depends on the increase in CFS concentration. This result indicates that CFS accelerated the rate of reactive oxygen species production and increased the accumulation of reactive oxygen species. In addition to ROS burst, upregulation of essential ROS synthesis genes (*Nox1*, *Nox1f*, *Nox3*) was observed universally at the very early phase of CFS processing, and the expression level was positively correlated with the concentration of CFS (Figure 4C). These results indicated that the ROS level raised by CFS might be closely related to the damage of hyphae.

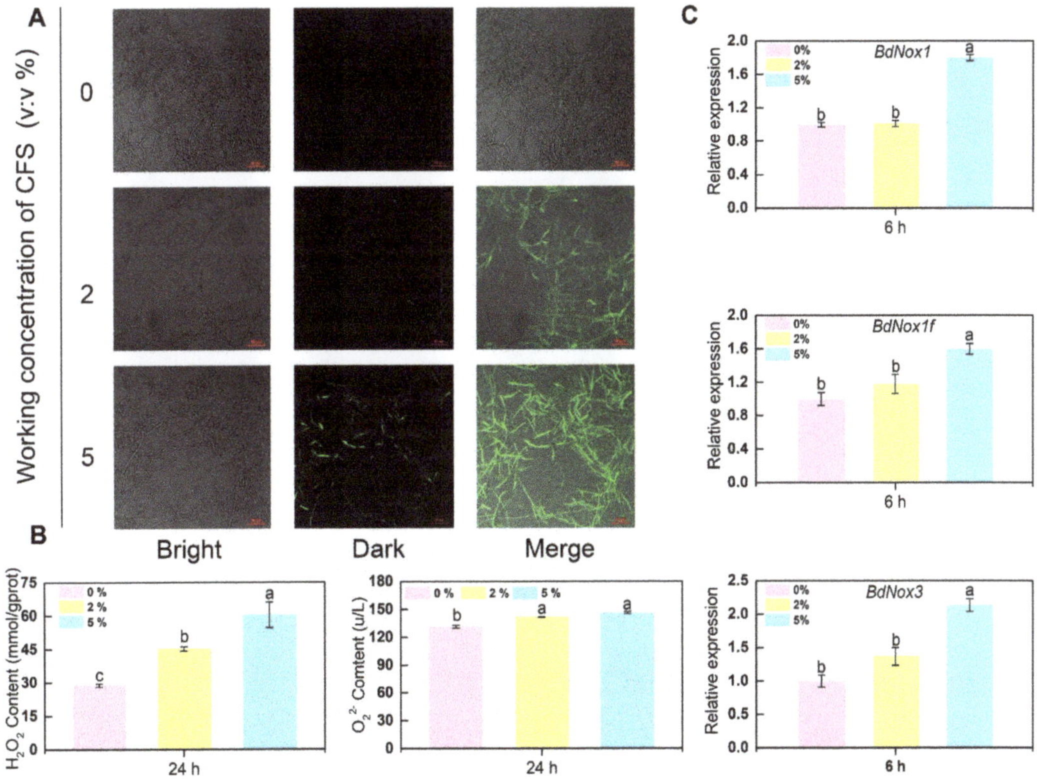

Figure 4. Effects of CFS on reactive oxygen species (ROS) accumulation in *B. dothidea*. (**A**) The images were recorded using confocal laser microscopy at 488 nm excitation and 520 nm emission. The hyphal cells treated with different concentrations of CFS were harvested after 24 h. Harvested hyphal cells were stained with the oxidant-sensitive probe, H2DCFDA, and then washed with PBS buffer for observation. (**B**) The contents of H2O2 and $O_2^{\cdot-}$ measured after 24 h treatment. (**C**) Gene expression levels of BdNox1, BdNox1f, and BdNox3 at 6 h after treatment. Different letters above the columns indicate significant differences within each group according to Duncan's multiple range test ($p < 0.05$).

3.5. Defense-Related Enzyme Activity and Corresponding Defense Enzyme Synthesis-Related Gene Expression in B. dothidea Mycelial Cells

The accumulation of ROS in cells is usually accompanied by the enhancement of scavenging capacity. The enzymatic scavenging system of ROS is largely dependent on the sequential cooperation of superoxide dismutase (SOD), catalase (CAT), and peroxidase (POD), which eventually catalyze ROS into harmless substances. Enzymatic activity was assessed in the hyphae with and without CFS after 24 h of culture. Indeed, a significant increase in scavenging enzyme activity was seen at diverse degrees (Figure 5A). The most pronounced increase in CAT and POD enzyme activity was observed in the group with 5% CFS application, about three times that of the control group. Moreover, the maximal activities of SOD among different CFS treatment groups exhibited similar behavior. The expression levels of genes encoding these ROS-scavenging enzymes were also assessed by qRT-PCR in the same samples. As Figure 5B shows, enzyme synthesis genes (*BdSOD*, *BdPOD*, *BdCAT*) were significantly upregulated at the early stage of CFS treatment. In conclusion, these results suggested that CFS induced ROS rapid accumulation in mycelial cells and leads to mycelial oxidative damage.

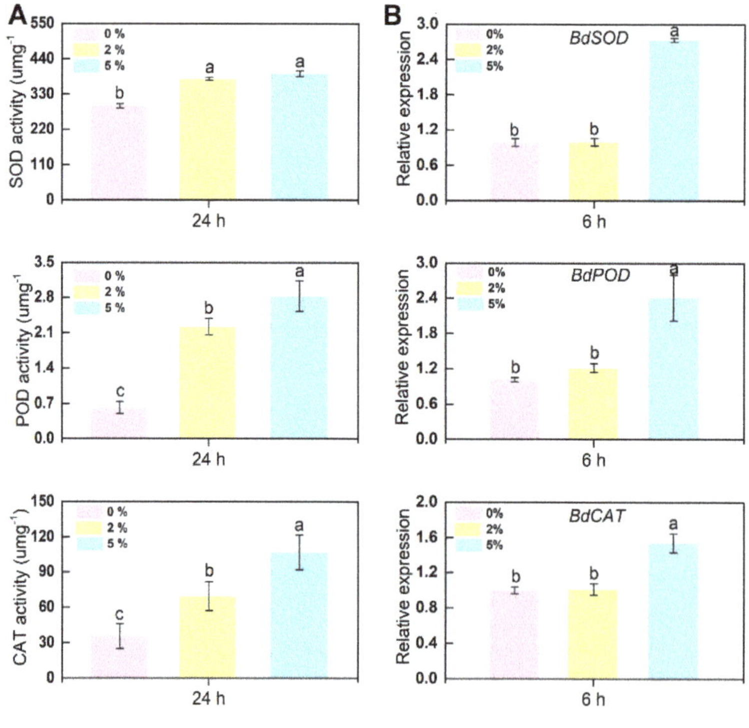

Figure 5. Effects of CFS on the enzyme activities involved in scavenging reactive oxygen species in *B. dothidea*. (**A**) The activities of ROS scavenging enzymes (POD, SOD, and CAT) were recorded in harvested hyphal cells. (**B**) Gene expression levels of *BdSOD*, *BdPOD*, and *BdCAT* at 6 h after treatment. Different letters above the columns indicate significant differences within each group according to Duncan's multiple range test ($p < 0.05$).

3.6. N-Acetylcysteine Alleviated the Burst of Reactive Oxygen Species in Hyphae of B. dothidea Induced by CFS

To quantify the oxidative accumulation of hyphae induced by CFS, an inhibition assay was performed. The broad-spectrum ROS inhibitor N-acetylcysteine (NAC) was resuspended in *B. dothidea* hyphae culture with CFS to evaluate the ROS fluorescent signal changes between the CFS-alone group and the CFS and NAC group. As shown in Figure 6, NAC effectively decreased the CFS-induced green fluorescence signal, indicating that NAC could remove various types of ROS generated by CFS. These data proved again from another perspective that CFS of *B. subtilis* induced a considerable accumulation of ROS in hyphae cells, leading to oxidative damage of hyphae.

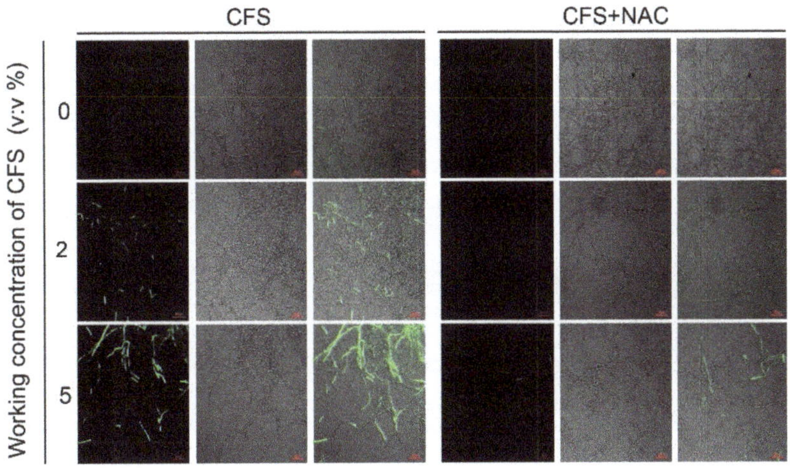

Figure 6. Effect of N-acetylcysteine (NAC) on the accumulation of reactive oxygen species (ROS) in hyphae. The culture medium was amended with different concentrations of CFS with or without 10 mM NAC, and the representative fluorescent pictures were recorded using confocal laser microscopy at 488 nm excitation and 520 nm emission.

3.7. *N-Acetylcysteine (NAC) Reduces Cell Death and Recovers the Weak Growth of B. dothidea Caused by CFS*

Considering the scavenging ability of NAC in terms of ROS, we next assessed the ability of NAC to restore the viability and growth of CFS-treated mycelium. Therefore, along with the original pathogen in vitro test, trypan blue staining test, and ROS accumulation staining test, NAC (10 mM) was applied to the CFS treatment groups to observe the status of NAC on hyphal cell death and mycelial growth. Hyphae swelling caused by CFS was hardly observed after NAC addition (Figure 7A), demonstrating that the changes in hyphal morphology induced by CFS were alleviated. In addition to the effect on hyphal morphology rescue, NAC was also effective in alleviating the hyphae death caused by CFS; the images clearly showed that the blue color of hyphae in NAC was much lighter than that of the CFS-treated group (Figure 7A). Referring to the NAC function on recovery hyphae growth under oxidative stress, we measured the CFS-damaged *B. dothidea* mycelia expansion capacity on the PDA plates with or without NAC. Figure 7B,C illustrate the image of the mycelia growth feature and statics of mycelia diameter after treatment with NAC. Regardless of the CFS concentration leveling the plate, the mycelia expansion became faster and more significant with NAC treatment, meaning NAC relieved the inhibition of CFS on mycelial growth. Taken together, these results suggested that the ROS scavenger NAC could rescue the growth disorder of mycelia treated with CFS by removing intracellular ROS.

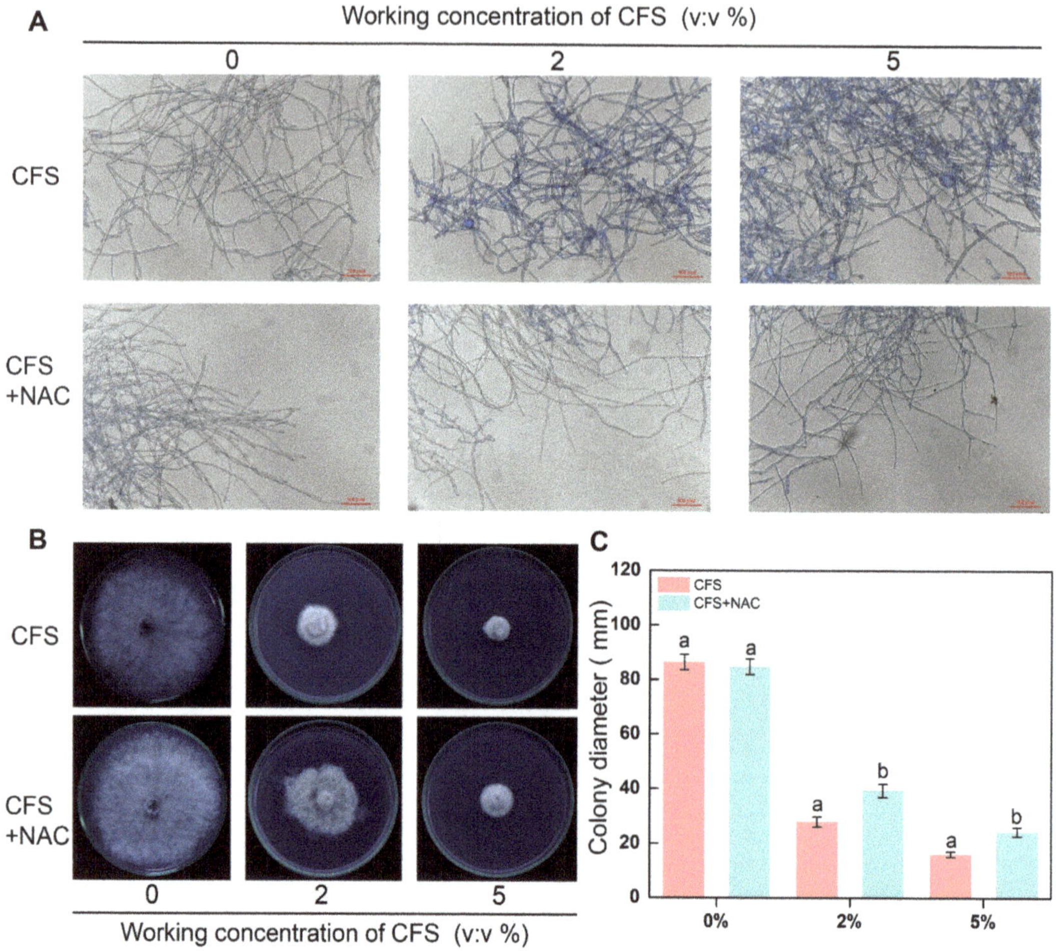

Figure 7. Effect of N-Acetylcysteine (NAC) on the activity and growth of mycelia treated with CFS. The cultures were amended with CFS (2% and 5% *v/v*) and 10 mM NAC, separately. (**A**) Hyphal cell death was determined by an accumulation of staining with trypan blue after the application of CFS and treatment with NAC for 24 h. Images were recorded using a microscope. (**B**) The mycelial growth diameter was measured, and images were recorded 5 days after inoculation. (**C**) Statistical analysis for colony diameters. Error bars indicate standard errors of the means of two repeated experiments. Treatments annotated with different letters were significantly different within a specific time point according to the Duncan's multiple range test. ($p < 0.05$).

4. Discussion

Fruit is an essential part of the diet, providing people with a rich source of vitamins and minerals, but postharvest losses of fruit amount to around 30% globally [32]. While biological control of postharvest fruit diseases has been subject to extensive attention and research, including the application of biocontrol microorganisms and natural substances from plants, this safe and effective control method remains a hot research topic. *Bacillus halotolerans* KLBC XJ-5 inhibits the growth of *Botrytis cinerea* mycelium and germination of conidia, effectively controlling postharvest strawberry grey mold [33]. Magnolol, a functional component of Magnolia officinalis, activates the autophagic activity of *B. cinerea* and substantially inhibits the mycelial growth of *B. cinerea* [23]. However, a great majority of postharvest biocontrol studies have concentrated on grey mold caused by the model fungus *B.cinerea*, with little attention on other diseases. In addition, studies on the biocontrol of pathogens have mainly focused on the effectiveness of control, but have seldom revealed their inhibition mechanism. These necessitate improvements to the research on the biocontrol of postharvest diseases.

Kiwifruit is a widely received fruit with a high vitamin C content and a rich nutritional profile. Nonetheless, the economic performance of kiwifruit is seriously affected by the infestation of several pathogens, with the soft rot of kiwifruit caused by *Botryosphaeria dothidea* being one of the leading causes affecting fruit production and quality [34]. This demonstrates the urgent need for effective, safe, and efficient disease control measures against the pathogen. Therefore, in this study, we preliminarily found that the *B. subtilis* strain BS-1 had a significant inhibitory effect on kiwifruit soft rot, and the interaction between BS-1 and the pathogen *B. dothidea* was further investigated to assess its potential biotic mechanisms against kiwifruit soft rot. Our results were similar to those of *B. subtilis* in primary postharvest diseases of ecological fruits such as strawberry, grapes, tomato, and cucumber, where treatment with the fermentation solution (CFS) of *B. subtilis* inhibited the expansion of the pathogen in vitro and weakened the pathogenicity of pathogens on the host in a dose-dependent manner [35–38]. These indicate that *B. subtilis* has a broad-spectrum inhibitory effect and is effective against kiwifruit soft rot. Previous studies have indicated three main mechanisms of biocontrol bacteria: competition for space and nutrients, production of inhibitory substances (cell-free supernatants, CFS and volatile substances), and induction of systemic resistance [39]. Considerable research suggests that inhibition substances releasing is the primary biocontrol method to reduce disease in postharvest fruit and vegetables [40,41]. The CFS of *B. subtilis* EA-CB0015 impedes the development of grey mold by inhibiting conidial germination, shoot tube growth, sporangial formation and tissue colonization of *Botrytis cinerea* [42]. The antifungal action of *B. subtilis* may be attributed to the ability to synthesize a wide range of bioactive secondary metabolites in CFS and volatile components against pathogens [43]. The secondary lipopeptide metabolites of *Bacillus amyloliquefaciens* S76-3 reduce the pathogenicity of *Fusarium graminearum* by disrupting mycelial cell structure and inhibiting mycelial growth and conidial germination [44]. Lipopeptide iturin A extracted from the CFS of *Bacillus subtilis* WL-2 induces cell membrane damage, oxidative stress, and mitochondrial dysfunction of Phytophthora infestans, leading to hyphal cell death [39]. Additionally, several investigations prove that the CFS of *B. subtilis* is highly resistant to a variety of stresses, including high temperature, UV irradiation and low temperature et al. [45,46]. Our previous study demonstrates that the CFS of *B. subtilis* strain BS-1 is tolerant to these stress conditions. Therefore, this study focused on the mechanism of CFS of BS-1 in the biocontrol of kiwifruit soft rot induced by *B. dothidea*.

The results we obtained showed that CFS inhibited the growth of *B. dothidea* in a dosage-dependent manner and effectively suppressed the development of soft rot in kiwifruit in both in vitro and antagonistic inhibition assays (Figure 1). Similarly, our data showed that the growth inhibition of *B. dothidea* was related to the morphological structure of the hyphae, which was damaged by the *B. subtilis* BS-1 CFS treatment, becoming rough, crumpled, and severely fractured (Figure 2B). Based on the observations, we hypothesized that the membrane permeability and intracellular contents of the treated hyphae cells were altered. In this work, the hypotheses were tentatively determined based on propidium iodide (PI) and trypan blue staining, the determination of malondialdehyde (MDA) and ergosterol, and the leakage of cellular contents (Figure 3A–F). It turned out that CFS could disrupt cellular membranes, leading to the leakage of nucleic acids and protein macromolecules from the cells and the alteration of cell permeability (Figure 3A–F). Changing the permeability of the hyphae cell membrane could conduct cell apoptosis or necrosis [43]. Therefore, our results suggest that CFS causes cell death in *B.dothidea* by interfering with the integrity of the cell membrane.

Reactive oxygen species (ROS) are one of the main factors that induce cellular damage and are a significant marker of oxidative stress [47]. When large amounts of ROS are produced, these will constantly attack and damage cellular tissues, exacerbating peroxidation of the plasma membrane, and leading to damage of the cellular membrane system, thereby destroying the structural and functional integrity of cellular tissues [48,49]. In our study, after 24 h of exposure to CFS, the hyphae ROS fluorescence signal became more robust with increasing CFS concentration. Hydrogen peroxide levels and superoxide anion production rates were significantly higher than those in controls, and ROS synthesis-associated genes (*BdNox1*, *BdNox1f*, *BdNox3*) were upregulated quickly after CFS treatment (Figure 4A–C). The NADPH oxidases (Nox) are *the* most prominent enzyme family with *the* primary function of highly regulated production of ROS, which involves *Nox* genes varying among fungi species [50]. Several biocontrol agents and biological metabolites have been identified as effective elicitors of *Nox* family genes, leading to the accumulation of ROS in pathogens and achieving the control effect. Treatment with *Bacillus velezensis* A4 CFS could upregulate the expression of *BcNoxA* and *BcNoxB* genes and stimulate the accumulation of ROS in the mycelium of *B. cinerea*, inducing oxidative damage and apoptosis in the mycelium [51]. The gene expression of the Nox complex subunit, *BcNoxB*, *BcNoxD*, and *BcNoxR* in the mycelia of *B. cinerea* was upregulated by methyl tartrate, and ROS in the cytoplasm of the mycelium were accumulated, eventually leading to mycelia death [52]. By comparing the homology of *BdNox1*, *BdNox1f*, and *BdNox3* with nox in *B. cinerea*, the homology of *BcnoxA* with *BdNox1* and *BdNox1f* is close to 70.75% and 67.45%, respectively, and the homology of *BcnoxB* with *BdNox3* is close to 66.48%. This suggests that the three *Nox* genes in *B. dothidea* perform functions similar to those genes of *noxA* and *noxB* in *B. cinerea*, as evidenced by our experimental results.

In order to scavenge excess ROS and reduce cell damage, the enzymes associated with ROS elimination (SOD, POD, CAT) and corresponding genes are induced and activated [17]. A 5% BS-1 CFS treatment for 6 h could significantly induce the expression of tested genes (*BdSOD*, *BdPOD*, *BdCAT*), while 2% CFS exhibited a mild induction effect. However, both 2% and 5% CFS resulted in significant increases in the activities of three enzymes in hyphae cells after 24 h of treatment (Figure 5). These results suggest that 2% CFS needs a relatively long time to induce gene expression, or CFS can induce the co-expression of several genes related to enzyme synthesis, which enhances the enzyme activity after CFS treatment at diverse amounts. N-acetylcysteine(NAC), an inhibitor of flavoenzymes such as NADPH oxidase, was introduced for further confirmation [53]; 10 mM NAC partially restored the growth of CFS-induced mycelium of *B. dothidea* and significantly weakened the accumulation of ROS and apoptosis of mycelial cells (Figures 6 and 7). Combined with the MDA content in the CFS condition (Figure 3C) [54], all the above results suggest that CFS of *B. subtilis* BS-1 can directly induce oxidative stress through crucial genes and disrupt the integrity of cell membranes, leading to cell collapse and death and reducing the pathogenicity of *B. dothidea*.

In addition, 1% CFS showed a positive inhibition on the pathogen in vitro and on kiwifruit soft rot in vivo (Figure 1), yet it could hardly affect the morphology and structure of the hyphae (Figure 2). These data implied that CFS of *B.subtilis* BS-1 could inhibit the activity of pathogens more than ROS-induced hyphal damage. Recent works have shown that *B. subtilis* MBI600 are able to stimulate the plant-induced systemic resistance (ISR) signal pathway in the plant, consequently inducing the disease resistance potential of plants [55]. In some cases, *B. subtilis* can reduce the growth of pathogens through nutritional competition, as *B. subtilis* (10-4, 26D) and *B. subtilis* KLBC BS6 [56,57]. The possible mechanism of CFS hampering soft rot caused by *B. dothidea* in our work might be the result of multiple causes. Further work is necessary to identify other potential pathways of CFS inhibition of *B. dothidea* to complete the panorama of the mechanism.

5. Conclusions

In conclusion, the cell-free supernatant (CFS) of *Bacillus subtilis* BS-1 exhibited a remarkable inhibitory effect on *B. dothidea* both in vitro and on kiwifruit. By inducing the accumulation of excessive ROS, CFS damaged the membrane of the hyphae cells, eventually leading to the death of hyphae cells, which threatened the normal growth and pathogenicity of *B. dothidea*. These results provide a theoretical basis for the subsequent biological application of *B. subtilis* and biocontrol of kiwifruit soft rot.

Author Contributions: Conceptualization, M.M., Y.F. and K.L.; methodology, Y.F., K.L., R.L. and J.G.; validation, M.M., Y.F. and K.L.; formal analysis, R.L., J.G., W.S. and H.Z.; investigation, Y.F., M.M., X.T. and Y.L.; resources, K.L; data curation, Y.F. and K.L.; writing—original draft preparation, Y.F.; writing—review and editing, M.M., X.T. and Y.L.; visualization, M.M., X.T. and Y.L.; supervision, M.M. and Y.L.; project administration, M.M.; funding acquisition, M.M. All authors have read and agreed to the published version of the manuscript.

Funding: This study was supported by the National Natural Science Foundation of China, grant numbers 31970345, 31972474 and 31701059. The Natural Science Foundation of Anhui Province, grant number 2208085MC66.

Institutional Review Board Statement: Not applicable.

Informed Consent Statement: Not applicable.

Data Availability Statement: Not applicable.

Conflicts of Interest: The authors declare no conflict of interest.

References

1. Maleki, S.; Maleki-Zanjani, B.; Gallego, P.P. Kiwifruit status in Iran: Management and production. *Acta Hortic.* **2018**, *12*, 39–44. [CrossRef]
2. Meng, F.B.; Gou, Z.Z.; Li, Y.C.; Zou, L.H.; Chen, W.J.; Liu, D.Y. The Efficiency of Lemon Essential Oil-Based Nanoemulsions on the Inhibition of Phomopsis sp. and Reduction of Postharvest Decay of Kiwifruit. *Foods* **2022**, *11*, 1510. [CrossRef] [PubMed]
3. Zhang, C.; Li, H.T.; Wu, X.M.; Su, Y.; Long, Y.Y. Co-Application of Tetramycin and Chitosan in Controlling Leaf Spot Disease of Kiwifruit and Enhancing Its Resistance, Photosynthesis, Quality and Amino Acids. *Biomolecules* **2022**, *12*, 500. [CrossRef] [PubMed]
4. Zhu, C.; Chou, O.; Lee, F.Y.; Wang, Z.; Barrow, C.; Dunshea, F.R.; Suleria, H.A.R. Characterization of Phenolics in Rejected Kiwifruit and Their Antioxidant Potential. *Processes* **2021**, *9*, 781. [CrossRef]
5. Li, L.; Pan, H.; Chen, M.Y.; Zhang, S.J.; Zhong, C.H. First Report of Nigrospora oryzae Causing Brown/Black Spot Disease of Kiwifruit in China. *Plant Dis.* **2018**, *102*, 243. [CrossRef]
6. Luo, A.; Bai, J.; Li, R.; Fang, Y.; Li, L.; Wang, D.; Zhang, L.; Liang, J.; Huang, T.; Kou, L. Effects of ozone treatment on the quality of kiwifruit during postharvest storage affected by Botrytis cinerea and Penicillium expansum. *J. Phytopathol.* **2019**, *167*, 470–478. [CrossRef]
7. Wang, Q.P.; Zhang, C.; Wu, X.M.; Long, Y.H.; Su, Y. Chitosan Augments Tetramycin against Soft Rot in Kiwifruit and Enhances Its Improvement for Kiwifruit Growth, Quality and Aroma. *Biomolecules* **2021**, *11*, 1257. [CrossRef]
8. Hur, J.; Oh, S.; Lim, K.; Jung, J.S.; Kim, J.; Koh, Y.J. Novel effects of TiO_2 photocatalytic ozonation on control of postharvest fungal spoilage of kiwifruit. *Postharvest Biol. Technol.* **2005**, *35*, 109–113. [CrossRef]
9. Li, W.; Long, Y.; Mo, F.; Shu, R.; Yin, X.; Wu, X.; Zhang, R.; Zhang, Z.; He, L.; Chen, T.; et al. Antifungal Activity and Biocontrol Mechanism of Fusicolla violacea J-1 against Soft Rot in Kiwifruit Caused by Alternaria alternata. *J. Fungi* **2021**, *7*, 937. [CrossRef]
10. Koh, Y.J.; Hur, J.; Jung, J.S. Postharvest fruit rots of kiwifruit (Actinidia deliciosa) in Korea. *N. Z. J. Crop Hortic. Sci.* **2005**, *33*, 303–310. [CrossRef]
11. Li, L.; Pan, H.; Chen, M.Y.; Zhang, S.J.; Zhong, C.H. Isolation and identification of pathogenic fungi causing postharvest fruit rot of kiwifruit (Actinidia chinensis) in China. *J. Phytopathol.* **2017**, *165*, 782–790. [CrossRef]
12. Zhang, C.; Long, Y.H.; Li, J.H.; Li, M.; Xing, D.K.; An, H.M.; Wu, X.M.; Wu, Y.Y. A Chitosan Composite Film Sprayed before Pathogen Infection Effectively Controls Postharvest Soft Rot in Kiwifruit. *Agronomy* **2020**, *10*, 265. [CrossRef]
13. Zhou, Y.; Gong, G.S.; Cui, Y.L.; Zhang, D.X.; Chang, X.L.; Hu, R.P.; Liu, N.; Sun, X.F. Identification of Botryosphaeriaceae Species Causing Kiwifruit Rot in Sichuan Province, China. *Plant Dis.* **2015**, *99*, 699–708. [CrossRef]
14. Wang, L.; Hou, H.; Zhou, Z.; Tu, H.; Yuan, H. Identification and Detection of Botryosphaeria dothidea from Kiwifruit (Actinidia chinensis) in China. *Plants* **2021**, *10*, 401. [CrossRef] [PubMed]
15. Liu, H.; Liu, M.; Zhu, H.; Zhong, J.; Liao, X.; Zhou, Q. Molecular characterization of a novel mitovirus from the plant-pathogenic fungus Botryosphaeria dothidea. *Arch. Virol.* **2021**, *166*, 633–637. [CrossRef] [PubMed]

16. Zhai, L.; Yang, M.; Zhang, M.; Hong, N.; Wang, G. Characterization of a Botybirnavirus Conferring Hypovirulence in the Phytopathogenic Fungus Botryosphaeria dothidea. *Viruses* **2019**, *11*, 266. [CrossRef]
17. Wang, Q.; Zhang, C.; Long, Y.; Wu, X.; Su, Y.; Lei, Y.; Ai, Q. Bioactivity and Control Efficacy of the Novel Antibiotic Tetramycin against Various Kiwifruit Diseases. *Antibiotics* **2021**, *10*, 589. [CrossRef]
18. Wang, F.; Zhang, R.; Yuan, Z.; Chen, P. Biological prevention and control of pitaya fruit canker disease using endophytic fungi isolated from papaya. *Arch. Microbiol.* **2021**, *203*, 4033–4040. [CrossRef]
19. Choi, H.W.; Ahsan, S.M. Biocontrol Activity of Aspergillus terreus ANU-301 against Two Distinct Plant Diseases, Tomato Fusarium Wilt and Potato Soft Rot. *Plant Pathol. J.* **2022**, *38*, 33–45. [CrossRef]
20. Ling, L.; Zhao, Y.; Tu, Y.; Yang, C.; Ma, W.; Feng, S.; Lu, L.; Zhang, J. The inhibitory effect of volatile organic compounds produced by Bacillus subtilis CL2 on pathogenic fungi of wolfberry. *J. Basic Microbiol.* **2021**, *61*, 110–121. [CrossRef]
21. Bu, S.W.; Munir, S.H.; He, P.F.; Li, Y.M.; Wu, Y.X.; Li, X.Y.; Kong, B.H.; He, P.B.; He, Y.Q. Bacillus subtilis L1-21 as a biocontrol agent for postharvest gray mold of tomato caused by Botrytis cinerea. *Biol. Control* **2021**, *157*, 104568. [CrossRef]
22. Hong, H.A.; Huang, J.M.; Khaneja, R.; Hiep, L.V.; Urdaci, M.C.; Cutting, S.M. The safety of Bacillus subtilis and Bacillus indicus as food probiotics. *J. Appl. Microbiol.* **2008**, *105*, 510–520. [CrossRef] [PubMed]
23. Cui, X.M.; Ma, D.Y.; Liu, X.Y.; Zhang, Z.Q.; Li, B.Q.; Xu, Y.; Chen, T.; Tian, S.P. Magnolol inhibits gray mold on postharvest fruit by inducing autophagic activity of Botrytis cinerea. *Postharvest Biol. Technol.* **2021**, *180*, 111596. [CrossRef]
24. Ben Khedher, S.; Kilani-Feki, O.; Dammak, M.; Jabnoun-Khiareddine, H.; Daami-Remadi, M.; Tounsi, S. Efficacy of Bacillus subtilis V26 as a biological control agent against Rhizoctonia solani on potato. *Comptes Rendus Biol.* **2015**, *338*, 784–792. [CrossRef]
25. Yan, Y.F.; Yang, C.J.; Shang, X.F.; Zhao, Z.M.; Liu, Y.Q.; Zhou, R.; Liu, H.; Wu, T.L.; Zhao, W.B.; Wang, Y.L.; et al. Bioassay-guided isolation of two antifungal compounds from Magnolia officinalis, and the mechanism of action of honokiol. *Pestic. Biochem. Physiol.* **2020**, *170*, 104705. [CrossRef]
26. Shu, C.; Zhao, H.; Jiao, W.; Liu, B.; Cao, J.; Jiang, W. Antifungal efficacy of ursolic acid in control of Alternaria alternata causing black spot rot on apple fruit and possible mechanisms involved. *Sci. Hortic.* **2019**, *256*, 108636. [CrossRef]
27. Engelking, B.; Flessa, H.; Joergensen, R.G. Shifts in amino sugar and ergosterol contents after addition of sucrose and cellulose to soil. *Soil Biol. Biochem.* **2007**, *39*, 2111–2118. [CrossRef]
28. Cai, J.H.; Chen, J.; Lu, G.B.; Zhao, Y.M.; Tian, S.P.; Qin, G.Z. Control of brown rot on jujube and peach fruits by trisodium phosphate. *Postharvest Biol. Technol.* **2015**, *99*, 93–98. [CrossRef]
29. Kai, K.; Bi, W.L.; Sui, Y.; Hua, C.Y.; Liu, Y.S.; Zhang, D.F. Curcumin inhibits Diaporthe phaseolorum and reduces postharvest decay in kiwifruit. *Sci. Hortic.* **2020**, *259*, 108860. [CrossRef]
30. Li, D.; Zhang, X.; Li, L.; Aghdam, M.S.; Wei, X.; Liu, J.; Xu, Y.; Luo, Z. Elevated CO(2) delayed the chlorophyll degradation and anthocyanin accumulation in postharvest strawberry fruit. *Food Chem.* **2019**, *285*, 163–170. [CrossRef]
31. Dou, Y.; Routledge, M.N.; Gong, Y.Y.; Godana, E.A.; Dhanasekaran, S.; Yang, Q.Y.; Zhang, X.Y.; Zhang, H.Y. Efficacy of epsilon-poly-L-lysine inhibition of postharvest blue mold in apples and potential mechanisms. *Postharvest Biol. Technol.* **2021**, *171*, 111346. [CrossRef]
32. Meitha, K.; Pramesti, Y.; Suhandono, S. Reactive Oxygen Species and Antioxidants in Postharvest Vegetables and Fruits. *Int. J. Food Sci.* **2020**, *2020*, 8817778. [CrossRef]
33. Wang, F.; Xiao, J.; Zhang, Y.; Li, R.; Liu, L.; Deng, J. Biocontrol ability and action mechanism of Bacillus halotolerans against Botrytis cinerea causing grey mould in postharvest strawberry fruit. *Postharvest Biol. Technol.* **2021**, *174*, 111456. [CrossRef]
34. Hua, C.Y.; Kai, K.; Bi, W.L.; Shi, W.; Liu, Y.S.; Zhang, D.F. Curcumin Induces Oxidative Stress in Botrytis cinerea, Resulting in a Reduction in Gray Mold Decay in Kiwifruit. *J. Agric. Food Chem.* **2019**, *67*, 7968–7976. [CrossRef] [PubMed]
35. Punja, Z.K.; Rodriguez, G.; Tirajoh, A. Effects of Bacillus subtilis strain QST 713 and storage temperatures on post-harvest disease development on greenhouse tomatoes. *Crop Prot.* **2016**, *84*, 98–104. [CrossRef]
36. Alijani, Z.; Amini, J.; Ashengroph, M.; Bahramnejad, B.; Mozafari, A.A. Biocontrol of strawberry anthracnose disease caused by Colletotrichum nymphaeae using Bacillus atrophaeus strain DM6120 with multiple mechanisms. *Trop. Plant Pathol.* **2021**, *47*, 245–259. [CrossRef]
37. Jiang, C.; Shi, J.; Liu, Y.; Zhu, C. Inhibition of Aspergillus carbonarius and fungal contamination in table grapes using Bacillus subtilis. *Food Control* **2014**, *35*, 41–48. [CrossRef]
38. Suryawanshi, K.T.; Sawant, I.S.; Sawant, S.D.; Shabeer, T.P.A.; Saha, S.; Pudale, A.; Dantre, R.K. Field evaluation of the bio-efficacy of Bacillus subtilis DR-39 formulation for enhancing pesticide degradation in grapes and optimisation of application dose. *Indian Phytopathol.* **2018**, *71*, 571–577. [CrossRef]
39. Wang, Y.Y.; Zhang, C.Y.; Wu, L.F.; Wang, L.; Gao, W.B.; Jiang, J.Z.; Wu, Y.Q. Inhibitory effect of Bacillus subtilis WL-2 and its IturinA lipopeptides against Phytophthora infestans. *bioRxiv* **2019**. [CrossRef]
40. Li, W.S.; Yuan, S.Z.; Li, Q.Q.; Sang, W.N.; Cao, J.K.; Jiang, W.B. Methyl p-coumarate inhibits black spot rot on jujube fruit through membrane damage and oxidative stress against Alternaria alternata. *Postharvest Biol. Technol.* **2018**, *145*, 230–238. [CrossRef]
41. Song, M.; Yun, H.Y.; Kim, Y.H. Antagonistic Bacillus species as a biological control of ginseng root rot caused by Fusarium cf. incarnatum. *J. Ginseng Res.* **2014**, *38*, 136–145. [CrossRef] [PubMed]
42. Arroyave-Toro, J.J.; Mosquera, S.; Villegas-Escobar, V. Biocontrol activity of Bacillus subtilis EA-CB0015 cells and lipopeptides against postharvest fungal pathogens. *Biol. Control* **2017**, *114*, 195–200. [CrossRef]

43. Palazzini, J.M.; Dunlap, C.A.; Bowman, M.J.; Chulze, S.N. Bacillus velezensis RC 218 as a biocontrol agent to reduce Fusarium head blight and deoxynivalenol accumulation: Genome sequencing and secondary metabolite cluster profiles. *Microbiol. Res.* **2016**, *192*, 30–36. [CrossRef]
44. Gong, A.D.; Li, H.P.; Yuan, Q.S.; Song, X.S.; Yao, W.; He, W.J.; Zhang, J.B.; Liao, Y.C. Antagonistic mechanism of iturin A and plipastatin A from Bacillus amyloliquefaciens S76-3 from wheat spikes against Fusarium graminearum. *PLoS ONE* **2015**, *10*, e0116871. [CrossRef] [PubMed]
45. Jin, M.; Yang, C.; Wei, L.; Cui, L.; Osei, R.; Cai, F.; Ma, T.; Wang, Y. Transcriptome analysis of Stipa purpurea interacted with endophytic Bacillus subtilis in response to temperature and ultraviolet stress. *Plant Growth Regul.* **2022**, *98*, 205–218. [CrossRef]
46. Xiang, Y.-Z.; Li, X.-Y.; Zheng, H.-L.; Chen, J.-Y.; Lin, L.-B.; Zhang, Q.-L. Purification and antibacterial properties of a novel bacteriocin against Escherichia coli from Bacillus subtilis isolated from blueberry ferments. *LWT* **2021**, *146*, 111456. [CrossRef]
47. Haddoudi, I.; Cabrefiga, J.; Mora, I.l.; Mhadhbi, H.; Montesinos, E.; Mrabet, M. Biological control of Fusarium wilt caused by Fusarium equiseti in Vicia faba with broad spectrum antifungal plant-associated Bacillus spp. *Biol. Control* **2021**, *160*, 104671. [CrossRef]
48. Ma, Q.; Xu, Y.Q.; Li, D.; Wu, X.W.; Zhang, X.C.; Chen, Y.P.; Li, L.; Luo, Z.S. Potential epigenetic regulation of RNA 5′-terminal NAD decapping associated with cellular energy status of postharvest Fragaria × ananassa in response to Botrytis cinerea invasion. *Postharvest Biol. Technol.* **2022**, *186*, 111840. [CrossRef]
49. Xu, Y.; Charles, M.T.; Luo, Z.; Mimee, B.; Tong, Z.; Veronneau, P.Y.; Roussel, D.; Rolland, D. Ultraviolet-C priming of strawberry leaves against subsequent Mycosphaerella fragariae infection involves the action of reactive oxygen species, plant hormones, and terpenes. *Plant Cell Environ.* **2019**, *42*, 815–831. [CrossRef] [PubMed]
50. Rossi, F.R.; Krapp, A.R.; Bisaro, F.; Maiale, S.J.; Pieckenstain, F.L.; Carrillo, N. Reactive oxygen species generated in chloroplasts contribute to tobacco leaf infection by the necrotrophic fungus Botrytis cinerea. *Plant J.* **2017**, *92*, 761–773. [CrossRef]
51. Zhao, H.L.; Liu, K.; Fan, Y.Z.; Cao, J.C.; Li, H.H.; Song, W.; Liu, Y.S.; Miao, M. Cell-free supernatant of Bacillus velezensis suppresses mycelial growth and reduces virulence of Botrytis cinerea by inducing oxidative stress. *Front. Microbiol.* **2022**, *13*, 980022. [CrossRef]
52. Ji, D.C.; Chen, T.; Ma, D.Y.; Liu, J.L.; Xu, Y.; Tian, S.P. Inhibitory effects of methyl thujate on mycelial growth of Botrytis cinerea and possible mechanisms. *Postharvest Biol. Technol.* **2018**, *142*, 46–54. [CrossRef]
53. Kim, H.J. Exploitation of reactive oxygen species by fungi: Roles in host-fungus interaction and fungal development. *J. Microbiol. Biotechnol.* **2014**, *24*, 1455–1463. [CrossRef]
54. Oloyede, H.O.B.; Ajiboye, H.O.; Salawu, M.O.; Ajiboye, T.O. Influence of oxidative stress on the antibacterial activity of betulin, betulinic acid and ursolic acid. *Microb. Pathog.* **2017**, *111*, 338–344. [CrossRef]
55. Samaras, A.; Roumeliotis, E.; Ntasiou, P.; Karaoglanidis, G. Bacillus subtilis MBI600 Promotes Growth of Tomato Plants and Induces Systemic Resistance Contributing to the Control of Soilborne Pathogens. *Plants* **2021**, *10*, 1113. [CrossRef]
56. Lastochkina, O.; Baymiev, A.; Shayahmetova, A.; Garshina, D.; Koryakov, I.; Shpirnaya, I.; Pusenkova, L.; Mardanshin, I.; Kasnak, C.; Palamutoglu, R. Effects of Endophytic Bacillus Subtilis and Salicylic Acid on Postharvest Diseases (Phytophthora infestans, Fusarium oxysporum) Development in Stored Potato Tubers. *Plants* **2020**, *9*, 76. [CrossRef]
57. Lu, Y.Y.; Ma, D.T.; He, X.; Wang, F.; Liu, Y.; Jiao, J.Y.; Deng, J. Bacillus subtilis KLBC BS6 induces resistance and defence-related response against Botrytis cinerea in blueberry fruit. *Physiol. Mol. Plant Pathol.* **2021**, *114*, 101599. [CrossRef]

Journal of *Fungi*

MDPI

Article

Isolation of Main Pathogens Causing Postharvest Disease in Fresh *Codonopsis pilosula* during Different Storage Stages and Ozone Control against Disease and Mycotoxin Accumulation

Bingyu Lv [1,†], Xi Yang [1,†], Huali Xue [1,*], Mina Nan [1], Yuan Zhang [1], Zhiguang Liu [1], Yang Bi [2] and Suqin Shang [3]

[1] College of Science, Gansu Agricultural University, Lanzhou 730070, China
[2] College of Food Science and Engineering, Gansu Agricultural University, Lanzhou 730070, China
[3] College of Plant Protection, Gansu Agricultural University, Lanzhou 730070, China
* Correspondence: xuehual@gsau.edu.cn; Tel.: +86-181-8954-1078
† These authors contributed equally to this work.

Abstract: *Codonopsis pilosula* is an important Chinese herbal medicine. However, fresh *C. pilosula* is prone to decay during storage due to microorganism infections, seriously affecting the medicinal value and even causing mycotoxin accumulation. Therefore, it is necessary to study the pathogens present and develop efficient control strategies to mitigate their detrimental effects on the herbs during storage. In this study, fresh *C. pilosula* was collected from Min County in Gansu Province, China. The natural disease symptoms were observed during different storage stages, and the pathogens causing *C. pilosula* postharvest decay were isolated from the infected fresh *C. pilosula*. Morphological and molecular identification were performed, and pathogenicity was tested using Koch's postulates. In addition, the control of ozone was examined against the isolates and mycotoxin accumulation. The results indicated that the naturally occurring symptom increased progressively with the extension of storage time. The mucor rot caused by *Mucor* was first observed on day 7, followed by root rot caused by *Fusarium* on day 14. Blue mold disease caused by *Penicillum expansum* was detected as the most serious postharvest disease on day 28. Pink rot disease caused by *Trichothecium roseum* was observed on day 56. Moreover, ozone treatment significantly decreased the development of postharvest disease and inhibited the accumulations of patulin, deoxynivalenol, 15-Acetyl-deoxynivalenol, and HT-2 toxin.

Keywords: *Codonopsis pilosula*; postharvest disease; morphological and molecular biology identification; ozone; mycotoxin accumulation

1. Introduction

Codonopsis pilosula, a perennial herb of the *Campanulaceae* family, is a traditional Chinese herb that is rich in bioactive compounds including polyacetylene, polyene, flavonoids, lignans, alkaloids, coumarins, terpenes, steroids, organic acids, and polysaccharides [1]. *C. pilosula* is widely used in the therapy of hyperlipidemia, asthma, bronchitis, tuberculosis, and dyspepsia [2], and also plays a key role in boosting the body's immunity [3], reducing blood glucose [4], promoting hematopoiesis [5], protecting the cardiovascular system [6], repairing damaged nerve cells, and regulating gastrointestinal function [7].

Owing to its unique climate, Gansu Province in China has more than 2000 years of history in cultivating *C. pilosula*. With its high commercial value and profitability, the cultivation of *C. pilosula* is continuously expanding. However, some soil-borne fungi, bacteria, and nematodes have adapted to the local host plants and natural environment in the region, thus, resulting in serious diseases that severely affect the yield and quality of *C. pilosula*. Research to date has focused on field disease of *C. pilosula*. For instance, Zhao et al. [8] suggested that *Fusarium oxysporum* is the predominant pathogen causing root rot of *C. pilosula* in Dingxi County, Gansu Province. Using molecular technology, Yu et al. [9] isolated and identified five pathogens in China: *Puccinia Campanumoeae* Pat causing rust disease, *Helicobasidium mompa* Tanaka causing violet root rot, *Sphaerotheca Codonopisi (Golov.)* Z.Y. Zhao causing powdery mildew, *Fusarium oxysporum* Schl. causing root rot, and *Septoria codonopsidis* Ziling causing blight of *Codonopsis* tangshen in Chongqing. Based on morphology and molecular characteristics, Wang et al. [10] isolated and identified *Septoria codonopsidis Ziling* causing leaf spot of *C. pilosula* in Gansu Province, while Chen et al. [11] proposed that *Botrytis cinerea* is the typical fungus causing gray mold in *C. pilosula*.

Modern research indicates that fresh Chinese herbs have higher active ingredients and pharmacological activities than those of dry products [12]. For instance, the contents of flavonoids, saponins, and polysaccharides in fresh *Astragalus* were 1.5 times more than those in dry products [13]. Because of this, the market for freshly harvested Chinese herbs has grown rapidly and is today more significant and well known than before. Unfortunately, the postharvest losses due to the disease of fresh Chinese herbs are quite severe. Nevertheless, no reports are available regarding the prevalence of pathogens during this period. In general, the herb is usually harvested in late autumn and allowed to dry naturally on the field for more than two months before being transported to the traders. The freshly harvested *C. pilosula* must be sufficiently dried (about 16% of water content), then stored for 3–6 years. If *C. pilosula* does not reach an optimum level of dry matter, its abundance storage of fat, starch, protein, sugar, and other organic substances may favor the development of some latent pathogens and ultimately result in serious postharvest disease. Fungal infections are thought to result in yearly losses of 15% to 25% [14]. Such postharvest losses cause enormous economic damage to *C. pilosula* processing industries, and more importantly, the product may completely lose its medicinal value, becoming contaminated with mycotoxins that have carcinogenic, teratogenic, and mutagenic toxicity [15]. It is therefore important to systematically study the postharvest diseases of freshly harvested *C. pilosula*, identify the pathogens present at various storage stages, and develop an efficient control strategy to mitigate their effects.

Ozone, an antioxidant compound, has been widely applied to manage the postharvest decay of fruits and vegetables. The efficacy of ozone in controlling postharvest disease is mainly ascribed to its strong inhibitory activity on pathogenic fungi. However, there are no reports on the influence of ozone on fresh Chinese herb medicine postharvest disease and mycotoxin production.

In this study, we collected fresh *C. pilosula* from Min County in Gansu Province, China, investigated the disease development of freshly harvested *C. pilosula* during storage, then isolated and identified pathogens causing *C. pilosula* postharvest disease based on morphological and molecular biological techniques during different storage stages. Finally, we examined the influence of ozone on fresh *C. pilosula* postharvest disease and mycotoxin production.

2. Materials and Methods

2.1. Sample

Samples of freshly harvested *C. pilosula* (cv. *Mindang*) were obtained from Min County (location 35° N and 104° E) in Gansu Province, China. Sample roots of similar size and without obvious pest or mechanical damage were selected in October 2020, and transported to the Chemical Biology Laboratory, College of Science, Gansu Agricultural University within 24 h after bagging, and stored at room temperature for further analysis.

2.2. Disease Development of Freshly Harvested C. pilosula during Different Storage Stages

Freshly harvested *C. pilosula* samples (without any processing treatment such as washing and drying) were placed directly in plastic bags (20 °C, 50 % RH) in darkness for 7, 14, 21, 28, 42, and 56 days. Subsequently, the naturally occurring symptoms were observed and described. Different pathogens cause different disease symptoms, for example, in the early stages of root rot, small brown spots appear on the surface of the lower fibrous or lateral roots, with mild decay. As the disease expands, it gradually spreads to the main roots. The roots gradually decay from the bottom upwards and become dark brown and waterlogged [9]. Three replicates were included in each treatment, each treatment consisted of 50 samples of C. *pilosulas*, with a total of 900 samples being included in the whole experiment (50 samples × 3 replicates × 6 time points).

2.3. *Isolation and Purification of Pathogens Causing Disease of Freshly Harvested C. pilosula during Different Storage Stages*

Pathogens were isolated and purified from C. *pilosula* based on typical decay symptoms (the appearance of mycelium and spores on the surface of C. *pilosula*) [16]. Fragments (5 mm × 5 mm) showing typical disease symptoms during different storage stages were excised with sterilized scalpels from the junction of the healthy and diseased tissue, and the fragments were disinfected using 1% NaClO for 3 min, and then rinsed with sterile water three times to remove any NaClO residue. The treated fragments were placed onto potato dextrose agar (PDA) medium and cultured in darkness at 25 °C for 5 to 7 d. Single colonies were picked and transferred to a new PDA medium with a puncher. After 4 to 5 cycles of purification, a single purified colony was obtained.

2.4. *Identification of Pathogens Causing Disease in Freshly Harvested C. pilosula during Different Storage Stages*

2.4.1. Morphological Identification

The preliminary identification was conducted based on colony morphology and macro and microconidia characteristics [17]. The spore suspension (1×10^6 spores/mL) was inoculated on a PDA plate, and after drying, a cover glass was inserted into the PDA medium at an angle of 45°, and then cultured at 25 °C for 2 to 5 d. Subsequently, to observe colony morphology and pigment production, a 2 μL spore suspension was inoculated centrally in the PDA medium at 25 °C for 7 to 9 d. Colony morphology and pigment production were recorded on day 9. Mycelia were transferred using the copper pick-in method [18]. All samples were sprayed gold with an ion sputtering apparatus (MSP-1S, Shenzhen Research Precision Instrument Co., Ltd., Shenzhen, China) under 220 V and 40 mA, then the morphology of the spores was observed using a scanning electron microscope (SEM) (JSM-5910LV, Japanese electronics company, Tokyo, Japan).

2.4.2. Molecular Identification

The morphologically identified pathogens were further subjected to molecular confirmation according to a previously used method [19] with some modifications. The pathogens were grown on PDA medium for 3 to 9 d, then the DNA of pathogenic mycelia was extracted using the CTAB method [20], and PCR amplification was performed using the primers *ITS1* (5′-TCCGTAGGTGAACCTGCGG-3′) and *ITS4* (5′-TCCTCCGCTTATTGATATGC-3′) for the 9 isolates. For isolates 21-1, 28-1, 28-2, and 56-1, the primers *Bt2a* (5′-GGTAACCAAATCGGTGCTGCTTTC-3′) and *Bt2b* (5′-ACCCTCAGTGTAGTGACCCTTGGC-3′) were employed for further PCR amplification. For the *Fusarium* strains (14-1, 14-2, 14-3), the special primers of *EF1*(5′-ATGGGTAAGGA (A/G)GACAAGAC-3) and *EF2*(5′-GGA(G/A)GTACCAGT(G/C)ATCATGTT-3′) were adopted [21,22]. Subsequently, the amplified products were detected using 2% agarose gel electrophoresis and clear bands were obtained.

The PCR amplification procedure was as follows: pre-denaturation at 94 °C for 5 min, denaturation at 94 °C for 10 s, annealing at 53 °C for 10 s, extension at 72 °C for 30 s, 3 cycles, and holding at 72 °C for 5 min. The amplified products were subject to electrophoresis with 2% agarose gel. The amplified fragments were sequenced by Beijing Bomede Biotechnology Co., Ltd., and the sequences were aligned to the NCBI (https://www.ncbi.nlm.nih.gov/ (accessed on 8 July 2021)) using BLAST for homology analysis. A phylogenetic tree was constructed by MEGA7.0 software (Molecular Evolutionary Genomics Analysis Version 7) using the neighbor-joining method.

2.5. Pathogenicity Test

The healthy and freshly harvested *C. pilosula* was thoroughly washed with tap water to remove any adhering soil and air dried. The herbs were surface disinfected with 0.1% NaClO for 15 min, rinsed with sterile water three times to remove excessive NaClO, and air dried again. Spore suspensions (1×10^6 spores/mL) of the above isolated pathogens were prepared [18] and artificially inoculated by spraying onto the surface of the fresh and healthy *C. pilosula*. The control group was inoculated with sterile water. After natural drying, the inoculated *C. pilosula* were kept in darkness (20 °C, 50% RH). After an incubation period of 28 days, the disease symptoms were recorded, and different pathogens led to different disease symptoms [23]. The pathogens were re-isolated again from the *C. pilosula*'s infected roots and stem, and their similarity to the original isolates' morphological characteristics was verified. The casual isolates that obeyed the criteria specified by Koch's postulates were conducted for further study.

2.6. Effect of Ozone Treatment on the Development of C. pilosula Postharvest Disease

Healthy *C. pilosula* were treated and inoculated according to the above Section 2.5. Sterile water spraying inoculation was regarded as control. Gaseous ozone was generated by the OSAN ozone generator (Aoshan Huanbao Technology Industry Co., Ltd., Dalian, China). The concentration of ozone (2 mg L^{-1}) was adjusted using an ozone detector [18]. Ozone fumigation was performed in a closed transparent bag (80 cm long × 60 cm wide) (25 °C, relative humidity 75%), and the inoculated samples were subjected to ozone treatment for 1 and 2 h each day, respectively. The treatment continued for 7 days. Subsequently, the treated tissue was stored for 56 days in plastic bags (22 ± 2 °C, 75–80%). The disease development was evaluated by statistically determining the disease index and natural incidence according to the report by Sha et al. [24] with minor modifications (Table 1). Each treatment contained three replicates, and one replicate included 50 samples.

Table 1. Disease classification standard.

Disease Rate	Symptom
0	No disease
1	The diseased area accounts for 1–5% of the total area of *C. pilosula*
2	The diseased area accounts for 6–25% of the total area of *C. pilosula*
3	The diseased area accounts for 25–50% of the total area of *C. pilosula*
4	The diseased area accounts for 51–75% of the total area of *C. pilosula*
5	The diseased area accounts for 76–100% of the total area of *C. pilosula*

Disease Index = [sum (class frequency × score of rating class)/[(Total number of plants) × (maximal disease index)] × 100;

Disease incidence = (Number of the infected plants/the number of plants sampled) × 100;

Class frequency: the number of diseased plants at each rate;

Score of rating class: the diseased value for each rate.

2.7. Effect of Ozone Treatment on the Mycotoxin Accumulation in the Rotten Tissue

For mycotoxins analysis, the samples treated by ozone fumigation were collected and the rotten tissue was excised from the diseased root and immediately kept in liquid nitrogen and stored at −80 °C until mycotoxin analysis. Each treatment contained three replicates, and one replicate included 50 samples.

A 5.0 g frozen sample was ground in liquid nitrogen, then was transferred to a 50 mL centrifuge tube with extraction solvent to extract mycotoxin. For different kinds of mycotoxin, various purification and detection methods were employed, patulin (PAT) purification and detection was carried out as described by the method [25]; trichothecene purification and detection was performed according to the method [26].

2.8. Statistical Analysis

The data of disease index, disease incidence, and mycotoxin accumulation were expressed as the means [standard error (±SE)] using Duncan's multiple range tests. The experiment of the effect of ozone on postharvest disease and mycotoxin accumulation of *C. pilosula* was performed at least three times. Statistical analyses were performed using SPSS v.17.0 (SPSS, Inc., Chicago, IL, USA), and Duncan's multiple range test ($p < 0.05$) was employed in this study.

3. Results

3.1. Disease Development of Freshly Harvested C. pilosula during Different Storage Stages

With the extension of storage time, the disease development of freshly harvested *C. pilosula* was more severe (Figure 1). When stored for 7 days, mild symptoms of disease were observed, and a small number of white hyphae and moldy spores were found on the main root and lateral root of *C. pilosula*. When stored for 14 days, hyphae were gradually diffused, and more white hyphae and moldy spores covered the root of *C. pilosula*. After 21 days of storage, colonies expanded and their color changed slightly from white to light green, with vigorous mycelium growth. When stored for 28 days, some yellow, red, and green hyphae appeared on the surface of the *C. pilosula*. When stored for 42 days, multiple colonies were distributed over the surface of *C. pilosula* and the tissue was damaged. After 56 days of storage, *C. pilosula* was seriously diseased, and the tissue was wrinkled, soft, and even rotten.

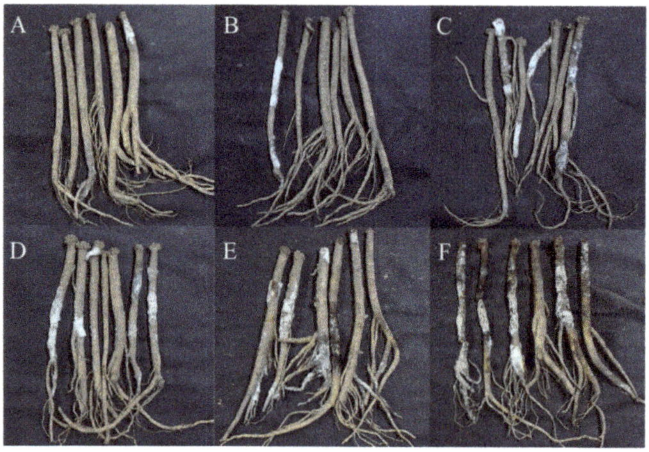

Figure 1. Disease development and naturally occurring symptoms of the fresh *C. pilosula* during different storage stages after harvest. (**A**) 7 d; (**B**) 14 d; (**C**) 21 d; (**D**) 28 d; (**E**) 42 d; (**F**) 56 d.

3.2. The Isolation of Pathogen from C. pilosula with Postharvest Disease during Different Storage Stages

During the whole storage period, a total of nine isolates of pathogenic fungi were isolated and purified. The nine different fungi were obtained using repeated plate streaking during different storage stages. For example, on the 7th day of storage, 2 isolates were obtained, which were named 7-1 and 7-2, respectively; on the 14th day of storage, 5 isolates were obtained, among them, 3 isolates were newly obtained, which were named 14-1, 14-2, and 14-3, respectively; on the 21st day of storage, 6 isolates were obtained, among them, a new isolate was named 21-1; on the 28th day of storage, 8 isolates were obtained, and two of them were newly isolated, named 28-1 and 28-2. There were no new isolates on the 42nd and 49th day. On the 56th day of storage, 9 isolates were obtained, and there was a new isolate named 56-1.

3.3. Morphological Identification of Pathogens at Different Storage Stages

After isolation and purification, the pathogens were cultivated on a PDA medium. Colony morphology, spore morphology, and sporangiophore morphology of the nine isolates of pathogenic fungi were observed.

For isolate 7-1, the mycelia grew fast at a rate of 24.57 mm/d on the PDA plate. The colony texture was flocculent with a white color, irregular edge, and no pigment (Figure 2A). Spores were spherical or nearly spherical in size at 4.8–5.6 μm × 5.0–6.2 μm (Figure 3A); sporangia were spherical with a diameter of 78–81 μm (Figure 4A). For isolate 7-2, the mycelia grew fast at a rate of 21.5 mm/d on the PDA plate. The colony texture was flocculent and white or gray-white with dense mycelia, a neat edge, and no pigment (Figure 2B); spores were spherical or subspherical and approximately 3.9–4.5 μm × 4.3–5.9 μm in size (Figure 3B); sporangia located at the end of hyphae were spherical, with a diameter of 59–63 μm (Figure 4B) (Table 2).

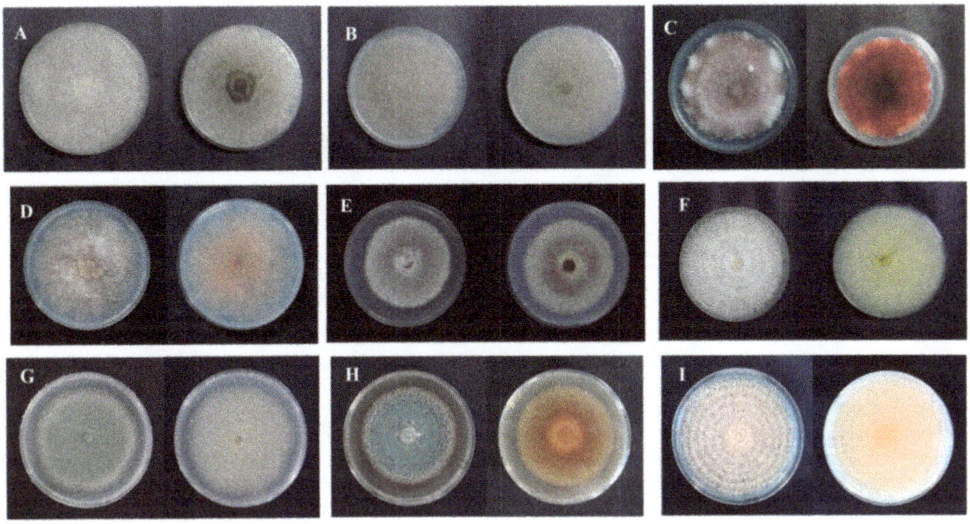

Figure 2. Colony morphology of isolates from fresh *C. pilosula* with postharvest disease during different storage stages. (**A**) *Actinomucor elegans*; (**B**) *Mucor hiemalis*; (**C**) *Fusarium acuminatum*; (**D**) *Fusarium equiseti*; (**E**) *Fusarium oxysporum*; (**F**) *Clonostachys rosea*; (**G**) *Penicillium expansum*; (**H**) *Penicillium aurantiogriseum*; (**I**) *Trichothecium roseum*.

For isolate 14-1, the mycelia grew at a rate of 7.39 mm/d on the PDA plate. The colony was round or nearly round with a fluffy texture, neat or wavy edges, and a thick and dense mycelium; the secreted pigment was cinnamon colored, the surface and back of the edge were rose-red (Figure 2C). The number of conidia was small, the shape was oval or fusiform with a size of 2.4–3.0 μm × 3.0–3.6 μm (Figure 3C); 2 to 4 septa were observed, and the conidiophores had branches (Figure 4C). For isolate 14-2, the mycelia grew at a speed of 4.75 mm/d on the PDA plate. The colony texture was villous or cotton floc, with a light pink color, and the colony edge was light pink or white, with dense mycelia in the middle of the colony and sparse mycelia at the border of the colony (Figure 2D); conidia were oblong or fusiform and 2.1–3.2 μm × 2.8–3.5 μm in size (Figure 3D); conidiophores were erect and branched (Figure 4D). For isolate 14-3, the mycelia grew at a speed of 4.57 mm/d on the PDA plate. The colony was round with neat edges, pale purple or dark purple in color, and the edges were white or grayish-white with flocculent or fluffy hyphae (Figure 2E). The conidia with 3 to 4 septa were elongated oval, oval, or slightly curved, measuring 2.2–3.3 μm × 2.8–3.5 μm (Figure 3E); conidiophores had branches (Figure 4E).

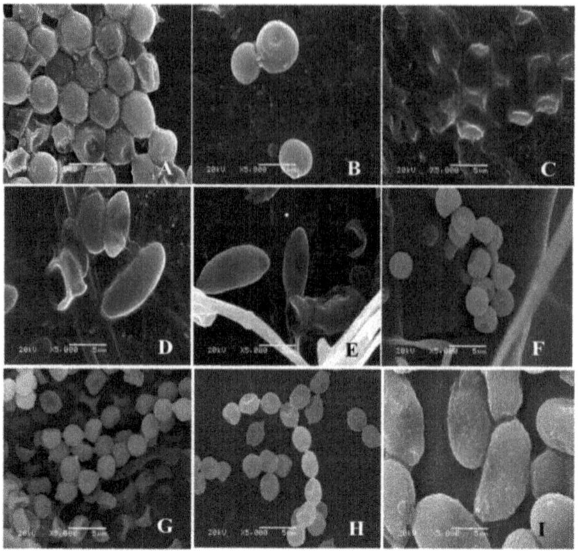

Figure 3. Morphology of conidium of isolates from fresh *C. pilosula* with postharvest disease during different storage stages. (**A**) *Actinomucor elegans*; (**B**) *Mucor hiemalis*; (**C**) *Fusarium acuminatum*; (**D**) *Fusarium equiseti*; (**E**) *Fusarium oxysporum*; (**F**) *Clonostachys rosea*; (**G**) *Penicillium expansum*; (**H**) *Penicillium aurantiogriseum*; (**I**) *Trichothecium roseum*.

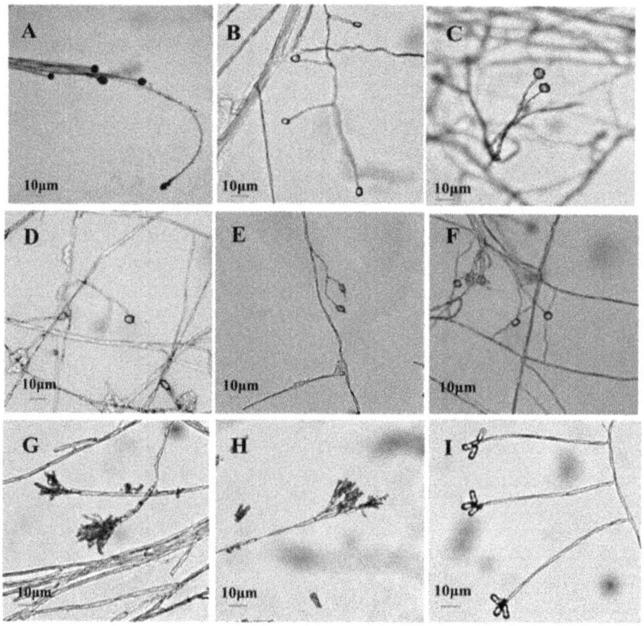

Figure 4. Morphology of conidiophore of isolates from *C. pilosula* with postharvest disease during different storage stages. (**A**) *Actinomucor elegans*; (**B**) *Mucor hiemalis*; (**C**) *Fusarium acuminatum*; (**D**) *Fusarium equiseti*; (**E**) *Fusarium oxysporum*; (**F**) *Clonostachys rosea*; (**G**) *Penicillium expansum*; (**H**) *Penicillium aurantiogriseum*; (**I**) *Trichothecium roseum*.

Table 2. Morphological characteristics of pathogens isolated at different storage periods.

	Colony Morphology				Microscopic Morphology	
Strain Number	Front Color	Back Color	Texture	Growth Speed (mm/d)	Conidium	Conidiophore
7-1	White	White	Flocculent	24.57	Spherical or near spherical	Sporangium
7-2	Off-white	White	Flocculent	21.5	Spherical or near spherical	Sporangium
14-1	Rose-red	Rose-red	Fluffiness	7.39	Spindle shaped	Erect and branch
14-2	Light pink	Light pink	Fluffiness	4.75	Spindle shaped	Erect and branch
14-3	Dark purple	Dark purple	Fluffiness	4.57	Spindle shaped	Erect and branch
21-1	White	Light yellow	Fluffiness	4.29	Spherical or near spherical	Erect and branch
28-1	Grey-green	White	Powdery or grainy	8.71	Spherical or flat spherical	Erect, broom
28-2	Blue-green	Tan	Grainy	6.50	Spherical or flat spherical Pear	Erect, broom
56-1	Orange	Orange	Grainy	10.14	shaped or obovate	Erect

For isolate 21-1, the mycelia grew at a speed of 4.29 mm/d on the PDA plate. The colony's front color was white, while the back was pale yellow; mycelia were creeping and looser with a neat edge (Figure 2F); conidia were spherical or nearly spherical with a size of 2.3–3.2 μm × 2.8–3.5 μm (Figure 3F), and the conidiophores had branches (Figure 4F).

For isolate 28-1, the mycelia grew at a rate of 8.71 mm/d on the PDA plate. The front color of the colony was gray-green with dense granular texture, and the reverse was white (Figure 2G); conidia were striate, with colorless monospores, globose or oblate, with a size of 2.1–3.4 μm × 3.4–4.2 μm (Figure 3G); conidiophores were erect, septate, colorless, apex branched, and broom shaped (Figure 4G). For isolate 28-2, the mycelia grew with a speed of 6.50 mm/d on the PDA plate. A blue-green, and yellowish-brown with white areas were observed at the front and back of the colony, respectively. The center of the colony had protrusions and was slightly flocculent at the central surface and the texture was fluffy and powdery, forming several granular concentric rings (Figure 2H); Conidia were striate, spherical or ellipsoidal, with a size of 2.4–2.8 μm × 2.8–3.6 μm (Figure 3H); conidiophores were erect, septate, colorless, apex branched, and broom shaped (Figure 4H).

For isolate 56-1, the mycelia grew with a speed of 10.14 mm/d on the PDA plate. An orange-pink, and pink with annual rings was observed on the front and back of the colony (Figure 2I); the conidia were loosely gathered at the top of the mycelium, and were obovate or pear shaped, colorless, bicellular, with a size of 5.8–7.0 μm × 10–14 μm (Figure 3I); conidiophores were erect, colorless, with an ultimate swelling, and 2.0–3.5 μm × 100–160 μm in size (Figure 4I).

3.4. Molecular Identification of Pathogens at Different Storage Stages

Based on the above morphological observation during different storage stages, the nine isolates were further characterized by molecular biological technology based on the method of Gloria et al. [27]. The length of *ITS* primer amplified sequences were: 638 base pair (bp) of 7-1, 625 bp of 7-2, 538 bp of 14-1, 516 bp of 14-2, 533 bp of 14-3, 545 bp and 563 bp of 28-1, 761 bp of 28-2, and 589 bp of 56-1 (Figure 5A). The lengths of the sequences amplified by *TEF* primers were 695 bp for 14-1, 686 bp for 14-2, and 685 bp for 14-3 (Figure 5B). The lengths of the sequences amplified by *Bt* primers were 331 bp for 21-1, 448 bp for 28-1, 452 bp for 28-2, and 332 bp for 56-1 (Figure 5C). The sequences of the nine pathogens were searched using BLAST in NCBI, and the appropriate sequences were selected. The phylogenetic trees of *ITS*, *TUB*, and *TEF* were, respectively, constructed by MEGA7 (Figure 6).

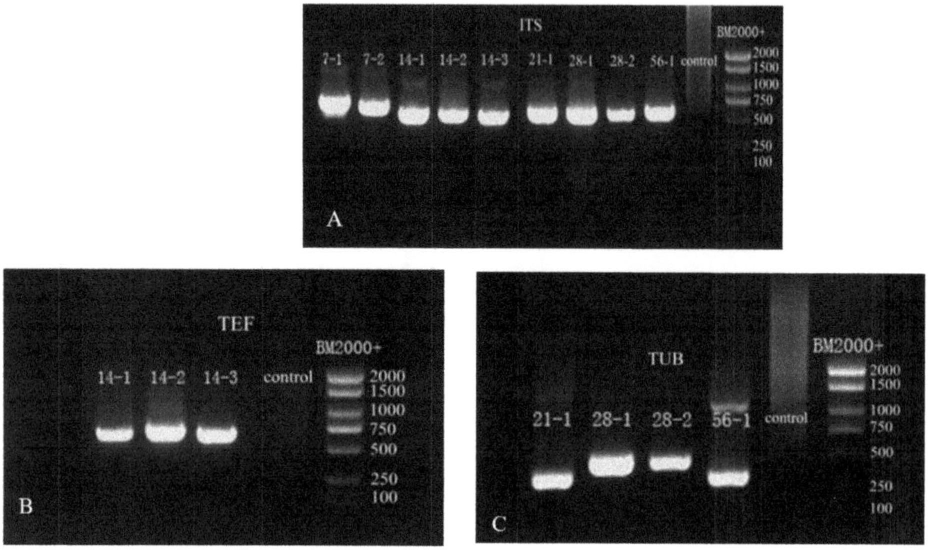

Figure 5. Gel electrophoresis images of PCR amplification products. (**A**) *ITS* gel electrophoresis images; (**B**) *TEF* gel electrophoresis images; (**C**) *TUB* gel electrophoresis images.

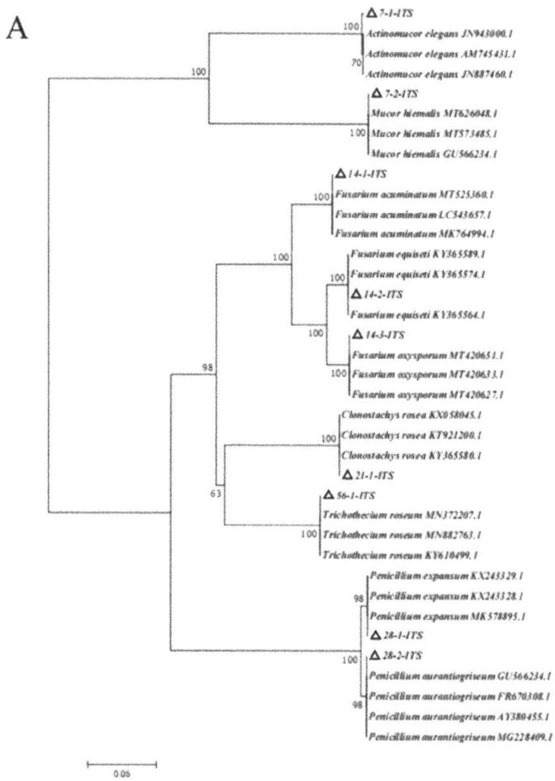

Figure 6. *Cont.*

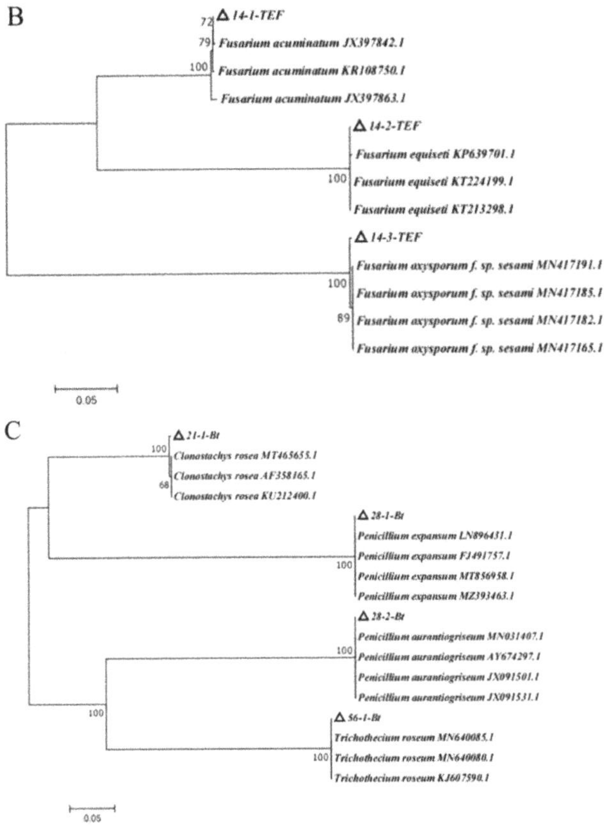

Figure 6. Phylogenetic tree of isolates based on analysis of different fungal genes. (**A**) *ITS* phylogenetic tree; (**B**) *TEF* phylogenetic tree; (**C**) *TUB* phylogenetic tree.

The phylogenetic tree of *ITS* analysis (Figure 6A) revealed that isolate 7-1 was closely related to the strains of *JN943000.1*, *AM745431.1*, and *JN887460.1*, which were in the same branch, with a homology of 100%. The combination of the morphology and biological characters, isolate 7-1, was identified as *Actinomucor elegans*. Isolate 7-2 shared 100% homology with the strains of *MT626048.1*, *MT573485.1*, and *GU566234.1*, which were located in the same branch; therefore, based on morphology and biological characters, isolate 7-2 was identified as *Mucor hiemalis*. The phylogenetic tree of *ITS* analysis showed that 14-1 shared 100% homology with *MT525360.1*, *LC543657.1*, and *MK764994.1* in the same branch; thus, 14-1 was preliminarily identified as *Fusarium acuminatum*. Isolate 14-2 shared 100% homology with *KY365589.1*, *KY365574.1*, and *KY365564.1* in the same branch; therefore, 14-2 was preliminarily identified as *Fusarium equiseti*. Isolate 14-3 was closely related to the strains of *MT420651.1*, *MT420633.1*, and *MT420627.1*, which were located in the same branch, with a homology of 100%; thus, 14-3 was preliminarily identified as *Fusarium oxysporum*. Isolate 21-1 was identified as *Clonostachys rosea* on the same branch as *KX058045.1*, *KT921200.1*, and *KY365580.1*, with a homology of 100%. Isolate 28-1 and *KX243329.1*, *KT243328.1*, *MK578895.1* were in the same branch, with a homology of 98%; therefore, combination morphology and biological characters, 28-1, was identified as *Penicillium expansum*. Isolate 28-2 and *GU566234.1*, *FR670308.1*, *AY380455.1*, and MG228409.1 were in the same branch, with a homology of 98%; therefore, based on the morphology and biological characters, isolate 28-2 was identified as *Penicillium aurantiogriseum*. Isolate 56-1 and *MN372207.1*, *MN882763.1*, and *KY610499.1* were in the same branch, with a homology of 100%; therefore, based on the morphology and biological features, isolate 56-1 was identified as *Trichothecium roseum* (Figure 6A).

Based on *ITS* phylogenetic tree analysis, 14-1, 14-2, and 14-3 were the *Fusarium* species. In order to more correctly characterize the *Fusarium* species, the special primer of *TEF* for *Fusarium* was employed, and the *TEF* phylogenetic tree was constructed (Figure 6B). *TEF* phylogenetic tree analysis suggested that isolate 14-1 was closely related to *JX397842.1*, *KR108750.1*, and *JX397863.1*, which were in the same branch, with a homology of 100%. Isolate 14-2 was related to *KP639701.0*, *KT224199.1*, and *KT213298.1*, which were located in the same branch, with a homology of 100%. Isolate 14-3 was related with *MN417191.1*, *MN417185.1*, *MN417182.1*, and *MN417165.1*, which were in the same branch, with a homology of 100%. With the combined analyses of *ITS* and *TEF* phylogenetic trees and morphology feature, isolates 14-1, 14-2, and 14-3 were, respectively, identified as *Fusarium acuminatum*, *Fusarium equiseti*, and *Fusarium oxysporum* (Figure 6A,B). According to the results of the *ITS* and *TUB* phylogenetic tree, isolates 21-1, 28-1, 28-2, and 56-1 were identified as *Clonostachys rosea*, *Penicillium expansum*, *Penicillium aurantiogriseum*, and *Trichothecium roseum*, respectively (Figure 6A,C).

3.5. Pathogenicity Test

Pathogenicity tests were used to verify the pathogenicity of the isolates causing postharvest disease according to Koch's postulates. During the whole incubation, for *C. pilosula* inoculated with 7-1 (*Actinomucor elegans*), dense white flocculent hyphae covered the main root on the 7th day, grew rapidly with time, and the expansion of colonies on the 28th day led to the decay of the epidermis of *C. pilosula* and leakage of sap (Figure 7A). For *C. pilosula* inoculated with 7-2 (*Mucor hiemalis*), gray flocculent hyphae appeared on the 7th day, grew rapidly on the 14th day, covering the whole tissue rotted on the 28th day (Figure 7B). For *C. pilosula* inoculated with 14-1 (*Fusarium acuminatum*): white dense mycelia were observed on the 7th day, the color of the dense mycelia gradually became rose-red on the 14th day; a rose-red glue was secreted and leaked from the *C. pilosula* tissue on 28th day (Figure 7C). For *C. pilosula* inoculated with 14-2 (*Fusarium equiseti*), there was less white velvet mycelial growth in the early stage, the colony became larger on the 14th day, and the color of *C. pilosula* epidermis turned pale pink or light yellow on the 28th day (Figure 7D). For *C. pilosula* inoculated with 14-3 (*Fusarium oxysporum*), there were white velvet colonies on the main root and lateral root on the 7th day, the color became light purple on the 14th day, and the color of *C. pilosula* epidermis turned dark purple on 28th day (Figure 7E). For *C. pilosula* inoculated with 21-1 (*Clonostachys rosea*), small white villous colonies were found on the surface of the tissue on the 7th day, the colony gradually expanded on the 14th day, and turned yellow and the tissue of *C. pilosula* produced yellowish glue on the 28th day (Figure 7F); For *C. pilosula* inoculated with 28-1 (*Penicillium expansum*), granular mold spots (2 to 3 mm in diameter) were observed on the surface of *C. pilosula* at the initial stage, a cluster of colonies gradually formed on the 14th day with a layer of blue powder on the surface; the plaque spread continuously, and the color of *C. pilosula* epidermis became gray and green, the tissue at the lesion developed soft rot and mildew on the 28th day (Figure 7G). For *C. pilosula* inoculated with 28-2 (*Penicillium aurantiogriseum*), small white granular mold spots appeared on the 7th day; the plaque expanded, and the number of colonies increased on the 14th day; the color of colonies turned blue and green, and *C. pilosula* developed serious mildew on the 28th day (Figure 7H).

For *C. pilosula* inoculated with 56-1 (*Trichothecium roseum*), white plaque with a diameter of 2 to 3 mm was initially observed on the 7th day; the orange-pink plaque expanded irregularly on the 14th day; the tissue was wrinkled with soft collapses and a dark color on the 28th day (Figure 7I). *C. pilosula* inoculated with the nine isolates had typical symptoms that were similar to the original natural symptoms. Based on the above typical and similar symptoms, the nine isolates were again identified through isolation, purification, and cultivation on PDA culture, and the same morphological and molecular biological characteristics were observed.

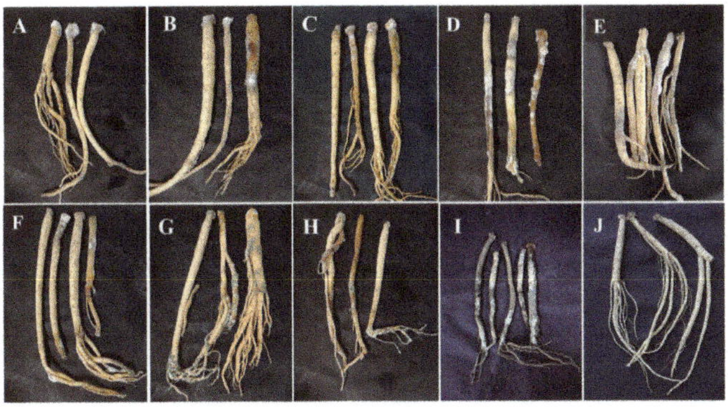

Figure 7. Pathogenicity testing of isolates from fresh *C. pilosula* with postharvest disease during different storage stages. (**A**) *Actinomucor elegans;* (**B**) *Mucor hiemalis;* (**C**) *Fusarium acuminatum;* (**D**) *Fusarium equiseti;* (**E**) *Fusarium oxysporum;* (**F**) *Clonostachys rosea;* (**G**) *Penicillium expansum;* (**H**) *Penicillium aurantiogriseum;* (**I**) *Trichothecium roseum.* (**J**) Control (no inoculation).

3.6. Ozone Fumigation Inhibited the Development of C. pilosula Postharvest Disease

The development of postharvest disease was effectively suppressed in *C. pilosula* inoculated with nine isolates after ozone fumigation treatment, and there was an ozone-exposure-time-dependent relationship with the inhibitory effect. For instance, the disease indexes in *C. pilosula* inoculated with *Actinomucor elegans* (7-1) after 1 and 2 h of ozone exposure were, respectively, 70% and 37% higher than those in control (Figure 8A). The disease incidences in *C. pilosula* inoculated with *F. acuminatum* (14-1) after 1 and 2 h ozone exposure were, respectively, 69% and 36% higher than those in control (Figure 8B). Similar results were obtained in the other isolates inoculated with *C. pilosula* after 1 and 2 h ozone exposure for 56 days of storage (Figure 8). It was obvious that with prolonged ozone treatment, the disease index and disease incidence dropped sharply.

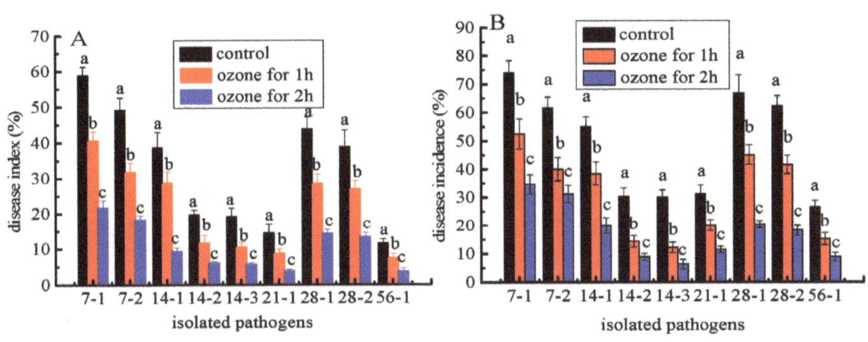

Figure 8. The effect of ozone treatment on disease index (**A**) and disease incidence (**B**) of fresh *C. pilosula* infected by the 9 isolates at 56 days of storage. (7-1: *Actinomucor elegans;* 7-2: *Mucor hiemalis;* 14-1: *Fusarium acuminatum;* 14-2: *Fusarium equiseti;* 14-3: *Fusarium oxysporum;* 21-1: *Clonostachys rosea;* 28-1: *Penicillium expansum;* 28-2: *Penicillium aurantiogriseum;* 56-1: *Trichothecium roseum*). The different letters indicate significant differences during the same storage period ($p < 0.05$).

3.7. Ozone Fumigation Inhibited the Mycotoxin Accumulation in the Rotten Tissue

More importantly, ozone treatment significantly inhibited mycotoxin accumulation in the rotten tissue of the inoculated *C. pilosula*. For instance, the content of patulin (PAT) of the rotten tissue in *C. pilosula* inoculated with *P. expansum* (28-1) was significantly ($p < 0.05$) inhibited by 38.9% and 53.0%, respectively, after 1 and 2 h of ozone exposure

(Figure 9B). Similarly, the concentrations of 15-ADON and HT-2 of the rotten tissue in *C. pilosula* inoculated with *F. acuminatum* (14-1) were markedly decreased by 35.1% and 59.9% (15ADON), respectively, and 33.2 % and 50.8 (HT-2), respectively, after 1 and 2 h ozone exposure (Figure 9A). Similar results were also found in *C. pilosula* inoculated with *T. roseum* (56-1) after ozone application (Figure 9C). For the control group and other pathogens infected with fresh *C. pilosula*, no mycotoxins were found (Figure 9).

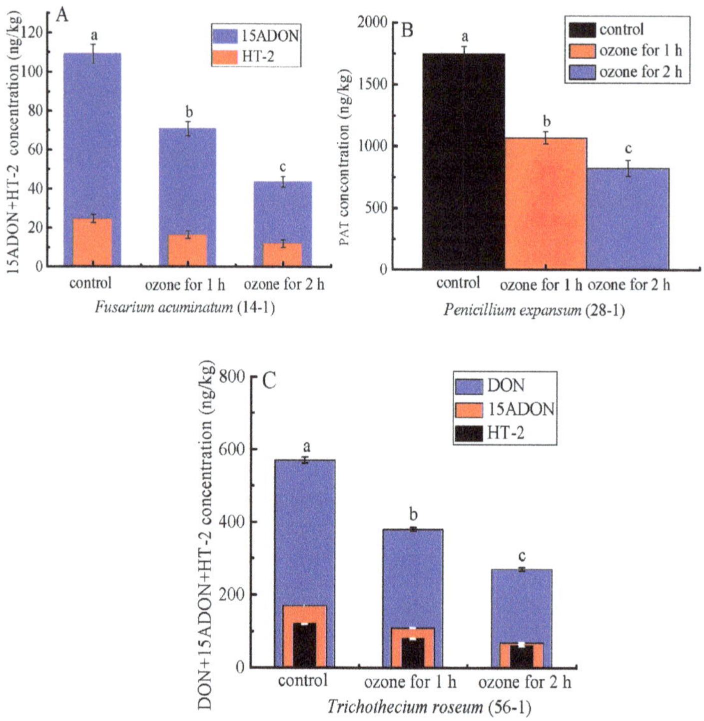

Figure 9. The effect of ozone treatment on mycotoxin accumulation of fresh *C. pilosula* infected by the isolates at 56 days of storage. (**A**) 15ADON and HT-2 toxin for *Fusarium acuminatum* (14-1); (**B**) Patulin (PAT) for *Penicillum expansum* (28-1); (**C**) DON, 15ADON, and HT-2 toxin for *Trichothecium roseum* (56-1); The different letters indicate significant differences during the same storage period ($p < 0.05$).

4. Discussion

At present, numerous studies have addressed the pathogens causing preharvest disease in Chinese herbs, these reports indicated that *Fusarium* species are the dominant pathogens that cause preharvest disease in the various regions [28–31]. However, there are limited reports on postharvest diseases of freshly harvested Chinese herbal medicines during storage. Of relevance, Chen et al. [32] isolated and identified *P. crustosum*, *P. viridicatum*, *P. aurantiogriseum*, and *P. brevicompactum* from *Angelica sinensis* and *C. pilosula* decoction pieces during storage. Nevertheless, there are significant differences between Chinese medicinal decoction pieces and freshly harvested Chinese herbs as experimental materials. Chinese medicinal decoction pieces are mostly obtained from the Chinese herbal medicine market, and these products are available in dry or dehydrated forms. On the other hand, freshly harvested Chinese herbs come from the field, without any processing and treatments after harvest such as drying or dehydration. Therefore, fresh *C. pilosula* is more susceptible to molds and decay, owing to the higher water content and the abundance of nourishing substances that allow the growth of pathogens.

To our knowledge, this is the first report to isolate and characterize the pathogens causing postharvest disease of *C. pilosula* during different storage periods. The results indicated that, with the extension of storage time, the symptoms of the disease were more severe. Moreover, a total of 9 isolates were isolated and characterized based on the morphology and biological characters during storage for 56 days. All the nine isolates could cause different postharvest disease by Koch's postulate. However, the results went beyond Koch's postulate because Koch's postulate is mainly applied to one pathogen. However, we initially observed in the postharvest disease that many pathogens were interacting for disease, and we isolated and identified the different pathogens, then inoculated the healthy *C. pilosula* with the obtained pathogen. Finally, we observed the different disease symptoms caused by different pathogens. Therefore, we thought Koch's postulate had limitations for the results of this study.

A. elegans and *M. hiemalis* were the main pathogens causing mucor rot of freshly harvested *C. pilosula* on the 7th day of storage. After 14 days of storage, *Fusarium* species of *F. acuminatum*, *F. equiseti*, and *F. oxysporum* were the major pathogenic fungi causing root rot in *C. pilosula*. After 21 days of storage, besides the above-isolated pathogens, a new pathogen of *C. rosea* was isolated and characterized. On the 28th day of storage, blue mold caused by *Penicillum* spp. was the typical postharvest disease, and two new pathogens of *P. expansum* and *P. aurantiogriseum* were isolated and identified. On the 56th day of storage, a new pathogen *T. roseum* was isolated and characterized.

From the above results, the first postharvest disease was caused by *Mucor* infection on the 7th day of storage, and the main pathogens were *A. elegans* and *M. hiemalis*. *A. elegans* grew rapidly on PDA plates, and it took around 3 days to cover the whole petri dish. The morphology of colonies, spores and sporangiophores were similar to those isolated from necrotic skin lesions in humans, and the predominant fungal pathogen causing invasive *mucormycosis* for humans [33]. In addition, *A. elegans*, which can result in sufu's white flake, was discovered in the sufu [34]. *Mucor hiemalis* had a fast growth speed on PDA medium, with a white colony initial color, which changed gradually to grayish brown (Figure 3B). The spore sacs were ellipsoid, and sporangiophores were erect and branched (Figure 4B), This observation was similar to the morphology of *M. hiemalis* causing mucor rot of mandarin fruit in California reported by Saito et al. [35]. The spores of the two Mucor species can survive in the air and infect hosts mainly through the air.

During storage of 14 days, *Fusarium* spp. were the main pathogens causing postharvest root rot of *C. pilosula*; among them, *F. acuminatum* had the strongest pathogenicity, and also produced mycotoxins, with nearly round colonies, neat edges, vigorous mycelia, flocculent, and dense (Figure 2C). Most conidia were fusiform and septate (Figure 3C), whose morphology was similar to *F. acuminatum* from garlic bulb rot in Serbia [36]. Moreover, *F. acuminatum* also metabolizes 15ADON and HT-2 toxins (Figure 9). The toxins 15-ADON and HT-2 are attributed to trichothecene with detrimental effects, including cytotoxicity, acute toxicity, immunotoxicity, and chronic toxicity, and pose a serious threat to human health. Tan et al. [37] suggested that DON and 3-ADON were detected in Fusarium head blight (FHB) of wheat infected by *F. acuminatum*. The colony of *F. equiseti* was white, and the mycelium had cotton-like flocculence (Figure 2D). Conidia were oval (Figure 3D), and the morphology was consistent with the observation by Afroz et al. [22], who mentioned that *F. equiseti* isolated from cabbage *Fusarium* wilt in Korea had a morphology of loosely floccose and whitish-brown aerial mycelia, and the pigmentation of pale orange on PDA medium; moreover, the size and morphology of macroconidia and chlamydospores were also in accordance with our observation. The *F. oxysporum* colony was initially white, then turned off-white later, to light purple-dark purple pigment. Hyphae were flocculent or villous and denser (Figure 2E), the conidia were elongated oval and slightly pointed at both ends (Figure 3E), which was consistent with the morphology of *F. oxysporum* isolated from sesame plants. Previous study revealed that *Fusarium* species is attributed to "latent infected pathogen" [38], the infection of *Fusarium* usually occur in the field (growth stage of plants) during the blooming or heading of plants. The spores adhere to the surface of the leaf, then transported into the calyx, finally colonize on the root of plants, and remain in a latent status until there are favorable environmental conditions (such as high temperature and humidity) [39]. Therefore, it is possible to manage plant disease more effectively during the pathogen infection time.

After 21 days of storage, *C. rosea* was isolated and identified, but its disease infection was not severe. The hyphae were white with neat edges, mycelium was creeping and loose (Figure 2F), and the morphology was similar to that isolated from root rot of *Astragalus membranaceus* in China [40]. *C. rosea* has been regarded as a typical biocontrol fungus to inhibit pathogenic fungal growth. The most serious postharvest disease was blue mold caused by *Penicillium* spp., including *P. expansum* and *P. aurantiogriseum*. *P. expansum* is a typical postharvest pathogen of pome fruits (such as apples and pears) causing blue mold and patulin contamination. The isolated *P. expansum* colony was grayish-green, and the edge was white, with a granular or velvet dense texture (Figure 2G). The microconidia were colorless with monospores that were globose or oblate. The number of conidia was high, and the conidiophores were erect, separated, and colorless (Figure 3G), whose morphology was consistent with the blue mold of apple fruits as reported by Bahri et al. [41]. In the present study, patulin (PAT) was detected in *C. pilosula* infected by *P. expansum*. Patulin is a secondary metabolite generated by *Aspergillus* and *Penicillium* species under favorable conditions, and usually found in blue mold disease of pome fruits and their corresponding products. The maximum limit of PAT in fruit juice was set as 50 μg/kg by the European Union [42]; however, no standard of PAT level limits is established in Chinese herbal medicines. *P. aurantiogriseum*, a kind of plant endophytic fungi, is also a casual pathogen causing fruit postharvest decay. Shim et al. [43] isolated the pathogen from pear fruit infected with blue mold in Korea, and Liu et al. [44] isolated *P. aurantiogriseum* from fresh-cut lettuce corruption. The cultivation characters and microscopic examination (Figures 2H, 3H and 4H) in our study were very similar to the above reports' morphological characteristics. *Penicillium* species are attributed to "wound pathogens" [38] that mainly infect host plants through wounds sustained by fruit cracking or mechanical damage during harvesting, transportation, handling, and storage. Therefore, to avoid pathogen infection, delicate operations should be recommended during packing and handling.

T. roseum was identified on the 56th day of storage, as the predominant fungus causing *C. pilosula* disease. *T. roseum* is a typical necuotrophic fungal pathogen that can infect various postharvest fruits and vegetables, leading to trichothecenes contamination. Sharma et al. [45] isolated *T. roseum* from postharvest pink rot of avocado in Israel. *T. roseum* is also known to cause pink rot in muskmelon and apple fruit. The cultivation characters and microscopic examination of the pathogen from the pink rot of fruit were consistent with our observation. Moreover, 15-ADON, DON, and HT-2 toxins were found in the rotten tissue of *C. pilosula* pink rot caused by *T. roseum*, which were trichothecenes. Tang et al. [46] also suggested T-2 toxin and neosolaniol (NEO) were detected in the core rot of apple fruit infected by *T. roseum*. *T. roseum* is attributed to "wound pathogen", which mainly infects host plants through wounds; however, it can also carry out latent infection when the plant grows in the field, then end its colonization and development, without showing any symptoms until the plant is harvested. Therefore, it is a challenging job to manage the disease caused by *T. roseum*. The combination application of preharvest spray and postharvest treatment is proposed.

Ozone, a strong oxidant compound, is widely used to control postharvest decay. In the present study, ozone treatment significantly reduced the development of postharvest disease of freshly harvested *C. pilosula*, especially, when the herbs were exposed to 2 h treatment compared to 1 h. The result was in accordance with the report that ozone exposure for 10 min was more effective than 5 min in controlling pear disease [47]. The inhibitory effect of ozone was attributed to oxidation. Ozone can attack the cell plasma of plant pathogenic fungus, leading to membrane lipid peroxidation, and cell membrane integrity destruction, and loss of pathogenicity to the plant host [18]. In addition, we found that, as the ozone exposure time increased, the control effect became more visible. The reason is maybe that there is a cumulative effect for ozone exposure, the cumulative effect of 2 h ozone exposure was more than that of 1 h ozone exposure.

More importantly, we found that ozone treatment greatly suppressed the mycotoxin accumulation in the rotten tissue of the infected *C. pilosula* (Figure 9). A similar result was documented by Xue et al. [48], who indicated ozone application significantly reduced NEO accumulation. Considering the two reasons, ozone treatment controlled postharvest disease by inhibiting fungi growth; on the other hand, ozone can act directly with the chemical structure of mycotoxin, thus destroying the structure of mycotoxin [48].

5. Conclusions

A total of nine isolates were identified and characterized by morphology and molecular technology from postharvest diseases of *C. pilosula* during different storage stages. The nine isolates, with varying morphologies of colony, spore, and conidiophore, could cause several symptomatic postharvest diseases during different storage stages. Postharvest diseases of *C. pilosula* during storage can also metabolize and produce mycotoxins, which pose health threats to humans. Ozone application not only controlled postharvest diseases of *C. pilosula* during storage but also reduced mycotoxin accumulation. Therefore, in order to effectively control postharvest decay, with different pathogens having various infection modes and infection periods, the incorporation application of preharvest spray and postharvest treatment should be proposed; on the other hand, ozone application needs to be widely recommended and the mechanism of ozone action on disease and mycotoxins suppression should be further studied.

Author Contributions: Conceptualization, B.L., X.Y. and H.X.; methodology, B.L.; software, X.Y.; validation, Y.Z., Z.L. and S.S.; formal analysis, Y.Z.; investigation, M.N.; resources, H.X. and Y.B.; data curation, B.L.; writing—original draft preparation, B.L. and X.Y.; writing—review and editing, H.X.; visualization, M.N.; supervision, Y.B.; project administration, H.X.; funding acquisition, H.X. All authors have read and agreed to the published version of the manuscript.

Funding: This research was funded by the Project of Enterprise Supporting Plan (2022CYZC-45, Education Department of Gansu Province), the Natural Science Foundation of China (32060566, National Natural Science Foundation of China), and the Science and Technology Project of Gansu Province (21JR7RA834, Science and Technology Department of Gansu Province) and the Universities of Gansu Province Foundation for Young Doctors (2022QB-078, Education Department of Gansu Province).

Institutional Review Board Statement: Not applicable.

Informed Consent Statement: Not applicable.

Data Availability Statement: The data presented in this study are included in the article; further inquiries can be directed to the corresponding author.

Conflicts of Interest: The authors declare no conflict of interest.

References

1. Du, Y.E.; Lee, J.S.; Kim, H.M.; Ahn, J.; Jung, I.H.; Ryu, J.H.; Choi, J.; Jang, D.S. Chemical constituents of the roots of *Codonopsis lanceolata*. *Arch. Pharm. Res.* **2018**, *41*, 1082–1091. [CrossRef] [PubMed]
2. Gao, S.M.; Liu, J.S.; Wang, M.; Cao, T.T.; Qi, Y.D.; Zhang, B.G.; Sun, X.B.; Liu, H.T.; Xiao, P.G. Traditional uses, phytochemistry, pharmacology and toxicology of *Codonopsis*: A review. *J. Ethnopharmacol.* **2018**, *219*, 50–70. [CrossRef] [PubMed]
3. Gao, Z.Z.; Zhang, C.; Jing, L.R.; Feng, M.; Li, R.; Yang, Y. The structural characterization and immune modulation activities comparison of *Codonopsis pilosula* polysaccharide (CPPS) and *Selenizing cpps* (sCPPS) on mouse in vitro and vivo. *Int. J. Biol. Macromol.* **2020**, *160*, 814–822. [CrossRef] [PubMed]
4. Jia, W.J.; Bi, Q.M.; Jiang, S.R.; Tao, J.H.; Liu, L.Y.; Yue, H.L.; Zhao, X.H. Hypoglycemic activity of Codonopsis pilosula (Franch.) Nannf. in vitro and in vivo and its chemical composition identification by UPLC-Triple-TOF-MS/MS. *Food Funct.* **2022**, *13*, 2456–2464. [CrossRef] [PubMed]
5. Guo, L.Z. Study on the pharmacological effects and clinical application of Dangshen Tonic. *China Health Stand. Manag.* **2015**, *22*, 130–131.
6. Zhong, Y.G.; Chen, L.Y.; Li, M.F.; Chen, L.; Qian, Y.F.; Chen, C.F.; Wang, Y.; Xu, Y.Z. Dangshen Erling decoction ameliorates myocardial hypertrophy via inhibiting myocardial inflammation. *Front. Pharmacol.* **2022**, *12*, 725816. [CrossRef]

7. Tang, L.J.; Chen, J.H.; Yin, J.; Fang, M.L. Screening of active components and key targets of *radix Codonopsis* in the treatment of gastric cancer. *J. Chem.* **2021**, *2021*, 6056636. [CrossRef]
8. Zhao, X.; Liang, Y.; Constantine, U.; Yang, L.; Yuan, T.; Zhao, H.J.; Zhou, Q.; Zhang, Y.B.; Wang, R.Y. First report of root rot caused by the *Fusarium oxysporum* species complex on *Codonopsis pilosula* in China. *Plant Dis.* **2021**, *105*, 3742. [CrossRef]
9. Yu, Z.L.; Lei, M.Y.; Pu, S.C.; Xiao, Z.; Cao, H.Q.; Yang, C.Q. Fungal disease survey and pathogen identification on *Codonopsis tangshen* in Chongqing. *J. Chin. Med. Mater.* **2015**, *38*, 1119–1122.
10. Wang, Y.; Zeng, C.Y.; Cheng, H.G.; Zhu, T.T.; Chen, X.R. Pathogenic fungi and biological characteristics of *Codonopsis pilosula* spot blight. *J. Plant Prot.* **2016**, *43*, 928–934. [CrossRef]
11. Chen, S.Z. Investigation on *Botrytis cinerea* of *Codonopsis pilosula* and its field control in Dingxi, Gansu Province. *Grassl. Lawn* **2017**, *37*, 94–97. [CrossRef]
12. Zhang, R.; Miao, M.S. Application characteristics of fresh traditional Chinese medicine. In Proceedings of the 2017 2nd International Conference on Biological Sciences and Technology (BST 2017), Zhuhai City, China, 17–19 November 2017; Atlantis Press: Dordrecht, The Netherlands, 2017; pp. 287–291.
13. Xu, H.; Zhou, Y.; Lei, T. Comparative analysis of volatile chemical constituents of fresh *Astragalus* and dried *Astragalus*. *Food Sci.* **2011**, *32*, 171–174.
14. Wang, X.; Yuan, Q.; Sun, K.; Guo, Z.X.; Chi, X.; Huang, L.Q. Population characteristics and threatened of wild *Angelica sinensis* in Gansu province. *Chin. J. Med.* **2019**, *44*, 2987–2995.
15. Han, Z.; Ren, Y.P.; Zhou, H.L.; Luan, L.J.; Cai, Z.X.; Wu, Y.J. A rapid method for simultaneous determination of zearalenone, α-zearalenol, β-zearalenol, zearalanone, α-zearalanol and β-zearalanol in traditional Chinese medicines by ultra-high-performance liquid chromatography-tandem mass spectrometry. *J. Chromatogr. B* **2011**, *879*, 411–420. [CrossRef]
16. Oh, S.Y.; Nam, K.W.; Yoon, D.H. Identification of *Acremonium acutatum* and *Trichothecium roseum* isolated from grape with white stain symptom in Korea. *Mycobiology* **2015**, *42*, 269–273. [CrossRef] [PubMed]
17. Gibert, S.; Edel-Hermann, V.; Gautheron, E.; Gautheron, N.; Sol, J.M.; Capelle, G.; Galland, R.; Bardon-Debats, A.; Lambert, C.; Steinberg, C. First report of *Fusarium avenaceum, Fusarium oxysporum, Fusarium redolens* and *Fusarium solani* causing root rot in pea in France. *Plant Dis.* **2022**, *106*, 1297. [CrossRef]
18. Li, L.; Xue, H.L.; Bi, Y.; Zhang, R.; Kouasseu, C.J.; Liu, Q.L.; Nan, M.N.; Pu, L.M.; Prusky, D. Ozone treatment inhibits dry rot development and diacetoxyscirpenol accumulation in inoculated potato tuber by influencing growth of *Fusarium sulphureum* and ergosterol biosynthesis. *Postharvest Biol. Technol.* **2022**, *2022*. *185*, 111796. [CrossRef]
19. Qin, P.W.; Xu, J.; Jiang, Y.; Hu, L.; van der Lee, T.; Waalwijk, C.; Zhang, W.M.; Xu, X.D. Survey for toxigenic *Fusarium* species on maize kernels in China. *World Mycotoxin J.* **2020**, *13*, 213–223. [CrossRef]
20. Edwards, K.; Johnstone, C.; Thompson, C. A simple and rapid method for the preparation of plant genomic DNA for PCR analysis. *Nucleic Acids Res.* **1991**, *19*, 1349. [CrossRef]
21. Duan, Y.H.; Qu, W.W.; Chang, S.X.; Li, C.; Xu, F.F.; Ju, M.; Zhao, R.H.; Wang, H.L.; Zhang, H.Y.; Miao, H.M. Identification of pathogenicity groups and pathogenic molecular characterization of *Fusarium oxysporum* f. sp. sesami in China. *Phytopathology* **2020**, *110*, 1093–1104. [CrossRef] [PubMed]
22. Afroz, T.; Je, S.; Choi, H.W.; Kim, J.H.; Assefa, A.D.; Aktaruzzaman, M.; Hahn, B.S.; Lee, H.S. First report of *Fusarium* wilt caused by *Fusarium equiseti* on cabbage (*Brassica oleracea var. capitate*) in Korea. *Plant. Dis.* **2020**, *105*, 1198. [CrossRef] [PubMed]
23. Wang, C.C.; Tang, Y.H.; Qiao, N.; Zhang, D.Z.; Chi, W.J.; Liu, J.; Pan, H.Q.; Li, J.T. First report of *Colletotrichum* black leaf spot on strawberry caused by *Colletotrichum siamense* in China. *J. Phytopathol.* **2022**, *170*, 279–281. [CrossRef]
24. Sha, C.; Yang, N.; Zhao, C.; Liu, J.; Han, C.; Wu, X. Diversity of *Fusarium* species associated with root rot of sugar beet in China. *J. Gen. Plant Pathol.* **2018**, *84*, 321–329. [CrossRef]
25. Jimdjio, C.K.; Xue, H.L.; Bi, Y.; Nan, M.N.; Li, L.; Zhang, R.; Liu, Q.L.; Pu, L.M. Effect of ambient pH on growth, pathogenicity, and patulin production of *Penicillium expansum*. *Toxins* **2021**, *13*, 550. [CrossRef] [PubMed]
26. Xue, H.L.; Bi, Y.; Wei, J.M.; Tang, Y.M.; Zhao, Y.; Wang, Y. New Method for the simultaneous analysis of types A and B trichothecenes by ultrahigh-performance liquid chromatography coupled with tandem mass spectrometry in potato tubers inoculated with *Fusarium sulphureum*. *J. Agric. Food Chem.* **2013**, *61*, 9333–9338. [CrossRef] [PubMed]
27. Vidal, G.S.; Hahn, M.H.; Pereira, W.V.; Pinho, D.B.; May-De-Mio, L.L.; Duarte, H.D.S.S. A molecular approach reveals Tranzschelia discolor as the causal agent of rust on plum and peach in Brazil. *Plant Dis.* **2021**, *6*, 1855. [CrossRef]
28. Li, J.S.; Yan, Z.Y.; Lan, Y.; Shen, X.F.; Wang, H.; He, D.M. Identification of pathogens causing root rot disease on *Ligusticum chuanxiong* in Sichuan. *J. Chin. Med. Mater.* **2015**, *38*, 443–446. [CrossRef]
29. Zhou, M.; Bai, R.Q. Isolation and identification of root rot pathogen of *Astragalus membranaceus* in Moqi, Inner Mongolia. *J. Northeast Agric. Sci.* **2021**, *46*, 52–55. [CrossRef]
30. Wu, X.L.; Wang, Y.; Liu, F.; Chen, D.X.; Li, L.Y. Identification of Coptis Chinensis root rot disease pathogenic *Fusarium* spp. fungi. *Chin. J. Med.* **2020**, *45*, 1323–1328. [CrossRef]
31. Chen, Y.P.; Chen, X.M.; Liu, X.; Xiao, R.F.; Liu, B. Pathogen identification of the new disease of *Pseudostellariae Radix* sour rot. *Acta Phytopathol. Sin.* **2021**, *51*, 464–468. [CrossRef]
32. Chen, J.; Gao, W.W.; Tang, D.; Cai, F.; Yang, M.H. Analysis of fungi in seven kinds of root medicines contaminated by Ochratoxin A. *Chin. Med. J.* **2010**, *35*, 2647–2651. [CrossRef]

33. Mahmud, A.; Lee, R.; Munfus-McCray, D.; Kwiatkowski, N.; Subramanian, A.; Neofytos, D.; Carroll, K.; Zhang, S.X. *Actinomucor elegans* as an emerging cause of mucormycosis. *J. Clin. Microbiol.* **2012**, *50*, 1092–1095. [CrossRef]
34. Yao, D.; Xu, L.; Wu, M.N.; Wang, X.Y.; Wang, K.; Li, Z.J.; Zhang, D.J. Microbial Community succession and metabolite changes during fermentation of BS Sufu, the fermented black soybean curd by *Rhizopus microsporus*, *Rhizopus oryzae*, and *Actinomucor elegans*. *Front. Microbiol.* **2021**, *12*, 665826. [CrossRef] [PubMed]
35. Saito, S.; Michailides, T.J.; Xiao, C.L. Mucor rot-an emerging postharvest disease of mandarin fruit caused by *Mucor piriformis* and other *Mucor* spp. in California. *Plant Dis.* **2016**, *100*, 1054–1063. [CrossRef] [PubMed]
36. Ignjatov, M.; Bjelic, D.; Nikolic, Z.; Milosevic, D.; Gvozdanovic-Varga, J.; Marinkovic, J.; Ivanovic, Z. First report of *Fusarium acuminatum* causing garlic bulb rot in Serbia. *Plant Dis.* **2017**, *101*, 1047–1048. [CrossRef]
37. Tan, D.C.; Flematti, G.R.; Ghisalberti, E.L.; Sivasithamparam, K.; Chakraborty, S.; Obanor, F.; Jayasena, K.; Barbetti, M.J. Mycotoxins produced by *Fusarium* spp. associated with *Fusarium* head blight of wheat in Western Australia. *Mycotoxin Res.* **2012**, *28*, 89–96. [CrossRef]
38. Mincuzzi, A.; Sanzani, S.M.; Palou, L.; Ragni, M.; Ippolito, A. Postharvest rot of pomegranate fruit in Southern Italy: Characterization of the main pathogens. *J. Fungi* **2022**, *8*, 475. [CrossRef]
39. Alisaac, E.; Behmann, J.; Rathgeb, A.; Karlovsky, P.; Dehne, H.; Mahlein, A. Assessment of *Fusarium* infection and mycotoxin contamination of wheat kernels and flour using hyperspectral imaging. *Toxins* **2019**, *11*, 556. [CrossRef]
40. Qi, H.X.; Duan, X.M.; Xu, W.H.; Zhou, Y.T.; Ma, H.X.; Ma, W.L.; Ma, G.H. First report disease of *Clonostachys rosea* causing root rot on *Astragalus membranaceus* in China. *Plant Dis.* **2022**, *106*, 1752. [CrossRef]
41. Bahri, B.A.; Mechichi, G.; Rouissi, W.; Ben Haj, J.I.; Ghrabi-Gammar, Z. Effects of cold-storage facility characteristics on the virulence and sporulation of *Penicillium expansum* and the efficacy of essential oils against blue mold rot of apples. *Folia Hortic.* **2019**, *31*, 301–317. [CrossRef]
42. Mahunu, G.K.; Zhang, H.Y.; Yang, Q.Y.; Li, C.L.; Zheng, X.F. Biological control of patulin by antagonistic yeast: A case study and possible model. *Crit. Rev. Microbiol.* **2016**, *42*, 643–655. [CrossRef] [PubMed]
43. Shim, J.O.; Choi, K.D.; Hahn, K.D.; Lee, J.H.; Hyun, I.H.; Lee, T.S.; Ko, K.; Lee, H.P.; Lee, M.W. Blue mold of pear caused by *Penicillium aurantiogriseum* in Korea. *Mycobiology* **2002**, *30*, 105–106. [CrossRef]
44. Liu, C.H.; Hu, W.Z.; Wang, Y.Y.; Tian, M.X.; Sun, L. Isolation and identification of spoilage molds from fresh-cut lettuce. *Sci. Technol. Food Ind.* **2016**, *37*, 135–138. [CrossRef]
45. Sharma, G.; Maymon, M.; Freeman, S. First detailed report of *Trichothecium roseum* causing post-harvest pink rot of avocado in Israel. *Plant Dis.* **2016**, *100*, 856. [CrossRef]
46. Tang, Y.M.; Xue, H.L.; Bi, Y.; Li, Y.C.; Wang, Y.; Zhao, Y.; Shen, K.P. A method of analysis for T-2 toxin and neosolaniol by UPLC-MS/MS in apple fruit inoculated with *Trichothecium roseum*. *Food Addit. Contam. Part A* **2015**, *32*, 480–487. [CrossRef]
47. Al-Haq, M.I.; Seo, Y.; Oshita, S.; Kawagoe, Y. Disinfection effects of electrolyzed oxidizing water on suppressing fruit rot of pear caused by *Botryosphaeria be-rengeriana*. *Food Res. Int.* **2002**, *35*, 657–666. [CrossRef]
48. Xue, H.L.; Bi, Y.; Hussain, R.; Wang, H.J.; Pu, L.M.; Nan, M.N.; Cheng, X.Y.; Wang, Y.; Li, Y.C. Detection of NEO in muskmelon fruits inoculated with *Fusarium sulphureum* and its control by postharvest ozone treatment. *Food Chem.* **2018**, *254*, 193–200. [CrossRef]

Journal of
Fungi

MDPI

Article

Ferric Chloride Controls Citrus Anthracnose by Inducing the Autophagy Activity of *Colletotrichum gloeosporioides*

Yuqing Wang [1,2,3,4,5,†], Xiaoxiao Wu [6,7,8,†], Yongqing Lu [1,2,3,4,5], Huimin Fu [6,7,8], Shuqi Liu [1,2,3,4,5], Juan Zhao [1,2,3,4,5] and Chaoan Long [1,2,3,4,5,*]

1 National Key Laboratory for Germplasm Innovation & Utilization of Horticultural Crops, Wuhan 430070, China
2 Key Laboratory of Horticultural Plant Biology of Ministry of Education, Wuhan 430070, China
3 National R&D Center for Citrus Preservation, Wuhan 430070, China
4 National Centre of Citrus Breeding, Wuhan 430070, China
5 College of Horticulture & Forestry Sciences of Huazhong Agricultural University, Wuhan 430070, China
6 Guangxi Laboratory of Germplasm Innovation and Utilization of Specialty Commercial Crops in North Guangxi, Guilin 541004, China
7 Guangxi Citrus Breeding and Cultivation Research Center of Engineering Technology, Guilin 541004, China
8 Guangxi Academy of Specialty Crops, Guilin 541004, China
* Correspondence: postharvest@mail.hzau.edu.cn
† These authors contributed equally to this work.

Abstract: *Colletotrichum gloeosporioides* causes citrus anthracnose, which seriously endangers the pre-harvest production and post-harvest storage of citrus due to its devastating effects on fruit quality, shelf life, and profits. However, although some chemical agents have been proven to effectively control this plant disease, little to no efforts have been made to identify effective and safe anti-anthracnose alternatives. Therefore, this study assessed and verified the inhibitory effect of ferric chloride ($FeCl_3$) against *C. gloeosporioides*. Our findings demonstrated that $FeCl_3$ could effectively inhibit *C. gloeosporioides* spore germination. After $FeCl_3$ treatment, the germination rate of the spores in the minimum inhibitory concentration (MIC) and minimum fungicidal concentration (MFC) groups decreased by 84.04% and 89.0%, respectively. Additionally, $FeCl_3$ could effectively inhibit the pathogenicity of *C. gloeosporioides* in vivo. Optical microscopy (OM) and scanning electron microscopy (SEM) analyses demonstrated the occurrence of wrinkled and atrophic mycelia. Moreover, $FeCl_3$ induced autophagosome formation in the test pathogen, as confirmed by transmission electron microscopy (TEM) and monodansylcadaverine (MDC) staining. Additionally, a positive correlation was identified between the $FeCl_3$ concentration and the damage rate of the fungal sporophyte cell membrane, as the staining rates of the control (untreated), 1/2 MIC, and MIC $FeCl_3$ treatment groups were 1.87%, 6.52%, and 18.15%, respectively. Furthermore, the ROS content in sporophyte cells increased by 3.6%, 29.27%, and 52.33% in the control, 1/2 MIC, and MIC $FeCl_3$ groups, respectively. Therefore, $FeCl_3$ could reduce the virulence and pathogenicity of *C. gloeosporioides*. Finally, $FeCl_3$-handled citrus fruit exhibited similar physiological qualities to water-handled fruit. The results show that $FeCl_3$ may prove to be a good substitute for the treatment of citrus anthracnose in the future.

Keywords: citrus; *Colletotrichum gloeosporioides;* ferric chloride; autophagy; ROS; membrane integrity

1. Introduction

Anthracnose caused by *Colletotrichum gloeosporioides* is a tree and fruit disease that seriously affects the quality and postharvest storage of citrus fruit and causes considerable economic losses [1,2]. Although the occurrence of this disease can be controlled through orchard management methods such as plant removal and crop rotation [3], chemical fungicides such as mancozeb and copper compounds, alone or in combination with fosetyl-Al, are the most effective and commonly used agents to manage this disease [4–6]. The continuous use of chemical fungicides not only promotes the occurrence of fungicide-resistant pathogenic fungi but also leads to the accumulation of chemical residues in the environment, thus threatening human and ecosystem health [7]. Therefore, developing novel green and efficient alternative methods to manage citrus anthracnose is crucial.

In this study, we investigated the inhibitory effects and mechanisms of $FeCl_3$ on *C. gloeosporioides*. Previous studies have found that iron chelates, iron ion complexes, and organic acid iron salts possess excellent antifungal properties and affect a series of growth and development processes [8,9]. Moreover, iron salts are generally considered safe and are commonly used in the food industry. In fact, some iron salts are used to fortify food, as previously reported by the National Health and Family Planning Commission of China, the "Standard for the Use of Food Nutritional Fortifiers" (GB 14880-2012), and the "Standard for the Use of Food Additives" (GB 2760-2014). In addition to its use as a food additive, $FeCl_3$ is also used in medicine [10].

Autophagy has been increasingly important in plant–pathogen interactions in recent years [11]. Macromolecular proteins, organelles, ribosomes, and other cell components are degraded by the process of autophagy in response to cellular stress circumstances such as inadequate energy supply, hunger, and environmental stress [12,13]. In addition to influencing and promoting programmed cell death [14], autophagy is essential for a number of regulatory processes, including growth and development [15–18].

We discovered that $FeCl_3$ can promote the autophagy of fungi, reducing the incidence of citrus anthracnose, in the control experiment of citrus anthracnose with $FeCl_3$. $FeCl_3$'s method of action against *C. gloeosporioides*, however, has not been documented. Through in vitro and in vivo tests, this work investigated the impact of $FeCl_3$ solution on the pathogenicity of *C. gloeosporioides* and its suppression on mycelia growth. The analysis of the spore germination rate, cell membrane integrity, autophagy structure, spore activity, and ROS buildup after treatment, as well as the outcomes of storage tests in producing regions, led to a discussion of the potential inhibitory mechanism of $FeCl_3$ on *C. gloeosporioides*.

2. Materials and Methods

2.1. Fungal Pathogens, Citrus, and Metal Salt

C. gloeosporioides was donated by professor Yanping Fu from the College of Plant Science & Technology of Huazhong Agricultural University (*C. gloeosporioides* is not currently particularly sensitive or resistant to fungicides). It was cultured on potato dextrose agar (PDA) medium at 28 °C for 7–10 days. 'Newhall' navel orange fruits (*C. sinensis* Osbeck) were harvested from the citrus orchard of Huazhong Agricultural University in Wuhan, China, and a commercial orchard in Zhijiang, Hubei, China. $FeCl_3$ was purchased from Shanghai Wokai Biotechnology Co., Ltd. (Shanghai, China).

2.2. Effect of $FeCl_3$ on C. gloeosporioides In Vitro

The antifungal activity of $FeCl_3$ on *C. gloeosporioides* was assessed as described by Liu et al. [19] with some modifications. $FeCl_3$ was mixed into PDA media to obtain final concentrations of 0.15, 0.3, 0.6, 1.2, and 2.4 g/L. The media were then poured into sterilized 90-mm-diameter Petri dishes. Once the media had cooled and solidified, a 6 mm lawn of fresh *C. gloeosporioides* was placed at the center and incubated at 28 °C for 6–7 d. The mycelial diameter of every plate was recorded via the interior extrapolation method at six days. The lowest concentration that inhibited mycelial growth after two days was defined as the minimum inhibitory concentration (MIC), and the lowest concentration that completely inhibited mycelial growth after four days was defined as the minimum fungicidal concentration (MFC). All experiments were conducted in triplicate.

2.3. Effect of $FeCl_3$ on C. gloeosporioides In Vivo

Citrus fruits were treated with sodium hypochlorite solution (2%, v/v) for 2 min, then washed with sterile water and allowed to air dry. As described by Cui et al. [20], a wound

(5 mm in diameter and 5 mm in depth) was made around the equator using a punch needle, after which the wound was inoculated with 10 μL of *C. gloeosporioides* spore suspension (1×10^6 spores mL^{-1}). After drying for 1–2 min, 10 μL of $FeCl_3$ solution at 10× MFC and 15× MFC was added to the wounds, respectively, and sterile water was used as a control. The treated fruit wounds were observed after storing the fruit in a plastic case with water at the bottom at room temperature for a week. Each treatment consisted of 15 fruits, and the experiments were conducted in triplicate.

2.4. Effect of $FeCl_3$ on *C. gloeosporioides* Spore Germination

The spore suspensions (1×10^6 spores mL^{-1}) of fresh *C. gloeosporioides* were prepared with 1 mL PDB medium in 1.5 mL sterile centrifuge tubes, after which $FeCl_3$ was added to the final concentrations of 1/2 MIC, MIC, and MFC, adding sterile water as a control. The suspensions were then cultivated at 28 °C and 180 rpm on a shaker for 12 h, after which the samples were examined with an optical microscope (OM). Each treatment contained 200 spores from at least three replicates, and spore germination rates were calculated as follows: Spore germination = (number of germination spores/total spores) × 100%.

2.5. Observation by OM, SEM, and TEM

OM: PDA media containing $FeCl_3$ at the MIC and without $FeCl_3$ were applied onto cellophane, over which a 6-mm lawn plate of fresh *C. gloeosporioides* was placed. All samples were cultured in an incubator at 28 °C for 6–7 d. Afterward, 5 × 5 mm pieces of cellophane covered with mycelium were cut and observed by OM.

SEM: The pieces of cellophane covered with *C. gloeosporioides* mycelium were prepared following the same method as for OM. However, all sample pieces were fixed in glutaraldehyde solution (3%, *v*/*v*) at 4 °C for 12 h and washed three times with phosphate buffer solution (PBS). Afterward, the samples were dehydrated two times with an ethanol gradient (30%, 50%, 70%, 95%, and 100%, *v*/*v*; 20 min per ethanol solution), after which they were dried at a critical point in liquid CO_2 and coated with a layer of gold. All samples were observed via SEM (JEOL, JSM-6390LV, Tokyo, Japan).

TEM: The fresh spore solution of *C. gloeosporioides* (1×10^4 conidia/mL) was added into 50 mL PDB medium, then cultivated at 28 °C and 180 rpm in a shaking incubator for 24 h. Next, $FeCl_3$ was added to reach a final concentration of 0 and MIC. Afterward, 3–5 mycelium pellets were selected and cultured for 12 h at 28 °C and 180 rpm. The pellets were then fixed in glutaraldehyde solution (3%, *v*/*v*) for 12 h and stored for sample preparation. The samples were postfixed again with osmic acid solution (1%, *v*/*v*) for 2 h. All specimens were washed three times with PBS and dehydrated two times with an ethanol gradient (30%, 50%, 70%, 95%, and 100%, *v*/*v*; 20 min per ethanol solution), followed by polymerization in 21-well silicate embedded plates at 60 °C for 48 h. Finally, 60-nm thin sections were obtained using an Ultratome Leica UC6, after which the slices were stained in uranyl acetate (2%) and lead citrate for 30 min and 10 min, respectively. Finally, the samples were observed via TEM (Hitachi H-7650, Tokyo, Japan).

2.6. Detection of $FeCl_3$ on the Cell Membrane Integrity of *C. gloeosporioides*

PI staining: Fresh *C. gloeosporioides* spore suspension (1×10^6 spore mL^{-1}) was prepared with 1 mL PDB staining medium in 1.5 mL sterile centrifuge tubes. Then, it was cultured at 28 °C and 180 rpm on a shaker for 3 h, centrifuged for 3 min at 4000 rpm and washed three times with PBS. The spores were collected and dyed with 200 mL propidium iodide (PI) (Coolaber Technology Co., Ltd., Beijing, China). All samples were stained at 37 °C for 5–10 min, after which they were rinsed twice with PBS to remove excess dye. The samples were then observed under a fluorescence microscope (Nikon Eclipse 90i).

2.7. Detection of Autophagic Structures, Conidial Viability and ROS Accumulation

A total of 1 mL of *C. gloeosporioides* conidial suspension (1×10^6 conidia/mL) was prepared, and $FeCl_3$ was added to reach a final concentration of 1/2 MIC and MIC, followed by incubation at 28 °C and 180 rpm on a shaker for 3 h before staining. Conidial viability was detected using fluorescein diacetate (FDA, 5 g L^{-1}). The samples were then incubated with the corresponding stains in the dark for 5 min, after which they were rinsed twice with PBS (pH = 7.0). The autophagosomes and intracellular ROS accumulation of the conidia were detected using a monodansylcadaverine assay kit (MDC, Solarbio, Beijing, China) and a Reactive Oxygen Species (ROS) assay kit (Solarbio, Beijing). All samples were observed under a fluorescence microscope (Nikon Eclipse 90i).

2.8. Physiological Qualities Test of $FeCl_3$-Treated Citrus Fruit

The physiological qualities of $FeCl_3$-treated citrus fruit were detected. For 30–45 s, 'Newhall' navel orange fruits (*C. sinensis* Osbeck) were immersed in 10× MFC $FeCl_3$ (=24 g/L), water, and chemicals. All fruits were transferred into a climate chamber for 30 days of storage. Total soluble solids (TSS) and titratable acid (TA) were tested using a Pocket Brix acidity meter (ATAGO Co., Ltd., Saitama, Japan) in accordance with the operation manual; 2,6-dichlorophenolindophenol was used to determine the VC content in fruit, there were five fruits in each group, and trials were repeated three times. Fruit weight losses were measured using an electronic scale, there were ten fruits in each group, and trials were repeated three times.

2.9. Statistical Analysis

All experiments were conducted in triplicate following a completely randomized design. Significant differences were determined via one-way ANOVA followed by Duncan's multiple range test (SPSS 26.0, $p < 0.05$).

3. Results

3.1. Inhibiting Effect In Vitro and In Vivo

The in vitro growth of *C. gloeosporioides* mycelia treated with $FeCl_3$ was significantly inhibited in the PDA medium (Figure 1A,B). Without $FeCl_3$ treatment, the colony center became dense and thickened, and the aging hyphae exhibited a greyish green coloration. This color disappeared when the $FeCl_3$ concentration reached 1.2 g/L, and mycelium growth was significantly inhibited. When the $FeCl_3$ concentration reached 2.4 g/L, colony growth was completely inhibited.

As illustrated in Figure 1C, the in vivo antifungal activity of $FeCl_3$ increased in a dose-dependent manner. When the concentration of $FeCl_3$ in citrus reached 15× MFC, the number of *C. gloeosporioides* cells in the treatment group was significantly lower than that in the control group after one week of inoculating the spore suspension.

3.2. Effect of $FeCl_3$ on the Spore Germination

The spore germination of *C. gloeosporioides* was remarkably inhibited by $FeCl_3$ (Figure 2A). When the $FeCl_3$ concentration reached 1/2 MIC and MIC, the germination rates of *C. gloeosporioides* were 55.77% and 4.83%, respectively, which represented a significant decrease compared with the control (88.23%, $p < 0.05$) (Figure 2B). Moreover, the length of the germ tubes in the $FeCl_3$-treated group was markedly shorter than that in the control group.

3.3. Effect of $FeCl_3$ on Mycelial Morphology and the Cell Membrane Integrity

OM and SEM: As observed by OM and SEM, normal hyphae exhibited a homogeneous and linear smooth surface, whereas those treated with MIC $FeCl_3$ were severely damaged, with irregularly contracted or expanded cell membranes, or were even bound together (marked with red arrows) (Figure 3A).

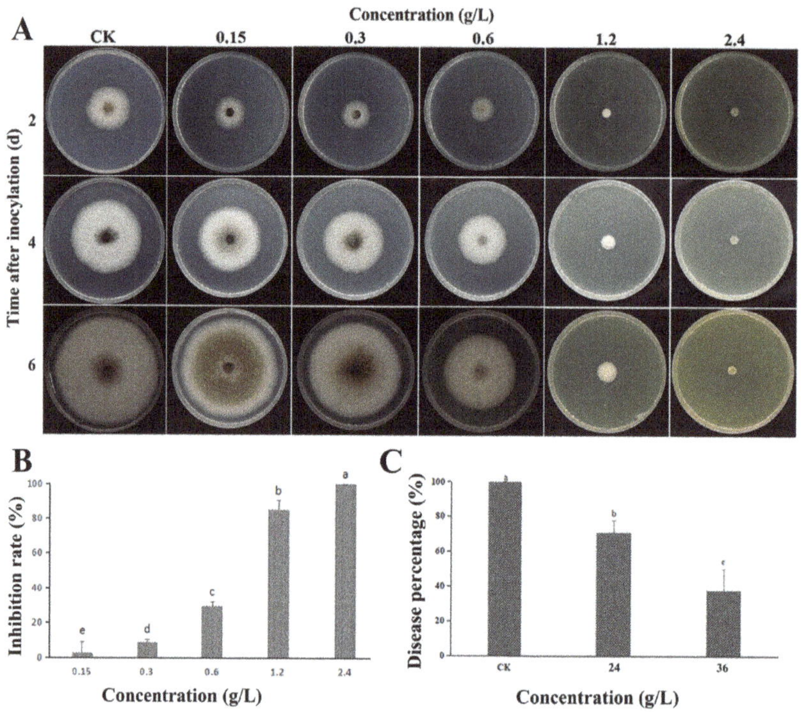

Figure 1. Inhibitory effect of $FeCl_3$ on *C. gloeosporioides* growth in vitro and in vivo. (**A**) $FeCl_3$ inhibits the mycelial growth of *C. gloeosporioides* in vitro; (**B**) Statistical analysis of the effects of different $FeCl_3$ concentrations on the mycelial growth of *C. gloeosporioides* (6 d); (**C**) $FeCl_3$ inhibits *C. gloeosporioides* virulence in vivo (7 d); the control group (CK) was treated with sterile water. All data are reported as the mean ± SD. Different letters indicate significant differences according to Duncan's multiple range test ($p < 0.05$).

PI staining: As shown by PI staining, the number of inactivated spores increased as the concentration of $FeCl_3$ increased. Almost all spores (98.13%) in the control group were not stained (Figure 3B), whereas substantially more *C. gloeosporioides* spores (6.52% and 18.15%) were stained in the 1/2 MIC and MIC treatment groups (Figure 3C), indicating that the membrane of the *C. gloeosporioides* spores was damaged.

3.4. Effect of $FeCl_3$ on Internal Structure and Autophagic Activity

Surprisingly, our TEM observations revealed the presence of transparent irregular-shaped vacuolar structures in the cytosol of the $FeCl_3$-treated conidia (Figure 4A). Therefore, MDC staining was conducted to identify these structures. As shown in Figure 4B, the $FeCl_3$-treated spores exhibited concentrated spots of MDC fluorescence, whereas the normal spores were uniformly dyed in MDC fluorescence. Apoptotic spores were also observed and appeared as stained particles of varying sizes and densities, indicating that their chromatin condensed, and the cell nucleus burst. Furthermore, these effects were more notable at higher $FeCl_3$ concentrations (as shown by the arrow).

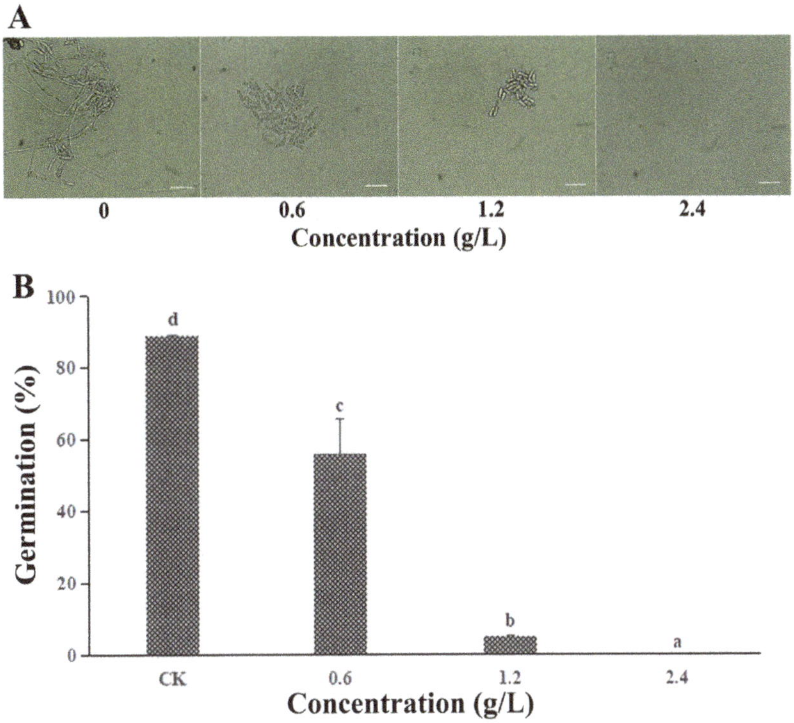

Figure 2. Effects of $FeCl_3$ treatments on *C. gloeosporioides* spore germination. (**A**) Inhibition efficacy of $FeCl_3$ on *C. gloeosporioides* spore germination (scale bar = 20 μm); (**B**) Spore germination rates. All data are reported as the mean ± SD. Different letters indicate significant differences according to Duncan's multiple range test ($p < 0.05$).

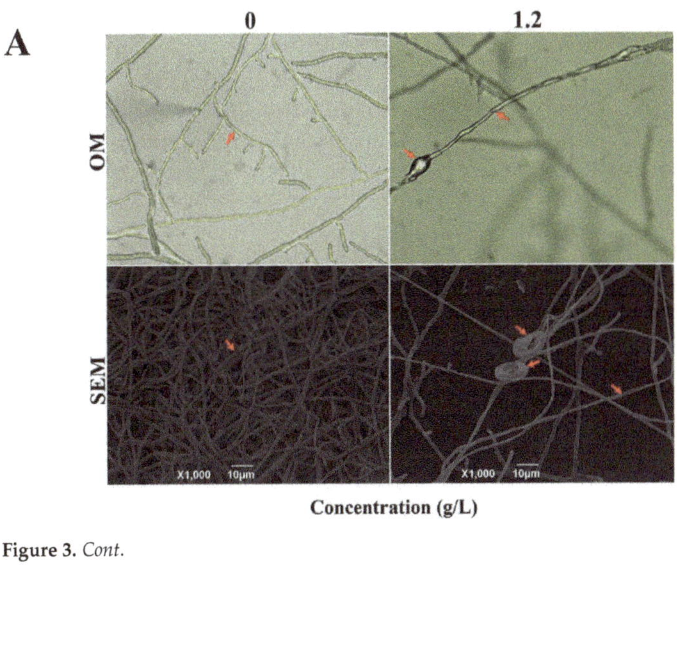

Figure 3. *Cont.*

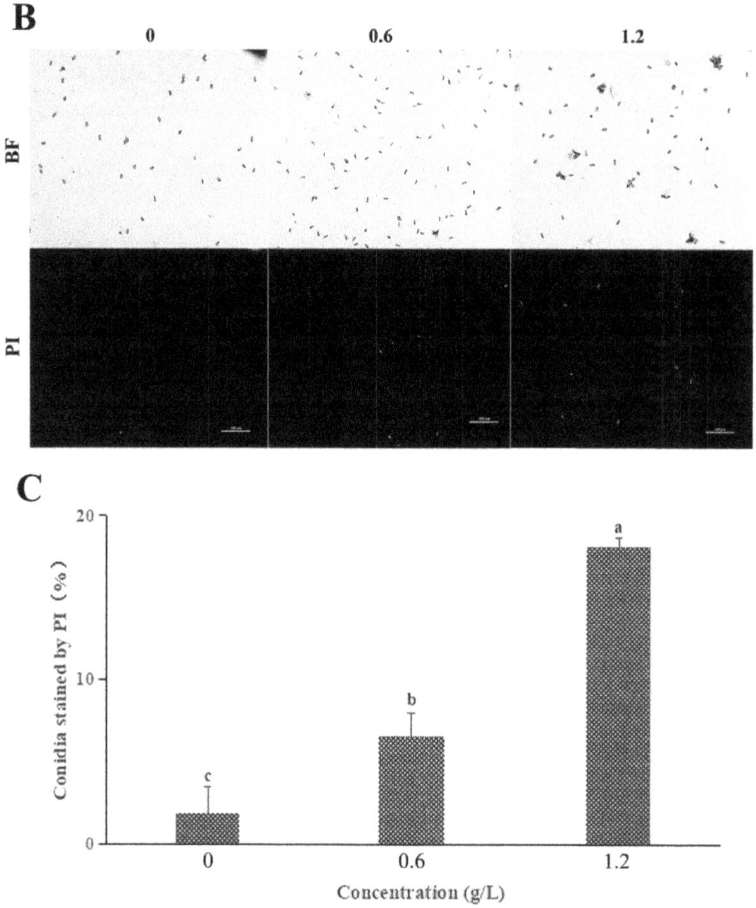

Figure 3. Effects of $FeCl_3$ on the cell membrane integrity of *C. gloeosporioides*. (**A**) Observation of mycelium morphology by OM and SEM (hyphal shrinkage, depression, coiling, and winding are indicated by arrows); (**B**) Observation of PI staining; (**C**) Statistical analysis for spores stained with PI. The error bars indicate the standard error of the mean of three replicate isolates. All data are reported as the mean ± SD. Different letters indicate significant differences according to Duncan's multiple range test ($p < 0.05$).

3.5. Effect of $FeCl_3$ on Vital Activity and ROS Accumulation

Overall, spores with high viability were detected by FDA green fluorescence. As shown in Figure 5A, there was no significant difference between the stained spores of the control and treated groups, indicating that $FeCl_3$ did not inactivate the spores. However, in the detection of reactive oxygen species, the control spores (3.61%) were rarely detected with high DCHF-DA fluorescence intensity labeling, whereas *C. gloeosporioides* spores cultured with 1/2 MIC and MIC $FeCl_3$ showed 29.27% and 52.33% staining, respectively (Figure 5B,C). This suggests that $FeCl_3$ induces the accumulation of ROS within the spores of *C. gloeosporioides*.

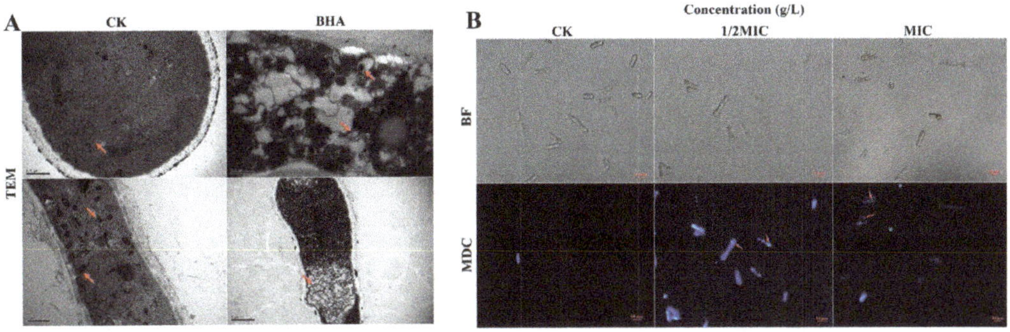

Figure 4. $FeCl_3$ triggers autophagosome formation. (**A**) Observation of internal structures by TEM (the arrows indicate the double-membrane vesicular structures (autophagosomes)); (**B**) MDC staining of *C. gloeosporioides* (the arrows indicate the condensed chromatin, i.e., autophagosomes).

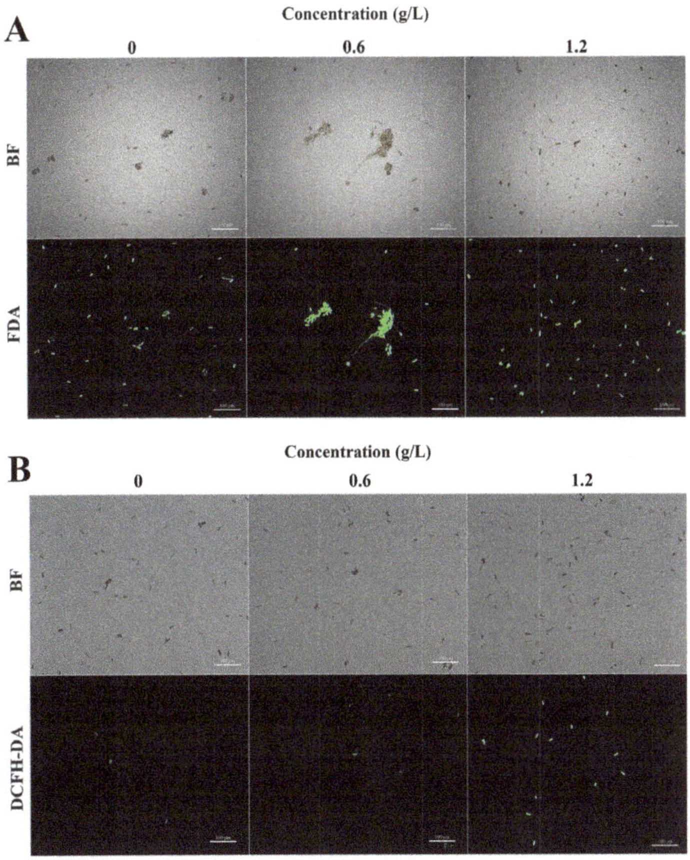

Figure 5. *Cont.*

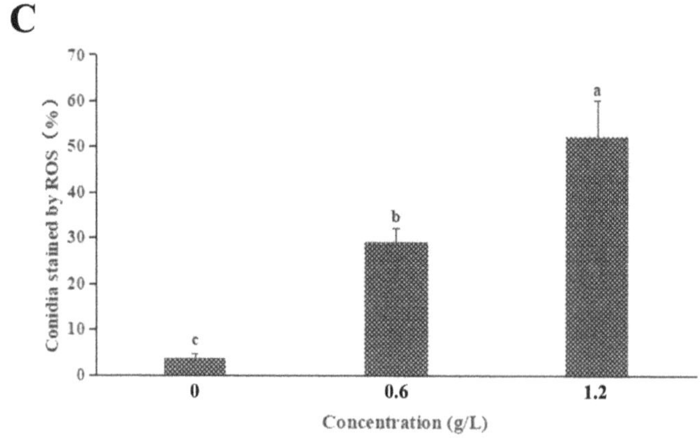

Figure 5. $FeCl_3$ decreases cell viability and increases intracellular ROS levels. (**A**) *C. gloeosporioides* conidia stained by FDA; (**B**) *C. gloeosporioides* conidia stained by DCFH-DA; (**C**) Statistical analysis of excessive ROS accumulation. All data are reported as the mean ± SD. Different letters indicate significant differences according to Duncan's multiple range test ($p < 0.05$).

3.6. Physiological Qualities Test of $FeCl_3$-Treated Citrus Fruit

After 30 days of storage, citrus fruit treated with $FeCl_3$ displayed TSS, TA, and VC values that were comparable to those of citrus fruit treated with water and a chemical fungicide (Figure 6A–C). These three groups all saw fruit weight decreases of roughly 2.5%. (Figure 6D). These findings demonstrated that $FeCl_3$ had no byproduct effects on citrus fruit.

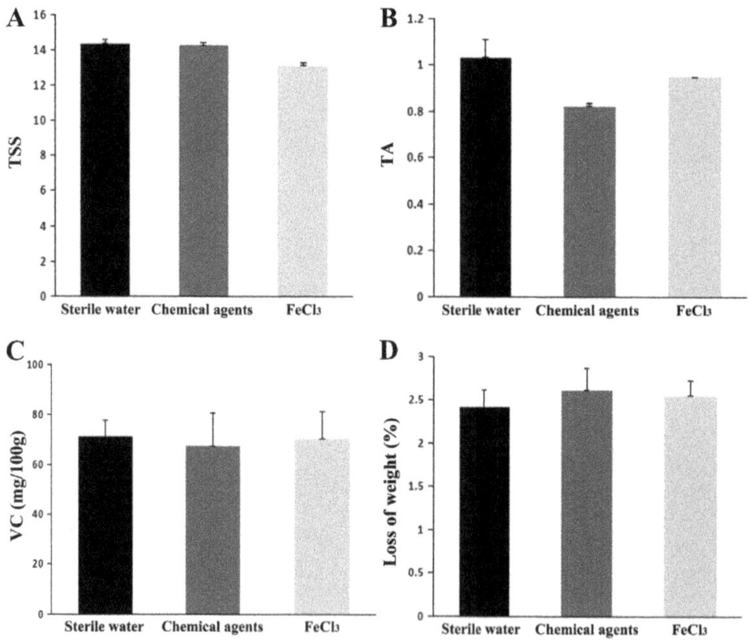

Figure 6. Physiological qualities test of $FeCl_3$-treated citrus fruit. (**A**) Total soluble solids (TSS). (**B**) Titratable acid (TA). (**C**) L-Ascorbic acid (VC). (**D**) Fruit weight loss. The fruits were stored in the ventilating chamber for 30 days.

4. Discussion

Anthracnose is one of the most prevalent diseases among a variety of fruit trees, particularly citrus, apple, mango, and other fruit trees [21–23]. As discussed above, mancozeb and copper compounds alone or coupled with fosetyl-Al are chemical fungicides that are commonly used in agriculture [4–6]. However, the uncontrolled use of these agents can promote the occurrence of fungicide-resistant pathogenic fungi, in addition to affecting human and environmental health [7,24]. Autophagy research has mainly focused on basic biology, whereas relatively few studies have been conducted on plants and plant pathogens. Nevertheless, the study of plant antibacterial mechanisms has recently garnered increasing attention [25]. Our studies demonstrated that autophagy could be exploited for the development of green disease prevention and treatment strategies in croplands.

TEM and MDC fluorescence staining are two commonly used methods to confirm the occurrence of autophagy [26]. In this study, our TEM and MDC fluorescence staining observations confirmed the formation of autophagy structures in $FeCl_3$-treated *C. gloeosporioides* spores (Figure 4A). Numerous studies have demonstrated that autophagy plays a crucial role in pathogen–plant interaction [27], which is consistent with the findings of this study.

Autophagy is rarely studied in fungi, and most of the research data are focused on biomedicine and plants. It has been reported that autophagy plays an important role in the life cycle of fungi, and most autophagy is beneficial to life itself. However, this viewpoint has not been reported for *C. gloeosporioides*. According to the experimental results in vitro and in vivo, $FeCl_3$-treated *C. gloeosporioides* not only produce autophagy phenomena but also have their growth significantly inhibited. Therefore, it is reasonable to assume that $FeCl_3$-induced autophagy of *C. gloeosporioides* is harmful to *C. gloeosporioides*.

The $FeCl_3$-treated *C. gloeosporioides* spores exhibited significant growth inhibition both in vitro and in vivo, suggesting that $FeCl_3$ exerted a potent antifungal effect (Figure 2). The results of the spore germination experiment showed that the germination rate of spores in the $FeCl_3$ treatment group decreased with higher $FeCl_3$ concentrations. Moreover, the diameter and length of the germinated spores in the treated group were smaller than in the control group. Therefore, MDC fluorescence staining was used to observe the internal changes of the spore cells (Figure 3B). SEM and TEM were used to observe the external morphological changes of hyphae and the changes in the spore cells of the treated groups (Figures 3A and 4A). Our findings demonstrated that, unlike the control group, the chromatin of the $FeCl_3$-treated spore cells was condensed, as demonstrated by the occurrence of concentrated bright spots in the MDC fluorescence staining experiments. In the control group, the staining was uniformly spread throughout the spore cells. Additionally, our SEM observations revealed that the hyphae of pathogenic fungi exhibited different degrees of coiling and shrinking. Our TEM observations revealed the occurrence of various double-membrane vesicular autophagic structures and countless vacuolated structures in the interior of the spore cells. Our findings suggest that the generation of autophagy structures affects spore germination and the formation of infective structures of pathogenic fungi, thereby reducing the invasiveness and pathogenicity of pathogenic fungi (Figure 7). Previous studies have demonstrated that autophagy can affect the growth, vitality, and pathogenicity of pathogenic fungi by inducing autophagy-related functions and apoptosis [20,28]. Therefore, we concluded that $FeCl_3$ triggers apoptosis by inducing autophagy in *C. gloeosporioides*.

Interestingly, our PI and FDA staining experiments demonstrated that although the $FeCl_3$-treated spores exhibited some degree of membrane damage, the cells did not die and were still viable (Figures 3B and 5A). These results were inconsistent with our original hypothesis that the induction of autophagy inhibits hyphal growth and triggers cell apoptosis in pathogenic fungi. Moreover, previous reports have demonstrated that autophagy promotes cell apoptosis in pathogenic fungi, which is also inconsistent with our findings [29]. After studying the inhibition mechanism of iron ion magnetic nanomaterials on model fungi, Qi Peng et al. concluded that the antifungal properties of iron ions could not be attributed to common plasma membrane damage or cell wall damage [30].

Therefore, we performed DCFH-DA fluorescent staining on the spores of the pathogenic fungi (Figure 5B,C). In the $FeCl_3$ treatment group, the DCFH-DA staining rate increased at higher $FeCl_3$ concentrations. Furthermore, our TEM observations revealed the occurrence of irreversible degradation of some organelles in the spore cells, such as mitochondria. The formation of autophagic structures leads to the degradation of the contents of the spore cells, which includes but is not limited to mitochondria.

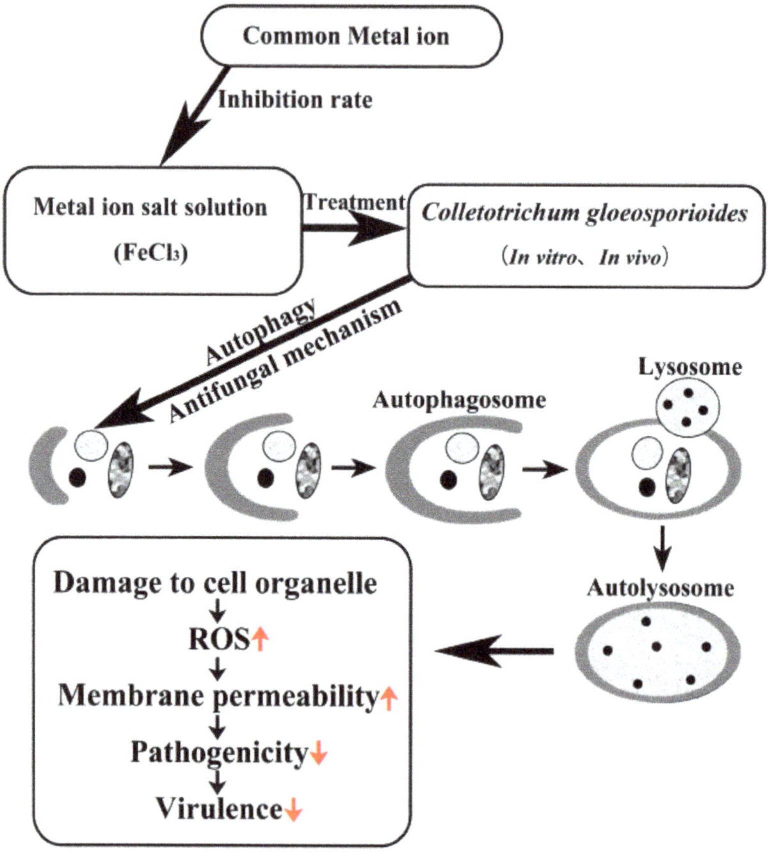

Figure 7. Hypothetic model of the effect of $FeCl_3$ treatment on *C. gloeosporioides* pathogenicity. ↑: up-regulated; ↓: down-regulated.

Combining the results of the above in vitro and in vivo experiments, we found that $FeCl_3$ significantly inhibited the growth of *C. goleosporioides*, both in mycelium growth and spore germination. Zhonghuan Tian et al. explored the physiological quality of fruits in terms of their application in producing areas [31]. In order to explore the application performance of $FeCl_3$ in actual production, we conducted a post-harvest storage and fresh-keeping experiment on 'Newhall' navel orange fruits (*C. sinensis* Osbeck) in a citrus industry orchard in Zhijiang (Hubei, China). The post-harvest treatment of fruit is simulated by soaking the fruit. After one month of positive storage, the results showed that the total soluble solid (TSS), titrable acid (TA), L-ascorbic acid (VC), and fruit weight loss were not significantly affected by water, $FeCl_3$, and chemical treatments. These results suggest that $FeCl_3$ has no by-product effect on citrus fruits [32]. In this regard, we believe that $FeCl_3$ can be used as a substitute for chemical fungicides to control anthracnose in producing areas in the future.

Therefore, our findings provided insights into the inhibitory mechanisms of $FeCl_3$ on *C. gloeosporioides* (Figure 7). Here, we conclude that $FeCl_3$ could induce autophagy for self-protection and increase the ROS content in pathogenic fungi. $FeCl_3$ could damage the internal structure of pathogenic fungi and induce the abnormal development of pathogenic fungi infection structures, thus decreasing the pathogenicity and virulence of *C. gloeosporioides*. Further assays are required to determine the controlling effect of anthracnose during citrus storage under general agronomic situations when $FeCl_3$ is applied, and research to detect its potential compatibility with other treatments is desired. $FeCl_3$, a type of iron salt with low cost and safety, offers a new option for replacing existing chemical fungicides.

5. Conclusions

Our findings demonstrate that $FeCl_3$ effectively inhibited the growth of *C. gloeosporioides* both in vitro and in vivo. $FeCl_3$ induced autophagosome formation and ROS accumulation, leading to organelle destruction and increased cell membrane permeability; however, it did not cause lethality. Our preliminary analysis of the mechanism of action and targets of $FeCl_3$ provides a theoretical basis for the future application of $FeCl_3$ to control citrus anthracnose. Additionally, our findings could serve as a guide for the future elucidation and characterization of invasion-related genes. However, additional studies are still needed to gain a more in-depth understanding of the molecular mechanisms through which $FeCl_3$ inhibits the growth of *C. gloeosporioides*. Based on the experiments in Zhijiang (Hubei, China), $FeCl_3$ is expected to replace chemical agents in the control of anthracnose in the future.

Author Contributions: Y.W. and X.W.: Data curation, investigation, validation, methodology, writing—original draft. Y.L., H.F., S.L. and J.Z.: investigation, methodology. C.L.: project administration, supervision, writing—review and editing. All authors have read and agreed to the published version of the manuscript.

Funding: This study was financially supported by the National Natural Science Foundation of China (Grant Nos. 31972122 & 32172255) and China Agriculture Research System of MOF and MARA.

Institutional Review Board Statement: Not applicable.

Informed Consent Statement: Not applicable.

Data Availability Statement: No new data were created in this study. Data sharing is not applicable to this article.

Conflicts of Interest: The authors declare that they have no known competing financial interest or personal relationship that could have appeared to influence the work reported in this paper.

References

1. Wang, W.; de Silva, D.; Moslemi, A.; Edwards, J.; Ades, P.; Crous, P.; Taylor, P.W.J. Colletotrichum Species Causing Anthracnose of Citrus in Australia. *J. Fungi* **2021**, *7*, 47. [CrossRef] [PubMed]
2. Zhang, J.; Timmer, L. Preharvest application of fungicides for postharvest disease control on early season tangerine hybrids in Florida. *Crop Prot.* **2007**, *26*, 886–893. [CrossRef]
3. Mehta, P.; Wiltse, C.; Rooney, W.; Collins, S.; Frederiksen, R.; Hess, D.; Chisi, M.; TeBeest, D.O. Classification and inheritance of genetic resistance to anthracnose in Sorghum Field. *Crops Res.* **2005**, *93*, 1–9. [CrossRef]
4. Giulio, P.; Raffaele, C.; Giancarlo, P.; Antonino, A.; Ernesto, L.; Dolores, F.; Alessandro, V. In vitro and in vivo activity of QoI fungicides against Colletotrichum gloeosporioides causing fruit anthracnose in Citrus sinensis. *Sci. Hortic.* **2018**, *236*, 90–95. [CrossRef]
5. Feng, L.; Wu, F.; Li, J.; Jiang, Y.; Duan, X. Antifungal activities of polyhexamethylene biguanide and polyhexamethylene guanide against the citrus sour rot pathogen Geotrichum citri-aurantii in vitro and in vivo. *Postharvest Biol. Technol.* **2011**, *61*, 160–164. [CrossRef]
6. Cia, P.; Pascholati, S.; Benato, E.; Camili, E.; Santos, C. Effects of gamma and UV-C irradiation on the postharvest control of papaya anthracnose. *Postharvest Biol. Technol.* **2007**, *43*, 366–373. [CrossRef]
7. Jeong, S.W.; Kim, H.G.; Park, S.; Lee, J.H.; Kim, Y.-H.; Kim, G.-S.; Jin, J.S.; Kwak, Y.-S.; Huh, M.R.; Lee, J.E.; et al. Variation in flavonoid levels in Citrus benikoji Hort. ex. Tan. infected by Colletotrichum gloeosporioides. *Food Chem.* **2014**, *148*, 284–288. [CrossRef]

8. Hernández-Monjaraz, W.S.; Caudillo-Pérez, C.; Ulises, P.; Salazar-Sánchez; Macías-Sánchez, K.L. Influence of iron and copper on the activity of laccases in Fusarium oxysporum f. sp. Lycopersici. *Braz. J. Microbiol.* **2018**, *49* (Suppl. S1), 269–275. [CrossRef]
9. Nazik, H.; Penner, J.C.; Ferreira, J.A.; Haagensen, J.A.; Cohen, K.; Spormann, A.M.; Martinez, M.; Chen, V.; Hsu, J.L.; Clemons, K.V.; et al. Effects of Iron Chelators on the Formation and Development of Aspergillus fumigatus Biofilm. *Antimicrob. Agents Chemother.* **2015**, *59*, 6514–6520, Erratum in: *Antimicrob. Agents Chemother.* **2015**, *59*, 7160. [CrossRef]
10. Food and Drug Administration. Health and human services. 2009, Pt.184: 73—FR—8607[S]. Available online: https://www.fda.gov/media/77832 (accessed on 9 January 2023).
11. Yang, M.; Ismayil, A.; Liu, Y. Autophagy in Plant-Virus Interactions. *Annu. Rev. Virol.* **2020**, *7*, 403–419. [CrossRef]
12. Yin, Z.; Pascual, C.; Klionsky, D. Autophagy: Machinery and regulation Microb. *Cell* **2016**, *3*, 588–596.
13. Klionsky, D.J. The molecular machinery of autophagy: Unanswered questions. *J. Cell Sci.* **2005**, *118 Pt 1*, 7–18. [CrossRef]
14. Bursch, W.; Ellinger, A.; Gerner, C.; Schulte-Hermann, R. Autophagocytosis and programmed cell death. In *Autophagy*; Klionsky Landes, D.J., Ed.; Bioscience: Georgetown, TX, USA, 2004; pp. 287–303.
15. Bergamini, E.; Cavallini, G.; Donati, A.; Gori, Z. The anti-ageing effects of caloric restriction may involve stimulation of macroautophagy and lysosomal degradation, and can be intensified pharmacologically. *Biomed. Pharmacother.* **2003**, *57*, 203–208. [CrossRef]
16. Longo, V.D.; Finch, C.E. Evolutionary medicine: From dwarf model systems to healthy centenarians? *Science* **2003**, *299*, 1342–1346. [CrossRef]
17. Melendez, A.; Tallóczy, Z.; Seaman, M.; Eskelinen, E.-L.; Hall, D.H.; Levine, B. Autophagy genes are essential for dauer development and life-span extension in C. elegans. *Science* **2003**, *301*, 1387–1391. [CrossRef]
18. Vellai, T.; Takacs-Vellai, K.; Zhang, Y.; Kovacs, A.L.; Orosz, L.; Muller, F. Influence of TOR kinase on lifespan in C. elegans. *Nature* **2003**, *426*, 620. [CrossRef]
19. Liu, S.Q.; Du, Y.J.; Zhang, D.Y.; Yang, F.; He, X.; Long, C.A. Aluminum sulfate inhibits green mold by inducing chitinase activity of Penicillium digitatum and enzyme activity of citrus fruit. *Food Control* **2022**, *136*, 108854. [CrossRef]
20. Cui, X.; Ma, D.; Liu, X.; Zhang, Z.; Li, B.; Xu, Y.; Chen, T.; Tian, S. Magnolol inhibits gray mold on postharvest fruit by inducing autophagic activity of Botrytis cinerea. *Postharvest Biol. Technol.* **2021**, *180*, 111596. [CrossRef]
21. Kim, Y.S.; Lee, Y.; Cheon, W.; Park, J.; Kwon, H.T.; Balaraju, K.; Kim, J.; Yoon, Y.J.; Jeon, Y. Characterization of Bacillus velezensis AK-0 as a biocontrol agent against apple bitter rot caused by Colletotrichum gloeosporioides. *Sci. Rep.* **2021**, *11*, 626. [CrossRef]
22. Fang, X.; Chai, W.; Li, S.; Zhang, L.; Yu, H.; Shen, J.; Xiao, W.; Liu, A.; Zhou, B.; Zhang, X. HSP17.4 mediates salicylic acid and jasmonic acid pathways in the regulation of resistance to Colletotrichum gloeosporioides in strawberry. *Mol. Plant Pathol.* **2021**, *22*, 817–828. [CrossRef]
23. Sudheeran, P.K.; Sela, N.; Carmeli-Weissberg, M.; Ovadia, R.; Panda, S.; Feygenberg, O.; Maurer, D.; Oren-Shamir, M.; Aharoni, A.; Alkan, N. Induced defense response in red mango fruit against Colletotrichum gloeosporioides. *Hortic. Res.* **2021**, *8*, 17. [CrossRef] [PubMed]
24. Liu, S.; Zhang, D.; Wang, Y.; Yang, F.; Zhao, J.; Du, Y.; Tian, Z.; Long, C. Dimethyl Dicarbonate as a Food Additive Effectively Inhibits Geotrichum citri-aurantii of Citrus. *Foods* **2022**, *11*, 2328. [CrossRef] [PubMed]
25. Zhu, X.M.; Li, L.; Wu, M.; Liang, S.; Shi, H.B.; Liu, X.H.; Lin, F.C. Current opinions on autophagy in pathogenicity of fungi. *Virulence* **2019**, *10*, 481–489. [CrossRef] [PubMed]
26. Cinque, L.; Forrester, A.; Bartolomeo, R.; Svelto, M.; Venditti, R.; Montefusco, S.; Polishchuk, E.; Nusco, E.; Rossi, A.; Medina, D.L.; et al. FGF signalling regulates bone growth through autophagy. *Nature* **2015**, *528*, 272–275. [CrossRef]
27. Hofius, D.; Li, L.; Hafrén, A.; Coll, N.S. Autophagy as an emerging arena for plant-pathogen interactions. *Curr. Opin. Plant Biol.* **2017**, *38*, 117–123. [CrossRef]
28. Talbot, N.; Kershaw, M. The emerging role of autophagy in plant pathogen attack and host defence. *Curr. Opin. Plant Biol.* **2009**, *12*, 444–450. [CrossRef]
29. Ma, D.; Cui, X.; Zhang, Z.; Li, B.; Xu, Y.; Tian, S.; Chen, T. Honokiol suppresses mycelial growth and reduces virulence of Botrytis cinerea by inducing autophagic activities and apoptosis. *Food Microbiol. J.* **2020**, *88*, 103411. [CrossRef]
30. Peng, Q.; Huo, D.; Li, H.; Zhang, B.; Li, Y.; Liang, A.; Wang, H.; Yu, Q.; Li, M. ROS-independent toxicity of Fe3O4 nanoparticles to yeast cells: Involvement of mitochondrial dysfunction. *Chem. Biol. Interact.* **2018**, *287*, 20–26. [CrossRef]
31. Tian, Z.; Chen, C.; Chen, K.; Liu, P.; Fan, Q.; Zhao, J.; Long, C. Biocontrol and the mechanisms of Bacillus sp. w176 against postharvest green mold in citrus. *Postharvest Biol. Technol.* **2020**, *159*, 111022. [CrossRef]
32. Tian, Z.; Li, Y.; Yang, Q.; Liu, P.; Du, Y.; Chen, C.; Long, C. Antifungal activities and the mechanisms of biocontrol agent WE-3 against postharvest sour rot in citrus. *Eur. J. Plant Pathol.* **2021**, *161*, 723–733. [CrossRef]

Journal of
Fungi

Article

Antifungal Activity of Perillaldehyde on *Fusarium solani* and Its Control Effect on Postharvest Decay of Sweet Potatoes

Chao Pan [1], Kunlong Yang [1], Famous Erhunmwunsee [1], Bo Wang [1], Dongjing Yang [2], Guoquan Lu [1,3], Man Liu [1], Yongxin Li [1] and Jun Tian [1,*]

[1] School of Life Science, Jiangsu Normal University, Xuzhou 221116, China
[2] Xuzhou Institute of Agricultural Sciences in Jiangsu Xuhuai District, Xuzhou 221131, China
[3] School of Agriculture and Food Science, Zhejiang A & F University, Hangzhou 311300, China
* Correspondence: tj-085@163.com; Tel.: +86-516-83403172

Abstract: Root rot caused by *Fusarium solani* is one of the major postharvest diseases limiting sweet potato production. Here, antifungal activity and the action mode of perillaldehyde (PAE) against *F. solani* were investigated. A PAE concentration of 0.15 mL/L in air (mL/L air) markedly inhibited the mycelial growth, spore reproduction and spore viability of *F. solani*. A PAE vapor of 0.25 mL/L in air could control the *F. solani* development in sweet potatoes during storage for 9 days at 28 °C. Moreover, the results of a flow cytometer demonstrated that PAE drove an increase in cell membrane permeability, reduction of mitochondrial membrane potential (MMP) and accumulation of reactive oxygen species (ROS) in *F. solani* spores. Subsequently, a fluorescence microscopy assay demonstrated that PAE caused serious damage to the cell nuclei in *F. solani* by inducing chromatin condensation. Further, the spread plate method showed that the spore survival rate was negatively correlated with the level of ROS and nuclear damage, of which the results indicated that PAE-driven ROS accumulation plays a critical role in contributing to cell death in *F. solani*. In all, the results revealed a specific antifungal mechanism of PAE against *F. solani*, and suggest that PAE could be a useful fumigant for controlling the postharvest diseases of sweet potatoes.

Keywords: root rot; storage; spore viability; mitochondria; reactive oxygen species

1. Introduction

The Food and Agriculture Organization of the United Nations reported that the yield of sweet potato (*Ipomoea batatas* Lam.) was 88.87 million tons all over the world in 2021, and about three quarters of this yield came from China [1]. Sweet potato is an important food crop around the world, especially in developing countries, because it can produce more dry matter, protein and minerals per unit area in comparison to cereals [2,3]. Due to containing a number of nutritional factors, such as carbohydrates, carotenes, vitamins, potassium et al., sweet potato roots have been developed into many foods and beverages [4]. However, because they have a high moisture and carbohydrate content, the roots are easily infected by *Fusarium solani*, resulting in postharvest deterioration [5,6].

The filamentous fungus *F. solani* is a plant and human pathogen belonging to the *Fusarium solani* species complex (FSSC), and its sexual state is *Nectria haematococca* [7]. *F. solani* is ubiquitously distributed in soil and decaying plant materials, where it acts as a decomposer [8]. However, *F. solani* is an important pathogen of a number of agriculturally important crops; for instance, soybean, potato and tomato, et al. [9]. *Fusarium* root rot caused by this fungus is one of the major postharvest diseases of sweet potato, particularly in the southeastern United States [5], and the main sweet potato-growing areas in China [3] and the Republic of Korea [10]. Muggy conditions, such as a high temperature (13 to 35 °C) and relative humidity (over 80%), contribute to this disease incidence [5]. It was reported that, in China in 2014 and South Korea in 2017, the incidence of Fusarium root rot on sweet potatoes caused by *F. solani* was 10–20% [10]. Further, *F. solani* is able to cause stem lesions and end rot of sweet potatoes in the process of cultivation, leading to a substantial reduction of sweet potato production [5]. On the other hand, *F. solani* was found to be an opportunistic pathogen causing human diseases, such as fungal keratitis or invasive mycoses [9]. Therefore, controlling the contamination of *F. solani* is a crucial approach to ensure food safety and human health.

Wounding is a prerequisite for an infection of sweet potatoes by *F. solani*. However, sweet potato roots are easily wounded due to a relatively thin and delicate skin during its harvest or transportation to market [6]. So far, some management strategies have been developed to reduce the microbial spoilage of sweet potato roots. Specifically, wound healing in large storage containers soon after harvest has served as an important strategy to minimize *Fusarium* root rot [5]. Nevertheless, due to a lack of large storage containers to implement the wound healing process, farmers prefer to use chemical fungicides to combat the sweet potato spoilage; for example, carbendazim, a cheap broad-spectrum fungicide which is widely used, especially in China [11]. However, the use of carbendazim for the storage of edible roots posed a potential risk to human health [12]. Hence, there is an urgent need to develop green preservatives for the storage of the edible part of plants [13].

Perillaldehyde (PAE), a natural monocyclic terpenoid, is a kind of essential oil (EOs) that is abundant in the perennial herb *Perilla* [14]. PAE has been used as a flavoring agent for foods such as baked goods, meat products and beverages et al. [15]. The foods have been certified as safe by the Food and Agriculture Organization of the United Nations (FAO) and United States Food and Drug Administration (FDA) [16]. Interestingly, PAE presents an effective antifungal activity against some pathogenic and food spoilage fungi, such as *Candida albicans* [17], *Ceratocystis fimbriata* [18] and *Aspergillus flavus* [19]. Thus, PAE holds promise as a novel antifungal agent used in food preservation [16]. However, the antifungal activity of PAE against *F. solani* remains unclear.

It is reported that apoptosis plays a key role in the fungistatic pathway executed by the activity of EOs [16], and the reactive oxygen species (ROS) produced by the mitochondria is a major key marker of the apoptotic process [20]. Our previous work revealed that PAE induced cell apoptosis of *A. flavus* via ROS accumulation [21]. According to our previous results of transcriptome sequencing, PAE drove an inhibition of glycometabolism resulting in the indirect suppression of glutathione synthesis in *A. flavus*, which contributed to reducing the ROS-scavenging capacity, ultimately leading to ROS accumulation [22]. *F. solani* was reported to be efficient in utilizing carbohydrates, by including a large number of multiple-copy coding genes for carbohydrate-active enzymes [7]. The results seemed to imply that *F. solani* may possess a high resistance to PAE due to its high carbohydrate-using capacity. In addition, in clinical practice, the infection of *F. solani* is difficult to treat because *Fusarium* spp. are highly resistant to most antifungals [23], such as amphotericin B and imidazoles [24]. However, whether PAE exhibits effective antifungal activity against *F. solani* in sweet potatoes remains unclear.

In the present study, the antifungal activity of PAE against *F. solani* and its preservative effect on sweet potatoes were estimated. First, the effects of PAE on mycelial growth, spore production and viability were examined. Second, the preservative effect of PAE on sweet potato roots was evaluated. Moreover, the mode of antifungal action of PAE was investigated by detecting the cell membrane integrity, mitochondrial membrane potential (MMP), ROS level and nuclear morphometry. With this information, the essential oil PAE can be recommended as a novel green preservative to limit the amount of postharvest loss of sweet potatoes due to *Fusarium* root rot.

2. Materials and Methods

2.1. Chemicals, Strain and Plant Materials

The PAE (CAS no. 18031-40-8, purity > 90.0%) was purchased from Tokyo Chemical Industry Co., Ltd. (Tokyo, Japan). The PAE was prepared as 10 × stock solutions in 0.1% (*v*/*v*) Tween 80 with an ultrasonic wave treatment for 30 min. *F. solani* X14011 was originally separated from the rot spot on sweet potato roots, and obtained from Hebei Academy of Agriculture and Forestry Sciences in China [25]. The fungus was cultured on potato dextrose agar (PDA; 20% potato, 2% dextrose, 1.5% agar) for 7 d at 28 °C. Sweet potato roots of a commercial cultivar Xushu 32 grown for about 150 d were obtained from the experimental station of the Xuzhou Institute of Agricultural Sciences in Jiangsu Xuhuai District in 2022. After harvest, these storage roots were placed into a storage facility, where they were cured at 29 °C for 7 days [5]. After wound healing, the roots were stored at 13 to 15 °C until inoculation.

2.2. Determination of Antifungal Activity

The effect of PAE on the mycelial growth of *F. solani* was tested using direct contact and vapor phase contact method [26]. In the direct contact method, 8-mm-diameter mycelial plugs were placed on the center of each PDA plate (9 cm diameter) supplemented with 0.01% Tween 80 and different PAE concentrations of 0.125, 0.25, 0.5, 0.75, 1, 1.25 and 1.5 mL/L. In the vapor phase contact method, mycelial plugs were inoculated on PDA plates containing 15 mL PDA and 80 mL air in the space of these dishes. The PAE was dissolved into methyl alcohol to gain different concentrations of stock solutions [27], and then aliquots of 100 μL stock solutions were pipetted on the inside of the lids of every plate to obtain various PAE concentrations of 0, 0.0125, 0.025, 0.05, 0.075, 0.1, 0.125, 0.15 and 0.175 mL/L in air (mL/L air) in the air space of plates. After that, all plates were sealed with a polythene preservative film, incubated for 9 d at 28 °C and colony diameters were measured every 24 h. To calculate the colony diameters, the following formula was used:

$$a = b - c \quad (1)$$

where a: colony diameter (cm); b: measured diameter of colony (cm); c: mycelial plug diameter (cm).

Further, the effect of PAE on the spore productivity of *F. solani* was tested [28]. After the colony diameters were measured, a certain volume of phosphate buffer solution (PBS, pH 7.0–7.2) was added to the plates and then both mycelia and spores were scraped off using a spreading rod. Following, the cells were transferred into 50 mL centrifuge tubes, and shaken violently with a vortex mixer. The spore number of each plate was counted using a hemocytometer under a light microscope.

The effect of PAE on spore viability was tested using a coating method via the contact method and vapor phase method [20,26]. In the contact approach, 100 μL spore suspension (2×10^3 spores/mL) was spread on PDA containing various PAE concentrations of 0, 0.25, 0.5, 1.0 and 1.5 mL/L. In the vapor phase assay, spore suspension was firstly spread on PDA plates, and then the plates were supplemented with 100 mL of PAE stock solutions dissolved in methanol to obtain various PAE concentrations of 0, 0.025, 0.05 0.10 and 0.15 mL/L air. After 5 d of inoculation, the number of colony forming units (CFUs) was counted. To calculate the percentage of conidial survival rate, the following formula was used:

$$a = b/c \times 100\% \quad (2)$$

where a: conidial survival rate (%); b: CFUs of PAE-treated group; c: CFUs of control group.

2.3. Determination of the Effect of PAE on Sweet Potato Preservation

The preventive effect of PAE on sweet potatoes [29,30] was detected using a vapor phase contact method [31], as shown in Figure 1. Firstly, 15 mL water agar (1.5% agar) was poured into 2 cm high petri dishes. After solidification, the dishes were placed inversely, and one sterile filter paper was placed on the inside of the lids of every plate. Afterwards, healthy sweet potato roots approximately four centimeters in diameter were washed, peeled and sterilized with 1% NaOCl for 10 min. Following this, 1 cm thick slices were cut from the equatorial region of the roots, and these slices were individually placed on the filter paper in each petri dish. Subsequently, 8 mm diameter mycelial plugs were placed on the center of each sweet potato slice. Finally, 100 mL of PAE stock solutions dissolved in methanol was pipetted on the filter paper resulting in a series of PAE concentrations of 0, 0.05, 0.1, 0.15, 0.2 and 0.25 mL/L air in the air in petri dishes. The dishes were incubated for 12 d at 28 °C, and lesion diameters were measured every 3 d. There were four replicates for each treatment. To calculate the percentage of the conidial survival rate, the following formula was used:

$$a = b - c \tag{3}$$

where a: lesion diameter (cm); b: measured lesion diameter (cm); c: mycelial plug diameter (cm).

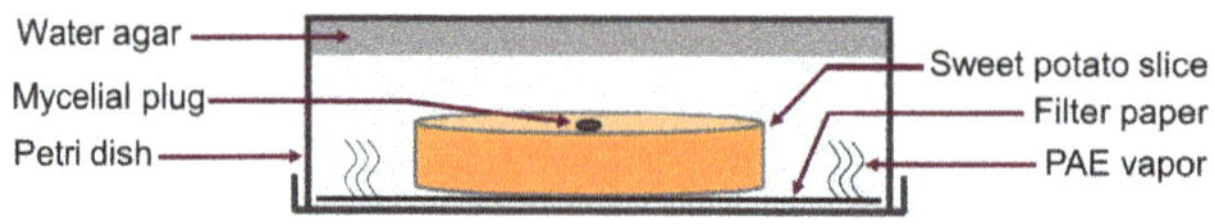

Figure 1. Schema for depicting the assay method determining the effect of PAE on sweet potato preservation.

2.4. Measurement of Cell Membrane Integrity

Cell membrane integrity was monitored by propidium iodide (PI; Solarbio, Beijing, China) [21]. A spore suspension of 5×10^6 spores/mL was incubated with 1.5 mL/L PAE in a rotary shaker for 2, 4 and 8 h at 28 °C. The spores without PAE treatment were regarded as a control group. After incubation, the spores were washed twice using PBS, and then stained with 1 mL of 10 mg/L PI for 30 min at 28 °C. Finally, the spores were washed three times with PBS, and analyzed using an Accuri™ C6 flow cytometer (BD Biosciences, San Jose, CA, USA).

2.5. Determination of MMP

MMP was detected by the fluorescent dye Rhodamine 123 (Rh123; Solarbio, Beijing, China) [18]. A spore suspension (5×10^6 spores/mL) was incubated with 1.5 mL/L PAE, for 0, 2, 4 and 8 h at 28 °C. After incubation, the spores were washed twice, and stained with 100 μg/L Rh123 for 30 min at 28 °C. The spores that were stained with distilled water instead of Rh123 were served as a dye-blank control. Finally, the spores were washed and analyzed by the flow cytometer.

2.6. Determination of ROS Level

The ROS production in *F. solani* spores was detected by fluorescent dye 2,7-Dichlorodihydrofluorescein diacetate (DCFH-DA; Sigma-Aldrich, St. Louis, MO, USA) [13]. Spore suspension of 5×10^6 spores/mL was incubated with 1.5 mL/L PAE for 4 and 8 h, 1.5 mL/L PAE plus 80 mM cysteine (Cys) for 4 and 8 h and 80 mM H_2O_2 for 8 h at 28 °C. Spores without PAE treatment were regarded as a control group. After treatment, the spores were washed and stained with 10 μM DCFH-DA for 30 min at 28 °C. Finally, the spores were analyzed by the flow cytometer. Meanwhile, spore suspension was serially diluted and spread on PDA. After an incubation for 5 d at 28 °C, the number of CFUs in each group was counted, and the conidial survival rate was calculated using Formula (2).

2.7. Determination of Nuclear Morphology

The effect of PAE on the nuclear morphology of *F. solani* was detected using 4′,6-diamidino-2-phenylindole (DAPI; Solarbio, Beijing, China) [18]. A spore suspension (5×10^6 spores/mL) was incubated with 0.75 and 0.15 mL/L PAE, 0.75 or 0.15 mL/L PAE plus 80 mM Cys and with 80 mM H_2O_2 for 12 h at 28 °C. After incubation, the spores were stained with 10 mg/L of DAPI for 30 min at 28 °C. After they were washed, these spores were placed on a glass slide and examined using a fluorescence microscope (Leica, Wetzlar, Germany). In addition, the spore suspension was diluted, and spread on PDA plates. After an incubation, the number of CFUs was counted, and the conidial survival rate was also calculated.

2.8. Statistical Analysis

The pathogenicity assay was carried out in quadruplicate, and the other assays were performed in triplicate. The results are expressed as mean ± standard deviations (SD), and the statistical significance was calculated by a one-way ANOVA with Duncan multiple range tests using SPSS 21 software (IBM, Chicago, IL, USA). Different letters indicated statistically significant differences at $p < 0.05$.

3. Results

3.1. Antifungal Activity of PAE against F. solani

The antifungal activity of PAE against *F. solani* was evaluated by detecting mycelial growth, spore production and viability. Diameters and spore productivity of the colonies that were treated with different concentrations of PAE for 9 d were measured via the contact method and vapor phase contact method. In the contact method, 0.125 mL/L PAE showed an inhibitory effect ($p < 0.05$) on mycelial growth and spore production (Figure 2A,B). The inhibitory effects of PAE on mycelial growth and spore production were in a dose-dependent manner. Notably, mycelial growth was completely inhibited as the PAE concentration reached 1.5 mL/L, and the spore production was also inhibited. This result indicated that the minimal inhibitory concentration (MIC) of the contact method against *F. solani* was 1.5 mL/L PAE. In the vapor phase method, mycelial growth and spore production were suppressed by PAE in a dose-dependent manner (Figure 2D,E). It is worth noting that 0.15 mL/L air PAE inhibited ($p < 0.05$) mycelial growth for 7 d, and 0.175 mL/L air PAE completely inhibited the mycelial growth of *F. solani* (Figure 2D), indicating that the MIC of the vapor phase method against *F. solani* was 1.5 mL/L PAE. Meanwhile, the spore production was also significantly inhibited ($p < 0.05$) under the PAE concentrations of 0.15 and 0.175 mL/L air (Figure 2E). Both assays above proved that PAE, especially PAE vapor, possesses effective inhibitory activity against the mycelial growth and spore production of *F. solani*.

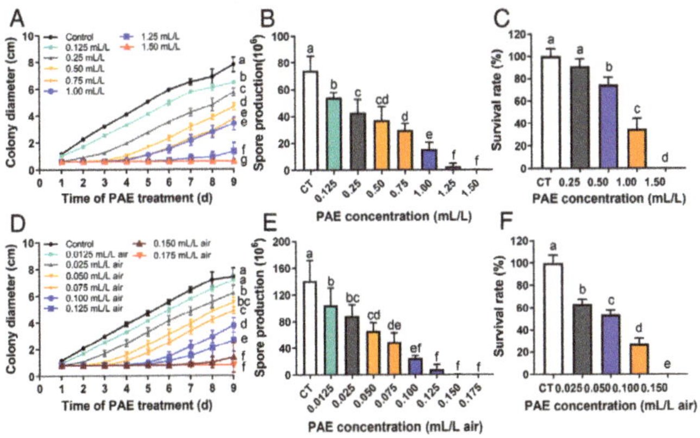

Figure 2. Effects of PAE on mycelial growth, spore production and viability of *F. solani*. Colony diameter (**A**) and spore production (**B**) analyzed by contact method after mycelial plugs were incubated with different concentrations of PAE for 9 d. Colony diameter (**D**) and spore production (**E**) evaluated via vapor phase method after mycelial plugs were incubated with indicated concentrations of PAE for 9 d. Spore survival rate assessed by contact method (**C**) and vapor phase method (**F**) after spores were incubated with different concentrations of PAE for 5 d. All data are presented as mean ± standard deviations (n = 3). Different letters indicate statistical significant differences ($p < 0.05$) calculated by one-way ANOVA.

Spore viability was tested using a coating method via the contact method and vapor phase contact method after the spores were treated with different concentrations of PAE for 5 d. In the contact method, the PAE concentration reached 0.5 mL/L which showed an obvious inhibitory effect ($p < 0.05$) on spore survival rate (Figure 2C). The inhibition degree increased as the PAE concentration increased. Exposure to 1.5 mL/L PAE completely inhibited the spore viability. In the vapor phase method, the PAE concentration reached 0.025 mL/L air, which showed an inhibitory effect ($p < 0.05$) on spore survival rate, and 0.15 mL/L PAE completely inhibited the spore viability (Figure 2F). Hence, PAE has a noteworthy inhibitory ability against spore viability in *F. solani*.

3.2. Effect of PAE on Sweet Potato Preservation

To evaluate of anti-decay effect of PAE vapor treatment, a pathogenicity assay of *F. solani* on sweet potatoes was carried out. The sweet potato slices inoculated with mycelial plugs were exposed to various concentrations of PAE for different days. The sweet potato slices without the PAE treatment decayed seriously, and a dark and sunken lesion appeared on their surface (Figure 3A). The result proved that the strain *F. solani* X14011 has a strong pathogenicity on sweet potato roots. A PAE concentration of 0.05 mL/L air exhibited a slight inhibitory effect on the expansion of the decayed diameter (Figure 3B). The inhibition degree increased as the PAE concentration increased. When the PAE concentration reached 0.25 mL/L air, sweet potato spoilage could be suppressed up to 9 d (Figure 3F,G). Therefore, PAE showed a notable preservative effect on sweet potato roots infected by *F. solani*.

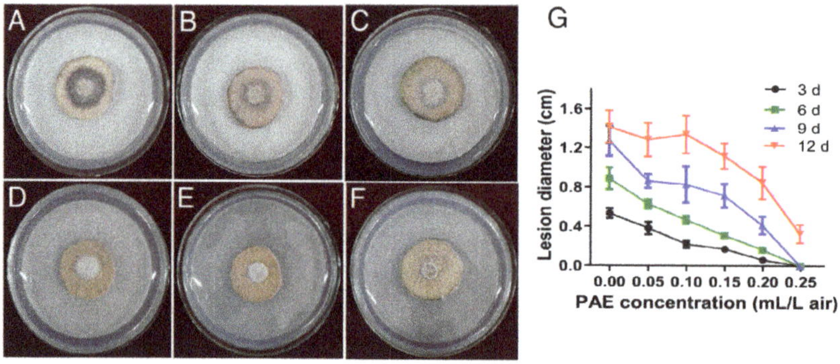

Figure 3. Effect of PAE on sweet potato spoilage caused by *F. solani*. The sweet potato slices inoculated with mycelial plugs were exposed to different PAE concentrations of 0 (**A**), 0.05 (**B**), 0.10 (**C**), 0.15 (**D**), 0.20 (**E**) and 0.25 (**F**) mL/L air for 12 d at 28 °C. Line chart of lesion diameter on these slices (**G**). Error bars represent the mean standard deviation of four replicates.

3.3. Effect of PAE on Cell Membrane Integrity

Cell membrane integrity was estimated using PI staining by a flow cytometer after *F. solani* spores were exposed to 1.5 mL/L PAE for different hours. In the control group without PAE treatment, the spores were not stained with PI, and their fluorescence intensity values were mainly about 10^3 (Figure 4A). After exposure to PAE for 2 h, the rate of stained spores rose from 21.1% in the control group (Figure 4A) to 38.2% (Figure 4B) with an 81.0% increase ($p < 0.05$). The rate of stained spores increased as the PAE exposure time extended (Figure 4E). As the treatment time of PAE reached 8 h, an obvious intensity peak appeared at the value of 10^4 on the horizontal axis, and the stained spore rate rose to 57.1% ($p < 0.05$, Figure 4E). This assay demonstrated that PAE caused obvious damage to cell membrane integrity in *F. solani*.

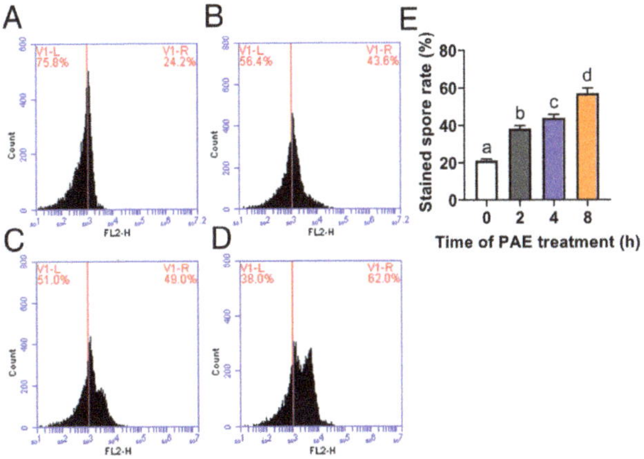

Figure 4. Distribution of fluorescence intensity values of cell membrane integrity in *F. solani* detected by PI staining, after spores exposed to 1.5 mL/L PAE for 0 (**A**), 2 (**B**), 4 (**C**) and 8 (**D**) h. Bar graph of rate of stained spores (**E**). Values are mean ± standard deviations (n = 3). Different letters indicate statistically significant differences ($p < 0.05$) calculated by one-way ANOVA.

3.4. *Effect of PAE on MMP*

MMP was detected using Rh123 staining by a flow cytometer after *F. solani* spores were exposed to 1.5 mL/L PAE for different hours. In the control group, there were two fluorescence intensity peaks (Figure 5B), of which the abscissa value of the first peak was approximately 10^3, similar to the value of the only fluorescence intensity peak in the dye-blank control (Figure 5A), and of which the abscissa value of another intensity peak that was produced by the stained spores was about 10^5, indicating that these spores had a high value of MMP. After treatment with PAE for 4 h, the peak area obviously decreased ($p < 0.05$), suggesting that PAE caused a reduction of MMP in *F. solani* spores. As the PAE treatment time reached 8 h, the second fluorescence intensity peak of the Rh123-stained spores almost disappeared, and the abscissa value of the first peak was similar to the fluorescence intensity peak in the dye-blank control (Figure 5A,E), implying that its MMP almost disappeared. So, the above result demonstrated that PAE destroyed MMP in *F. solani*.

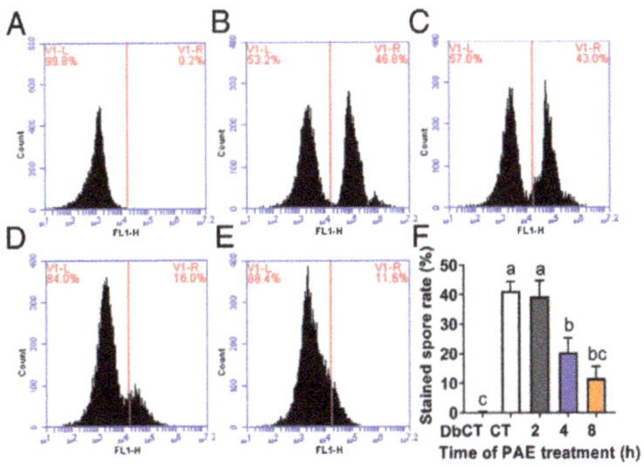

Figure 5. Distribution of fluorescence intensity values of mitochondrial membrane potential (MMP) in *F. solani* detected by Rh123 staining. Spores were exposed to 1.5 mL/L PAE for 0 as control (CT) (**B**), 2 (**C**), 4 (**D**) and 8 (**E**) h. The spores that were stained with water instead of Rh123 were served as a dye-blank control (DbCT) (**A**). Bar graph of stained spore rate (**F**). Values are mean ± standard deviations (n = 3). Different letters indicate statistically significant differences ($p < 0.05$) calculated by one-way ANOVA.

3.5. Effect of PAE on ROS Accumulation

The ROS level was measured using DCFH-DA staining by a flow cytometer after the spores of *F. solani* were exposed to 1.5 mL/L PAE for different hours. In the control group without PAE treatment, there was only one fluorescence intensity peak (Figure 6A). After incubation of PAE for 4 h, the rate of stained spores significantly ($p < 0.05$) increased (Figure 6B). The rate of stained spores increased as the time of PAE exposure extended (Figure 6G). After incubation of PAE for 8 h, there was a new-appeared fluorescence intensity peak that was located near 10^5 on the horizontal axis (Figure 6C), of which the abscissa value was similar to the abscissa value of the second fluorescence intensity peak in the H_2O_2-treated group (Figure 6D), indicating that PAE obviously drove ROS accumulation in *F. solani* spores. On the other hand, in the groups with spores that were treated by 1.5 mL/L PAE plus 80 mM Cys for 4 and 8 h (Figure 6E,F), there was just one fluorescence intensity peak located near 10^3 on the horizontal axis, similar to the non-PAE treated group (Figure 6A), proving that the antioxidant Cys had eliminated the intracellular excess ROS induced by PAE. Hence, the above result demonstrated that PAE was a key driver of ROS accumulation in *F. solani* spores.

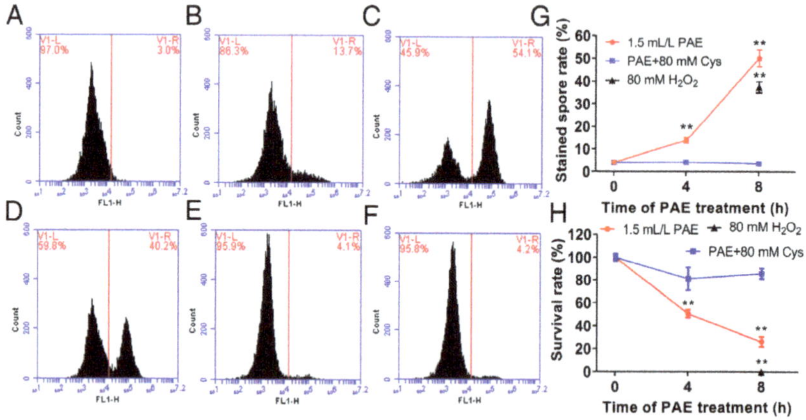

Figure 6. Effect of PAE on ROS level in *F. solani* detected by DCFH-DA staining. Spores were exposed to 1.5 mL/L PAE for 4 (**B**) and 8 (**C**) h, 80 mM H_2O_2 for 8 (**D**) h, and to 1.5 mL/L PAE plus 80 mM cysteine (Cys) for 4 (**E**) and 8 (**F**) h. Spores without PAE treatment were served as control (**A**). Line charts of rate of stained spores (**G**) and survival rate of spores in every experimental group (**H**). Values are mean ± standard deviations ($n = 3$). ** indicates $p < 0.01$ compared with the control group calculated by one-way ANOVA.

In order to analyze the role of ROS accumulation in PAE antifungal action, the survival rate of spores in every experimental group was further measured by the coating method. Figure 6H showed that PAE exhibited an ability to kill the *F. solani* spores in a concentration-dependent manner, and a treatment of 80 mM H_2O_2 for 8 h completely killed all the spores ($p < 0.05$). However, Cys supplementation in the PAE-treated group recovered the spore survival rate up to the level of the control group without PAE treatment. The above results strongly indicated that ROS accumulation induced by PAE plays a key role in killing *F. solani* spores.

3.6. Effect of PAE on Nuclear Morphometry

Nuclear morphometry was observed by DAPI staining after the spores incubated with different concentrations of PAE for 12 h. As shown in the control group (Figure 7A), the spore nuclei without PAE treatment was big and bright. However, the fluorescence intensity of the nuclei treated by 0.75 mL/L PAE became weaker and smaller than the control group (Figure 7A,B). When the PAE concentration reached 1.5 mL/L, the nuclear morphology disappeared (Figure 7C), similar to the H_2O_2-treated group (Figure 7D), proving that PAE drove severe nuclear damage in *F. solani* spores. Moreover, in order to estimate the relationship between nuclear damage and cell death caused by PAE, the survival rate of the spores was further quantified by the coating method. The PAE-treated concentration of 0.75 and 1.5 mL/L killed about 50 and 100% of the spores ($p < 0.01$), respectively (Figure 7G); this result was consistent with the changes of the nuclear morphology (Figure 7B,C). On the other hand, supplementation of the antioxidant Cys in the PAE-treated group was able to maintain a nuclear morphometry big and bright similar to the control group (Figure 7A,E,F), proving that the elimination of excess PAE-induced ROS contributed to keeping the nuclear architecture. Additionally, Cys supplementation had an ability to contribute to the restoration of the spore survival rate under PAE stress (Figure 7G). The above results proved that PAE treatment drove nuclear damage in the *F. solani* spores, and PAE-induced ROS accumulation acted as a critical cause of nuclear damage and spore death in *F. solani*.

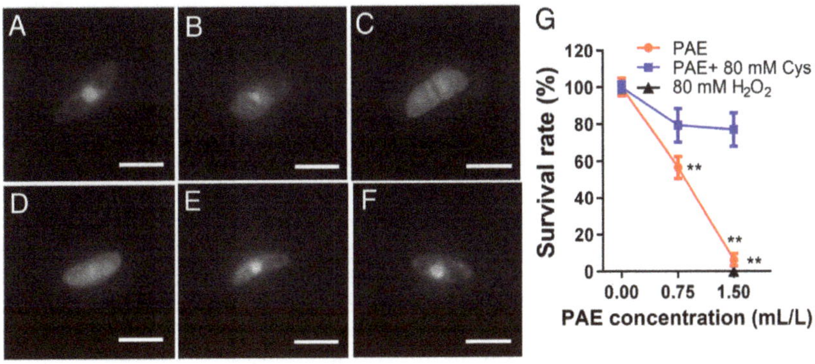

Figure 7. Effect of PAE on nuclear morphometry measured by DAPI staining. Spores were exposed to 0.75 (**B**) and 1.5 (**C**) mL/L PAE, 80 mM H_2O_2 (**D**), and to 0.75 (**E**) or 1.5 (**F**) mL/L PAE plus 80 mM Cys for 12 h. Spores without PAE treatment were served as control (**A**). Bar = 50 μm. Line chart of survival rate of spores in every experimental group (**G**). Values are mean ± standard deviations ($n = 3$). ** indicates $p < 0.01$ compared with the control group calculated by one-way ANOVA.

4. Discussion

Because of having a high moisture and free sugar content, sweet potato storage roots are easily infected by *F. solani,* leading to postharvest spoilage [5,6]. However, the use of chemical fungicides such as carbendazim for the storage of this edible crop posed a potential risk to human health [32]. As a promising green preservative, the antimicrobial effects of PAE have been reported [13,33]. Although eugenol, carvacrol or cinnamaldehyde had a stronger antifungal activity than PAE in traditional contact methods [34], PAE was reported to exhibit a higher antibacterial activity against airborne microbes than these three Eos, which was tested by the vapor method using an air washer [33]. The reason why PAE exhibited a higher antimicrobial activity may depend on its high volatility. Therefore, in order to develop PAE as a novel antifungal agent used in food preservation, the antifungal activity of PAE on *F. solani* was estimated through a vapor treatment, and its preservation effect on sweet potato roots was also evaluated in the present study.

In clinical practice, *F. solani* was reported to have a high resistance to some antifungals, such as amphotericin B [24]. In order to know the antifungal activity of PAE against *F. solani*, the effect on mycelial growth was firstly evaluated in the forms of a contact and vapor treatment. In mycelium assay, the MIC against the mycelial growth of *F. solani* in the direct contact method and vapor contact method was 1.5 mL/L PAE (Figure 2A) and 0.175 mL/L air (Figure 2D), respectively. According to our previous publications, 1 mL/L of PAE could completely suppress the growth of *Aspergillus niger*, which was determined using a micro-well dilution method [35]. In this regard, PAE also exhibited effective antifungal activity against the mycelial growth of *F. solani*. Mycelia are responsible for the spoilage of fruits and vegetables. Therefore, PAE could be considered as an antifungal candidate to protect sweet potato roots from *Fusarium* root rot.

The fungus *F. solani* is widely distributed in the environment, and produces a large numbers of spores which are the major source of infectious diseases in crops and food spoilage [9]. So, the effect of PAE on the spore production of *F. solani* was further evaluated. As shown in Figure 2B and E, MICs of 1.5 mL/L PAE in the contact method and 0.175 mL/L PAE in the vapor method could also completely inhibit spore production. On the other hand, spore viability was reported to be a main cause of root colonization in plants [36]. Thus, the effect of PAE on spore viability was further tested. Additionally, PAE showed an ability to suppress the spore viability of *F. solani*. PAE concentrations reached 1.5 mL/L in the contact method and 0.15 mL/L air in the vapor method, completely inhibiting the viability of the *F. solani* spores (Figure 2C,F). Interestingly, the dose of PAE against the spore viability of *F. solani* in the vapor method was as low as one-tenth of the dose in the contact method. For instance, in the group treated with 0.1 mL/L air PAE in vapor method was 27.36% (Figure 2C), the spore survival rate was lower than 34.65% of the group treated with 1 mL/L PAE in the contact method (Figure 2F), indicating that using PAE via the vapor phase method exhibited a more efficient inhibition of spore viability than the contact method. The result was consistent with our previous research on the antifungal effect of PAE on *A. niger* (Wang et al., 2015). Hence, PAE, specially via vapor treatment, has noteworthy antifungal activity against the mycelia and spores of *F. solani*.

F. solani can cause not only stem lesions and end rot in sweet potatoes in the process of cultivation [37], but also, *Fusarium* root rot during the storage period [5]. *Fusarium* root rot is a kind of dry decay, and its typical lesions on sweet potato roots are circular, with light and dark brown concentric rings, causing significant economic loss during storage [5,29]. Therefore, the effect of PAE on the control of *F. solani* infection of sweet potato roots in storage was further evaluated. In this study, slices cut from the equatorial region of sweet potato roots were used to perform the pathogenicity test [3,29]. The method can ensure uniform inoculation to the central area of each sweet potato root instead of the outer layer of sweet potato flesh with varying degrees of cell development [38]. On the other hand, the slices were inoculated with mycelial plugs; such a large inoculum would magnify the extent of pathogen infection compared with the natural infection caused by fungal spores [39]. In the control group without PAE treatment (Figure 3A), the sweet potato slices seriously decayed with an occurrence of dark and sunken lesions on its surface, a symptom which is consistent with previous reports [5], whereas PAE had an ability to inhibit the decay and spoilage of sweet potato slices (Figure 3G). Therefore, the preservative assay indicated that the PAE vapor treatment contributed to controlling the *F. solani* infection of sweet potato roots.

The cell membrane is commonly regarded as an important target of EOs [40]. The cell membrane is important in maintaining cell physiology, and its main function is to protect the internal substances from not leaking to the cell exterior. The cell membrane of fungi is enriched with diverse lipids, and due to the hydrophobic character of EO constituents, the cell membrane is easily attacked by EOs resulting in an increase of its permeability [40]. PI, a kind of DNA-binding fluorescent dye, cannot enter the cell until the cell membrane structure is disrupted [41]. A flow cytometer assay showed that the spores in the control group were not stained by PI (Figure 4A); however, the spores exposed to PAE led to an obvious PI influx (Figure 4B–D), indicating that cell membrane integrity had been disrupted. The result demonstrated that PAE had the ability to damage the cell membrane function of *F. solani*, which is consistent with previous reports of *A. flavus* [21].

Mitochondria are organelle composed of two layers of plasma membrane and are reported to be another important target of EOs [34]. Mitochondria play a key role in generating cellular ATP through oxidative phosphorylation depending on the MMP, and MMP is an important indicator of changes in mitochondrial function [41]. Rh123 is a specific fluorescent probe for mitochondria in living cells, and this dye preferentially enters mitochondria based on a high value of MMP [21]. In the control group, the spores without PAE treatment were stained with Rh123, while PAE treatment led to an obvious reduction of stained spore rate (Figure 5), indicating MMP being damaged by PAE. The result indicated that PAE disrupted the mitochondrial function in *F. solani*. This similar phenomenon was reported in research on *C. fimbriata* [18].

Dysfunctional mitochondria caused by fungistats could result in the generation of excess intracellular ROS, and subsequently in cell death due to its oxidative damage to cellular macromolecules [20]. DCFH-DA is a kind of ROS-sensitive probe used for ROS detection [13]. Furthermore, the effect of PAE on the ROS level and cell viability of *F. solani* was analyzed. PAE drove ROS accumulation in cells (Figure 6B,C) and meanwhile resulted in spore death in a concentration-dependent manner (Figure 6H). It is well known that Cys is a classical antioxidant to eliminate ROS in cells. As expected, Cys supplementation restored the ROS level to a low value similar to the non-PAE treated group, and also protected cell death induced by PAE (Figure 6E–H). Therefore, the combined use of DCFH-DA staining and a spore survival test demonstrated that ROS driven by PAE plays a critical role in contributing to cell death in *F. solani*. This result was consistent with previous reports of *C. fimbriata* [18].

Usually, excessive intracellular ROS is regarded as a molecular signal to initiate apoptosis, and nuclear damage is a late marker of apoptosis [21]. In order to confirm whether nuclear damage occurred, the nuclear morphology was further observed via DAPI staining. DAPI, a DNA-specific fluorescent probe, is commonly used to detect nuclear morphology changes. A DAPI assay showed that one half of the PAE MIC caused chromatin condensation in the PAE-treated spores, and PAE MIC resulted in the disappearance of nuclear morphology in *F. solani* spores (Figure 7B,C); this result was consistent with the reduction of spore viability induced by the corresponding PAE concentration (Figure 7G). In addition, Cys supplementation could contribute to maintaining nuclear morphology, and to restoring spore viability under PAE stress (Figure 7E–G). Although previous research has proved that PAE can disrupt the nuclear morphology in *A. flavus* [21,42], this result further demonstrates that ROS driven by PAE plays a critical role in killing *F. solani* spores by causing nuclear damage.

5. Conclusions

In summary, this study demonstrated that PAE shows notable inhibitory effects on *F. solani* and on sweet potato decay. A PAE vapor concentration of 0.15 mL/L air markedly inhibited the mycelial growth, spore reproduction and spore viability of *F. solani*. Furthermore, a PAE vapor of 0.25 mL/L air could control the *F. solani* development in sweet potato roots during storage for 9 days at 28 °C. The possible mode of antifungal action of PAE may be dependent on its ability to induce excess ROS generation. PAE firstly disrupts the barrier property of the cell membrane leading to an increase of cell membrane permeability, and subsequently drives a decrease in MMP. Consequently, excess ROS is derived from the dysfunctional mitochondria. Finally, the excess intracellular ROS plays a critical role in triggering cell death via inducing nuclear damage characterized by chromatin condensation. This study indicates that PAE vapor possesses an effective inhibitory activity against *F. solani*. In order to develop PAE as a novel fumigant used in food preservation, more research will be needed to develop the application strategies based on its volatile features.

Author Contributions: Conceptualization, J.T. and G.L.; methodology, C.P. and B.W.; validation, K.Y.; writing–original draft, F.E.; resources, D.Y.; formal analysis, M.L.; supervision, Y.L.; writing—review and editing; funding acquisition, J.T. All authors have read and agreed to the published version of the manuscript.

Funding: This work was supported by the National Natural Science Foundation of China (31972171, 32202180), 333 High-level Personnel Training Project of Jiangsu Province, Natural Science Foundation by Xuzhou City (KC21160), the earmarked fund for CARS-10-Sweetpotato, the Priority Academic

Program Development (PAPD) of Jiangsu Higher Education Institutions, and the International Cooperation Joint Laboratory of Jiangsu Province Colleges and Universities.

Institutional Review Board Statement: Not applicable.

Informed Consent Statement: Not applicable.

Data Availability Statement: All data sets in the current study are available from the corresponding author upon reasonable request.

Conflicts of Interest: The authors declare no conflict of interest.

References

1. Huang, M.; Cheng, J.; Chen, P.; Zheng, G.; Wang, D.; Hu, Y. Efficient production of succinic acid in engineered *Escherichia coli* strains controlled by anaerobically-induced *nirB* promoter using sweet potato waste hydrolysate. *J. Environ. Manag.* **2019**, *237*, 147–154. [CrossRef] [PubMed]
2. Oke, M.O.; Workneh, T.S. A review on sweet potato postharvest processing and preservation technology. *Afr. J. Agric. Res.* **2013**, *8*, 4990–5003. [CrossRef]
3. Xie, S.Y.; Ma, T.; Zhao, N.; Zhang, X.; Fang, B.; Huang, L. Whole-genome sequencing and comparative genome analysis of *Fusarium solani-melongenae* causing *Fusarium* root and stem rot in sweetpotatoes. *Microbiol. Spectr.* **2022**, *10*, e0068322. [CrossRef] [PubMed]
4. El Sheikha, A.F.; Ray, R.C. Potential impacts of bioprocessing of sweet potato: Review. *Crit. Rev. Food Sci. Nutr.* **2017**, *57*, 455–471. [CrossRef] [PubMed]
5. Scruggs, A.C.; Quesada-Ocampo, L.M. Etiology and epidemiological conditions promoting *Fusarium* root rot in sweetpotato. *Phytopathology* **2016**, *106*, 909–919. [CrossRef]
6. Ray, R.C.; Ravi, V. Post harvest spoilage of sweetpotato in tropics and control measures. *Crit. Rev. Food Sci. Nutr.* **2005**, *45*, 623–644. [CrossRef]
7. Coleman, J.J.; Rounsley, S.D.; Rodriguez-Carres, M.; Kuo, A.; Wasmann, C.C.; Grimwood, J.; Schmutz, J.; Taga, M.; White, G.J.; Zhou, S.; et al. The genome of *Nectria haematococca*: Contribution of supernumerary chromosomes to gene expansion. *PLoS Genet.* **2009**, *5*, e1000618. [CrossRef]
8. Zhang, N.; O'Donnell, K.; Sutton, D.A.; Nalim, F.A.; Summerbell, R.C.; Padhye, A.A.; Geiser, D.M. Members of the *Fusarium solani* species complex that cause infections in both humans and plants are common in the environment. *J. Clin. Microbiol.* **2006**, *44*, 2186–2190. [CrossRef]
9. Coleman, J.J. The *Fusarium solani* species complex: Ubiquitous pathogens of agricultural importance. *Mol. Plant Pathol.* **2016**, *17*, 146–158. [CrossRef]
10. Yang, J.-W.; Nam, S.-S.; Lee, H.-U.; Choi, K.-H.; Hwang, S.-G.; Paul, N.C. Fusarium root rot caused by *Fusarium solani* on sweet potato (*Ipomoea batatas*) in South Korea. *Can. J. Plant Pathol.* **2017**, *40*, 90–95. [CrossRef]
11. Xu, X.M.; Chen, J.Y.; Li, B.R.; Tang, L.J. Carbendazim residues in vegetables in China between 2014 and 2016 and a chronic carbendazim exposure risk assessment. *Food Control* **2018**, *91*, 20–25. [CrossRef]
12. Fan, R.; Zhang, W.; Li, L.; Jia, L.; Zhao, J.; Zhao, Z.; Peng, S.; Yuan, X.; Chen, Y. Individual and synergistic toxic effects of carbendazim and chlorpyrifos on zebrafish embryonic development. *Chemosphere* **2021**, *280*, 130769. [CrossRef]
13. Yang, K.; Geng, Q.; Luo, Y.; Xie, R.; Sun, T.; Wang, Z.; Qin, L.; Zhao, W.; Liu, M.; Li, Y.; et al. Dysfunction of FadA-cAMP signalling decreases *Aspergillus flavus* resistance to antimicrobial natural preservative perillaldehyde and AFB1 biosynthesis. *Environ. Microbiol.* **2022**, *86*, 326. [CrossRef] [PubMed]
14. Tian, J.; Zeng, X.B.; Zhang, S.; Wang, Y.Z.; Zhang, P.; Lu, A.J.; Peng, X. Regional variation in components and antioxidant and antifungal activities of *Perilla frutescens* essential oils in China. *Ind. Crops Prod.* **2014**, *59*, 69–79. [CrossRef]
15. Hobbs, C.A.; Taylor, S.V.; Beevers, C.; Lloyd, M.; Bowen, R.; Lillford, L.; Maronpot, R.; Hayashi, S.M. Genotoxicity assessment of the flavouring agent, perillaldehyde. *Food Chem. Toxicol.* **2016**, *97*, 232–242. [CrossRef]
16. Erhunmwunsee, F.; Pan, C.; Yang, K.; Li, Y.; Liu, M.; Tian, J. Recent development in biological activities and safety concerns of perillaldehyde from *Perilla* plants: A review. *Crit. Rev. Food Sci. Nutr.* **2022**, *62*, 6328–6340. [CrossRef] [PubMed]
17. Mcgeady, P.; Wansley, D.L.; Logan, D.A. Carvone and perillaldehyde interfere with the serum-induced formation of filamentous structures in *Candida albicans* at substantially lower concentrations than those causing significant inhibition of growth. *J. Nat. Prod.* **2002**, *65*, 953–955. [CrossRef] [PubMed]
18. Tian, J.; Pan, C.; Zhang, M.; Gan, Y.Y.; Pan, S.Y.; Liu, M.; Li, Y.X.; Zeng, X.B. Induced cell death in *Ceratocystis fimbriata* by pro-apoptotic activity of a natural organic compound, perillaldehyde, through Ca^{2+} overload and accumulation of reactive oxygen species. *Plant Pathol.* **2019**, *68*, 344–357. [CrossRef]
19. Li, Y.X.; Erhunmwunsee, F.; Liu, M.; Yang, K.; Zheng, W.; Tian, J. Antimicrobial mechanisms of spice essential oils and application in food industry. *Food Chem.* **2022**, *382*, 132312. [CrossRef]
20. Helmerhorst, E.J.; Troxler, R.F.; Oppenheim, F.G. The human salivary peptide histatin 5 exerts its antifungal activity through the formation of reactive oxygen species. *Proc. Natl. Acad. Sci. USA* **2001**, *98*, 14637–14642. [CrossRef]

21. Tian, J.; Wang, Y.; Lu, Z.; Sun, C.; Zhang, M.; Zhu, A.; Peng, X. Perillaldehyde, a promising antifungal agent used in food preservation, triggers apoptosis through a metacaspase-dependent pathway in *Aspergillus flavus*. *J. Agric. Food Chem.* **2016**, *64*, 7404–7413. [CrossRef] [PubMed]
22. Pan, C.; Li, Y.X.; Yang, K.; Famous, E.; Ma, Y.; He, X.; Geng, Q.; Liu, M.; Tian, J. The molecular mechanism of perillaldehyde inducing cell death in *Aspergillus flavus* by inhibiting energy metabolism revealed by transcriptome sequencing. *Int. J. Mol. Sci.* **2020**, *21*, 1518. [CrossRef] [PubMed]
23. Walther, G.; Stasch, S.; Kaerger, K.; Hamprecht, A.; Roth, M.; Cornely, O.A.; Geerling, G.; Mackenzie, C.R.; Kurzai, O.; von Lilienfeld-Toal, M. *Fusarium* keratitis in Germany. *J. Clin. Microbiol.* **2017**, *55*, 2983–2995. [CrossRef] [PubMed]
24. Venditti, M.; Micozzi, A.; Gentile, G.; Polonelli, L.; Morace, G.; Bianco, P.; Avvisati, G.; Papa, G.; Martino, P. Invasive *Fusarium solani* infections in patients with acute leukemia. *Rev. Infect. Dis.* **1988**, *10*, 653–660. [CrossRef] [PubMed]
25. Wang, R.Y.; Gao, B.; Li, X.H.; Ma, J.; Chen, S.L. First report of *Fusarium solani* causing *Fusarium* root rot and stem canker on storage roots of sweet potato in China. *Plant Dis.* **2014**, *98*, 160–161. [CrossRef] [PubMed]
26. Wang, Y.; Zeng, X.; Zhou, Z.; Xing, K.; Tessema, A.; Zeng, H.; Tian, J. Inhibitory effect of nerol against *Aspergillus niger* on grapes through a membrane lesion mechanism. *Food Control* **2015**, *55*, 54–61. [CrossRef]
27. Gulluce, M.; Sokmen, M.; Daferera, D.; Agar, G.; Ozkan, H.; Kartal, N.; Polissiou, M.; Sokmen, A.; Sahin, F. *In vitro* antibacterial, antifungal, and antioxidant activities of the essential oil and methanol extracts of herbal parts and callus cultures of *Satureja hortensis* L. *J. Agric. Food Chem.* **2003**, *51*, 3958–3965. [CrossRef]
28. Gong, Q.W.; Zhang, C.; Lu, F.X.; Zhao, H.Z.; Bie, X.M.; Lu, Z.X. Identification of bacillomycin D from *Bacillus subtilis* fmbJ and its inhibition effects against *Aspergillus flavus*. *Food Control* **2014**, *36*, 8–14. [CrossRef]
29. da Silva, W.L.; Clark, C.A. Infection of sweetpotato by *Fusarium solani* and *Macrophomina phaseolina* prior to harvest. *Plant Dis.* **2013**, *97*, 1636–1644. [CrossRef]
30. Pan, C.; Yang, K.; Erhunmwunsee, F.; Li, Y.-X.; Liu, M.; Pan, S.; Yang, D.; Lu, G.; Ma, D.; Tian, J. Inhibitory effect of cinnamaldehyde on *Fusarium solani* and its application in postharvest preservation of sweet potato. *Food Chem.* **2023**, *408*, 135213. [CrossRef]
31. Li, X.; Liu, M.; Huang, T.; Yang, K.; Zhou, S.; Li, Y.; Tian, J. Antifungal effect of nerol via transcriptome analysis and cell growth repression in sweet potato spoilage fungi *Ceratocystis fimbriata*. *Postharvest. Biol. Technol.* **2021**, *171*, 111343. [CrossRef]
32. Perreault, S.D.; Jeffay, S.; Poss, P.; Laskey, J.W. Use of the fungicide carbendazim as a model compound to determine the impact of acute chemical exposure during oocyte maturation and fertilization on pregnancy outcome in the hamster. *Toxicol. Appl. Pharmacol.* **1992**, *114*, 225–231. [CrossRef] [PubMed]
33. Sato, K.; Krist, S.; Buchbauer, G. Antimicrobial effect of *trans*-cinnamaldehyde, (−)-perillaldehyde, (−)-citronellal, citral, eugenol and carvacrol on airborne microbes using an airwasher. *Biol. Pharm. Bull.* **2006**, *29*, 2292–2294. [CrossRef]
34. Burt, S. Essential oils: Their antibacterial properties and potential applications in foods–a review. *Int. J. Food Microbiol.* **2004**, *94*, 223–253. [CrossRef] [PubMed]
35. Tian, J.; Wang, Y.; Zeng, H.; Li, Z.; Zhang, P.; Tessema, A.; Peng, X. Efficacy and possible mechanisms of perillaldehyde in control of *Aspergillus niger* causing grape decay. *Int. J. Food Microbiol.* **2015**, *202*, 27–34. [CrossRef]
36. Druille, M.; Cabello, M.N.; Omacini, M.; Golluscio, R.A. Glyphosate reduces spore viability and root colonization of arbuscular mycorrhizal fungi. *Appl. Soil. Ecol.* **2013**, *64*, 99–103. [CrossRef]
37. Clark, C.A. End rot, surface rot, and stem lesions caused on sweet potato by *Fusarium solani*. *Phytopathology* **1980**, *70*, 109–112. [CrossRef]
38. Ravi, V.; Chakrabarti, S.; Makeshkumar, T.; Saravanan, R. Molecular regulation of storage root formation and development in sweet potato. *Hortic. Rev.* **2014**, *42*, 157–208. [CrossRef]
39. Xing, K.; Li, T.J.; Liu, Y.F.; Zhang, J.; Zhang, Y.; Shen, X.Q.; Li, X.Y.; Miao, X.M.; Feng, Z.Z.; Peng, X.; et al. Antifungal and eliciting properties of chitosan against *Ceratocystis fimbriata* in sweet potato. *Food Chem.* **2018**, *268*, 188–195. [CrossRef]
40. Di Pasqua, R.; Betts, G.; Hoskins, N.; Edwards, M.; Ercolini, D.; Mauriello, G. Membrane toxicity of antimicrobial compounds from essential oils. *J. Agric. Food Chem.* **2007**, *55*, 4863–4870. [CrossRef]
41. Kong, J.; Zhang, Y.; Ju, J.; Xie, Y.; Guo, Y.; Cheng, Y.; Qian, H.; Quek, S.Y.; Yao, W. Antifungal effects of thymol and salicylic acid on cell membrane and mitochondria of *Rhizopus stolonifer* and their application in postharvest preservation of tomatoes. *Food Chem.* **2019**, *285*, 380–388. [CrossRef] [PubMed]
42. Liu, P.; Cai, Y.; Wang, R.; Li, B.; Weng, Q. Effect of Ethylenediaminetetraacetic acid (EDTA) on perillaldehyde-mediated regulation of postharvest *Aspergillus flavus* growth on peanuts. *LWT Food Sci. Technol.* **2022**, *154*, 112826. [CrossRef]

Journal of
Fungi

MDPI

Article

Biocontrol of Diseases Caused by *Phytophthora capsici* and *P. parasitica* in Pepper Plants

Mila Santos [1,*], Fernando Diánez [1], Brenda Sánchez-Montesinos [2], Victoria Huertas [1], Alejandro Moreno-Gavira [1], Belén Esteban García [3], José A. Garrido-Cárdenas [3] and Francisco J. Gea [4]

1 Departamento de Agronomía, Escuela Superior de Ingeniería, Universidad de Almería, 04120 Almería, Spain
2 Departamento de Agronomía, División Ciencias de la Vida, Campus Irapuato-Salamanca, Universidad de Guanajuato, Irapuato 36500, Guanajuato, Mexico
3 Departamento de Biología y Geología, Edificio CITE IIB, Universidad de Almería, 04120 Almería, Spain
4 Centro de Investigación, Experimentación y Servicios del Champiñón (CIES), Quintanar del Rey, 16220 Cuenca, Spain
* Correspondence: msantos@ual.es; Tel.: +34-628188339

Abstract: The main objective of this study was to evaluate the ability of *Trichoderma aggressivum* f. *europaeum*, *T. longibrachiatum*, *Paecilomyces variotii*, and *T. saturnisporum* as biological control agents (BCAs) against diseases caused by *P. capsici* and *P. parasitica* in pepper. For this purpose, their antagonistic activities were evaluated both in vitro and in vivo. We analysed the expression patterns of five defence related genes, *CaBGLU*, *CaRGA1*, *CaBPR1*, *CaPTI1*, and *CaSAR8.2*, in leaves. All BCAs showed a high in vitro antagonistic activity, significantly reducing the mycelial growth of *P. capsici* and *P. parasitica*. The treatments with *T. aggressivum* f. *europaeum*, *T. longibrachiatum*, and *P. variotii* substantially reduced the severity of the disease caused by *P. capsici* by 54, 76, and 70%, respectively, and of the disease caused by *P. parasitica* by 66, 55, and 64%, respectively. *T. saturnisporum* had the lowest values of disease reduction. Reinoculation with the four BCAs increased the control of both plant pathogens. Markedly different expression patterns were observed in the genes *CaBGLU*, *CaRGA1*, and *CaSAR8.2*. Based on the results, all four BCAs under study could be used as a biological alternative to chemicals for the control of *P. capsici* and *P. parasitica* in pepper with a high success rate.

Keywords: *Trichoderma*; *Paecilomyces*; biological control; root rot; blight; *Phytophthora*

1. Introduction

The genus *Phytophthora* includes a group of devastating plant pathogenic species that economically affect important crops worldwide. This genus has long been included in the family *Pythiaceae*, within the group of oomycetes, but was ultimately included in the family *Peronosporaceae* after ribosomal analysis [1,2]. Advances in molecular analysis have enabled the elucidation of these issues and the description of new genera such as *Phytopythium* [3–7]. Currently, a total of 365 species and subspecies have been described (www.mycobank.org, accessed on 2 May 2022) in the genus *Phytophthora*, and this number continues to increase [5,8–12]. These species are classified into 12 phylogenetic clades [13–15], and new species of *Phytophthora* hybrids have been recently identified [16–19].

A number of species in this genus have been characterised as pathogenic in plants, which have a wide range of hosts. *P. capsici* Leonian and *P. parasitica* Dastur (syn. *P. nicotianae* Breda de Haan) are the most important pathogenic species of the genus for pepper (*Capsicum annuum*) crops in Spain [20,21]. *Phytophthora capsici* causes root rot, crown rot, foliar blight, and fruit rot in pepper [22]; *P. parasitica* is a causal agent of root and crown rot [20]. Due to the similarities of the symptoms on the roots and crown, both species may cause diagnostic confusion. The symptoms considerably vary according to the host, areas of infection, and environmental conditions, such as soil, air, and water temperature [23,24]. Methods of *Phytophthora* control include cultivation practices, fungicide application, and the use of resistant or tolerant varieties [25,26]. Control in many vegetable-growing areas in Spain has been based on the use of chemical soil disinfectants, many of which have been banned [27]. Currently, only a few fungicides are authorised for *Phytophthora* control, and their effectiveness is not guaranteed. In addition, they often generate resistance; for example, the resistance of *P. capsici* to metalaxyl [26,28,29]. Furthermore, the ability of *Phytophthora* to overcome the genetic resistance of plants owing to its genetic variability creates the need for alternative control methods for both diseases. Different cultivation techniques, such as grafting on resistant rootstocks [30–34] or nonchemical disinfection methods [35–43], have been used as alternatives. Crop rotation is a key component in the integrated management of diseases caused by *Phytophthora*; nevertheless, the survival of oospores, even in the absence of hosts, limits the effectiveness of these methods [21,44].

In recent years, different studies have been conducted by combining biological and chemical control agents (BCAs and CCAs, respectively) and/or combining the techniques mentioned above [45]. Reduced doses of fungicide stress and weaken the pathogen, increasing its susceptibility to attack by the antagonist [46]. BCAs are alternatives or complements to CCAs. CCAs are also adversely affected by the application of microbial antagonists because these antagonists are harmed by the application of pesticides, such that their effectiveness is sometimes weakened [47]. Biodisinfection and the subsequent incorporation of antagonistic bacteria and/or fungi may increase the benefits of this practice. For example, Wang et al. [48] demonstrated that combining biofumigation with the addition of *Bacillus amyloliquefaciens* controlled the disease caused by *P. capsici* by 40% to 90% in peppers. Other authors also described the benefits of the combined action of biofumigation and microbial incorporation [49].

Many microorganisms are growth inhibitors of *P. capsici* and *P. parasitica*, including *Streptomyces* spp. [50–57], *Bacillus* spp. [58–65], *Paenibacillus* spp. [66], *Pseudomonas* spp. [67,68], *Rhizobium* spp. [69], *Serratia* spp. [63,70], *Trichoderma* spp. [68,71–77], *Aspergillus* spp. [78,79], *Penicillium* spp. [80], *Curvularia* spp. [81], *Clitocybe nuda* [82], *Cladobotryum mycophilum* [83], *Fusarium solani* [68], *Aureobasidium pullulans* [84], *Rhodotorula mucilaginosa* [85], *Muscodor albus* [86], mycorrhizal fungi [87], and mixtures of microorganisms [88–94]. The mechanisms used for their control include the production of lytic enzymes, volatile and nonvolatile active metabolites, mycoparasitism, competition for nutrients and space, and host resistance induction [75,95–101]. Similarly, soil bioactivation through the incorporation of microorganisms could reduce pathogen counts through the indirect effect of an optimised soil microbiome that improved the nonbiological factors of the soil [45]. Moreover, the rhizosphere microbiome plays a substantial role in reprogramming the defence responses of plants [102].

Plants recognise the presence of pathogens through interactions with receptors known as pathogen- and microbe-associated molecular patterns (PAMPs and MAMPs, respectively), inducing a local defence response termed PAMP-triggered immunity [103–105]. Some pathogens, including oomycetes, can suppress this response [105,106], which can be counteracted by cytoplasmic receptors (resistance proteins). These receptors, in turn, trigger a defence response termed effector-triggered immunity, which generates a hypersensitive response [105]. In addition to triggering local responses, plant pathogens induce systemic responses or systemic acquired resistance (SAR), such as *Fusarium oxysporum* fsp. *lycopersici* [107] and nonhost *Phytophthora nicotianae* [108], both of which protect pepper plants from subsequent infection with *P. capsici*. In addition to pathogens, numerous beneficial microorganisms trigger these immune responses [109,110]. Accordingly, systemic resistance against different *Phytophthora* species can be induced by *Trichoderma* [111–115], *Bacillus velezensis* [116], *B. subtilis* [115,117], *B. thuringiensis* [118], *B. vallismortis* [119], *B. amyloliquefaciens* [120], *Burkholderia* sp. [121], and the microorganisms present in aqueous compost extracts [122], among many others. The differential expression of genes involved in plant defence mechanisms allows us to compare how plants defend themselves against attack by different pathogens. Other control systems against *Phytophthora* include the inhibitory effect of extracellular self-DNA, which acts as a damage-associated molecular pattern (DAMP) on the pathogen, affecting the germination rate of *P. capsici* zoospores, thereby protecting the plant [123]. This type of technology should be studied in depth for subsequent applications in agriculture.

Pepper (*Capsicum annuum* L.) is the most important vegetable crop in Almeria, southeastern Spain, covering 12,310 ha of cultivated land. In the 2020/2021 crop year, the total pepper production was 1,508,168 t, reflecting a 63.5% increase in the area of land cultivated with peppers in the last 10 years [124]. Biological pest control is performed on 96.3% of this cultivated land area. Biocontrol, using an antagonist, represents a potentially attractive disease management approach to reduce the side effects of fungicides as environmental pollutants. Therefore, the main objectives of this study were to determine (a) the potential of different BCAs against *P. capsici* and *P. parasitica* in vitro and in vivo; (b) the effect of volatile and non-volatile antifungal metabolites in vitro; (c) the effect of different BCAs on the development of diseases caused by both plant pathogens in vivo; and (d) the differential expression of the genes involved in plant defence responses, *CaBGLU*, *CaRGA1*, *CaBPR1*, *CaPTI1*, and *CaSAR8.2*, during the onset of marked symptoms in plants inoculated with both pathogens.

2. Materials and Methods

2.1. Fungal Isolates

The following BCAs were selected in this study: *Trichoderma aggressivum* f. *europaeum* Samuels & W. Gams (TA) [125,126], *T. longibrachiatum* Rifai (TL) [126], *Paecilomyces variotii* Bainier (PAE) [127], and *T. saturnisporum* Hammil (TS) [72]. TS has been previously described as a BCA for *P. parasitica* and *P. capsici* [72,126], and was used as a reference to compare the efficacy of the fungal isolates TA, TL, and PAE tested in this study (Figure 1). All isolates were deposited in the Phytopathology laboratory of the Department of Agronomy, Universidad de Almería, (UAL), Spain.

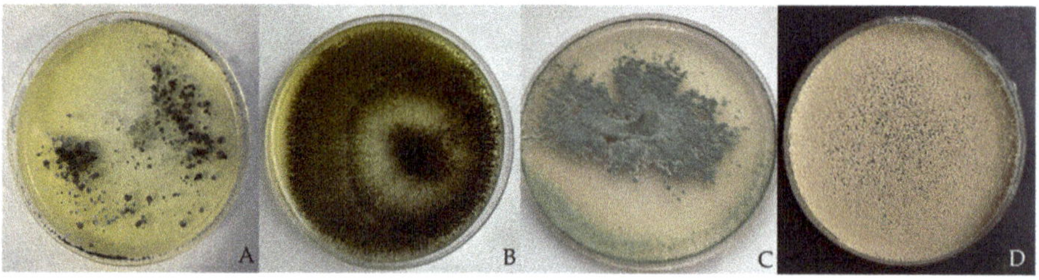

Figure 1. Fungal isolates used in this assay: (**A**) *T. saturnisporum*; (**B**) *T. longibrachiatum*; (**C**) *T. aggressivum* f. *europaeum*; (**D**) *P. variotii*.

Plants infected with *P. parasitica* and *P. capsici* were collected from pepper crops in the province of Almeria (Spain). Stem sections with active lesions were cut and tissue pieces from the boundaries between healthy and discoloured areas were disinfected with 2% NaOCl for 2 min and then abundantly washed with sterile distilled water. These fragments were dried on sterile paper and subsequently placed on potato dextrose agar (PDA, Cultimed Panreac EU). Once the isolates were obtained, pathogenicity tests were carried out on pepper plants (*Capsicum annuum* L. cv. Acorde) using the methods of Diánez et al. [72].

BCA isolates were grown on PDA for 5 or 10 days at 25–27 $\pm$ 2 °C under dark conditions and were maintained on PDA at 4 °C. *Phytophthora parasitica* and *P. capsici* were maintained on V8 agar.

2.2. Dual Culture Bioassays

TA, TS, TL, and PAE were screened for their antagonism in vitro against *P. parasitica* and *P. capsici*. The antagonism assay was performed on PDA in Petri dishes using the dual culture method [128]. Plugs of 0.5 cm of mycelia of all fungi were cut from the growing edge of 10-day-old cultures with active growth of each isolate. The plugs were placed at the ends of Petri dishes with a distance of 8 cm between the two fungi and antagonist–phytopathogen. The antagonistic fungus was placed in the Petri dish 24 h before the pathogen. All plates were sealed with Parafilm® and incubated in the dark at 25 °C until the controls reached the edge of the Petri dish. Radial fungal colony growth was measured daily. Results were transformed into percentages of mycelium growth inhibition (PIRM). These tests were carried out in quintuplicate.

2.3. Antifungal Volatile and Non-Volatile Metabolite Bioassay

The antifungal activity of volatile organic compounds (VOCs) of TA, TS, TL, and PAE against *P. parasitica* and *P. capsici* was assessed using the procedure described by Phoka et al. [97] and bi-compartment dishes. All fungi were grown in PDA medium for 5 days at 25 °C in the dark. A 0.5-cm-wide plug of each antagonist fungus was placed 0.5 cm from the edge of the plate in a compartment. Similarly, in the other compartment of the plate, the pathogen was placed as described above, 24 h later. The plates were sealed with 3 layers of parafilm and incubated at 25 °C in the dark for 5 days. The fungal diameter was measured and compared with control plates (without antagonist). The experiment was performed in quintuplicate and repeated twice.

To determine the antifungal activity of non-volatile organic compounds (N-VOCs) of TA, TS, TL, and PAE, Erlenmeyer flasks (500 mL) containing 100 mL of PDB medium (Cultimed Panreac EU) were inoculated with two 0.5-cm-wide plugs of each antagonistic fungus. The flasks were incubated without stirring at 25 °C in the dark for 7, 14, 21, and 30 days. Mycelia were harvested by filtration through two layers of cheesecloth and the supernatant was filtered through sterile Millipore membranes (pore size 0.22 μm) and collected in sterile tubes. Filtrates were then incorporated and mixed with the cooled PDA at 5, 10, or 15% (*v:v*) and immediately poured into 50 mm Petri dishes [129]. The plates without filtrate served as control. A mycelial disc of 5 mm diameter of *P. parasitica* and *P. capsici* was put in the centre of the Petri plates. The cultures were incubated at 25 °C for 7 days. The colony diameter was measured and the percentage inhibition of the radial growth was calculated. Each assay was performed in quintuplicate.

2.4. Effects of TA, TS, TL, and PAE Isolates on the Severity of Phytophthora Blight in Pepper

TA, TS, TL, and PAE isolates were tested for biocontrol of *Phytophthora* blight (*P. capsici* and *P. parasitica*) in pepper plants (*Capsicum annuum* L., cv. Largo de Reus). The experiment was performed in two phases, one phase in a nursery and another in a greenhouse. Two independent experiments were conducted using completely randomised block designs.

Seedling was performed according to the procedure described by Sanchez-Montesinos et al. [130]. Pepper seeds were sown in 96-cell, commercial peat mix-filled, nursery polystyrene planting trays (70 mL volume) and covered with vermiculite. Trays were relocated to a greenhouse and rinsed with sterile distilled water (control) or a 5 mL (TA, TS, TL, or PAE) spore suspension per cell at 10^5 spores per plant, after a 4-day period in a germination room (relative humidity (RH) = 95%; 25 °C). Two trays of seedlings for each treatment were cultivated under standard nursery culture conditions (18–28 °C; 75.4 ± 6.7% RH). After 45 days at the commercial nursery, 240 plants were transferred to pots (1 L capacity) containing peat moss, 40 plants of each antagonistic isolate, 40 control plants for each pathogen, and 40 plants for non-pathogen control. After transplanting, 50% of the plants were reinoculated with the same dose of the antagonist (R). After 7 days, all plants (except non-pathogen controls) were then inoculated with 5 mL of the zoospore suspension (10^4 zoospores·mL^{-1}) using a sterile micropipette, as described by Diánez et al. [72]. Symptom severity was rated periodically and final disease severity index was estimated according to the following scale [72]: 1, healthy plant; 2, symptoms beginning; 3, moderate symptoms; 4, severely affected plant; and 5, dead plant (Figure 2).

Figure 2. Disease severity scale for *Phytophthora capsici*; the same scale is used for *P. parasitica*. (**1**) Healthy plant; (**2**) symptoms beginning; (**3**) moderate symptoms; (**4**) severely affected plant; (**5**) dead plant.

2.5. Effect of Antagonists on the Chlorophyll Content of Peppers

Chlorophyll content from the fourth leaf was determined using a SPAD 502 Plus Chlorophyll Meter (Konica Minolta, Inc., Ramsey, NJ, United States). The SPAD values were converted to chlorophyll using the formula described by Ling et al. [131]. The experiment was carried out in triplicates, with 10 plants measured at 15 and 45 days after transplanting (DATs).

2.6. RNA Extraction and Real-Time Polymerase Chain Reaction (RT-PCR)

The differential expression of the genes *CaBGLU* (*C. annuum* β-1,3-glucanase), *CaRGA1* (blight resistance protein), *CaBPR1* (basic PR protein 1), *CaPTI1* (ethylene responsive factor), and *CaSAR8.2* (Systemic Acquired Resistance 8.2) was determined using real-time PCR for all the treatments when the plants showed symptoms at stage 2 (62–65 days after the first application of the BCAs and 7–10 days after pathogen inoculation) and, similarly, in healthy plants inoculated with different BCAs (without pathogen). Gene expression was compared with the controls without inoculation. Leaves of a similar developmental stage were collected, frozen in liquid nitrogen, and kept at −80 °C until processing.

Total RNA was extracted from samples of pepper leaves using a commercial RNA PureLink RNA Mini Kit (Invitrogen), following the manufacturer's manual. The samples were reduced to a smaller size and homogenised prior to the extraction with FastPrep-24 5G (MP Biomedical) for 40 s at a speed of 6 m/s. The quality and concentration of RNA was quantified by Nanodrop 2000 (Thermo Fisher Scientific, Waltham, MA, USA). In all cases, RNA concentrations were higher than 100 ng/μL and RNA extracts were stored at −20 °C. The high-capacity cDNA Reverse Transcription Kit (Applied Biosystems, by Thermo Fisher Scientific) was used to obtain cDNA from 1 μg of RNA. The cDNA was used as a template for the subsequent RT-PCR.

Quantitative RT-PCR was performed on a MyGo Pro® RealTime PCR System using the SYBR Green fluorophore with the specific primers shown in Table 1. The SYBR Green reactions were performed in a 20 μL reaction mix comprising 1.5 ng of DNA, 10 μL of the SensiFAST SYBR No-ROX Kit (Bioline), and 2 μL of each of the primers (2 μM). The *ACT* gene was used as the housekeeping gene for data normalisation. In all reactions, amplifications were carried out under the following conditions: an initial hold step of 95 °C for 5 min and 45 PCR cycles of 95 °C for 15 s and 60 °C for 1 min. All *Ct* (cycle threshold) values were considered positive in the 18–35 range. Double delta *Ct* (ΔΔCT) analysis was used for determining relative expression [132] and the measurement of each gene was normalised with respect to the *ACT* gene. For each pair of primers, the melting curve was analysed to evaluate the specificity of the amplification, with high specificity in all cases.

The visualisation of a single peak in the melting curve indicated a single specific fragment, the absence of primer dimers, and the lack of nonspecific products. For every experiment, mean values of six replicates are given for every concentration of samples tested, and their standard deviations are represented as error bars in figures.

Table 1. PCR primers used in SYBR Green assays.

Target	Primers	Sequences (5′→3′)	Reference
CaBGLU	CABGLU-F CABGLU-R	ACAGGCACATCTTCACTTACC CGAGCAAAGGCGAATTTATCC	[107]
CaRGA1	CARGA-F CARGA-R	ATGAGAAGGGAATAGGACGAG ACATCCAATGGCAGGAAACT	[133]
CaBPR1	CaBPR1-F CaBPR1-R	GTTGTGCTAGGGTTCGGTGT CAAGCAATTATTTAAACGATCCA	[99]
CaPTI1	CapPI1-F CapPI1-R	TTTGAAACGGCCGAAGAAGC TGCACGATTCTGTCTTAGCGT	[98,134]
CaSAR8.2	CapSAR8.2-F CapSAR8.2-R	TGTTGCCAGGGAGATGACTTC ACAACGGCCATGACAAGTTT	[135]
ACT	ACT-F ACT-R	TGTTATGGTAGGGATGGGTC TTCTCTCTATTTGCCTTGGG	[136]

2.7. Statistical Analysis

The experimental results are presented as mean values (±standard deviation) for the different replicates. Mean separation was carried out using Fisher's least significant difference (LSD) test. The data were tested by one-way analysis of variance (ANOVA) or Student's *t*-test, with significance defined as *p*-values less than 0.05 ($p < 0.05$). Statgraphics Centurion 18 software was utilised for statistical analysis.

3. Results

3.1. Dual Culture Bioassays

All isolates showed high antagonistic activity against both *Phytophthora* species. *P. variotii* inhibition peaked at 83 and 87% for *P. parasitica* and *P. capsici*, respectively, at 7 days of incubation. The BCAs TA, TS, and TL showed similar high antagonistic activity values of approximately 88 and 82% for *P. parasitica* and *P. capsici*, respectively, and their activity peaked after 3 days of incubation (Figure 3).

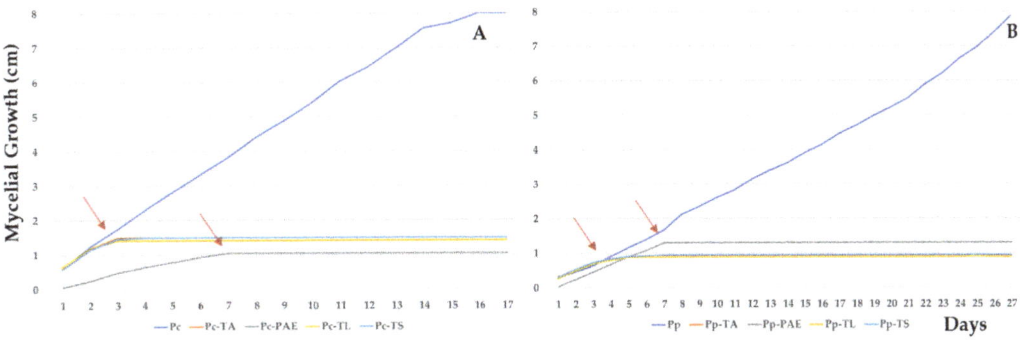

Figure 3. Mycelial growth of *P. capsici* (Pc) (**A**) and *P. parasitica* (Pp) (**B**) against *T. aggressivum* f. *europaeum* (TA), *P. variotii* (PAE), *T. longibrachiatum* (TL), and *T. saturnisporum* (TS), which peaked at 17 and 27 days, respectively.

3.2. Antifungal Volatile and Non-Volatile Metabolite Bioassay

The in vitro antifungal activity of VOCs (Figure 4) and N-VOCs (Table 2) produced by the isolates of TA, TS, TL, and PAE was tested against *P. parasitica* and *P. capsici*. The VOCs of TL and TA showed the highest percentages of growth inhibition for both plant pathogens of all isolates tested in this study, reaching approximately 50 and 20% inhibition, respectively. Conversely, the VOCs of PAE showed a weak growth inhibition effect against *P. parasitica* (5.43%) and no effect against *P. capsici*. Similarly, *P. capsici* growth was not affected by the VOCs of TS.

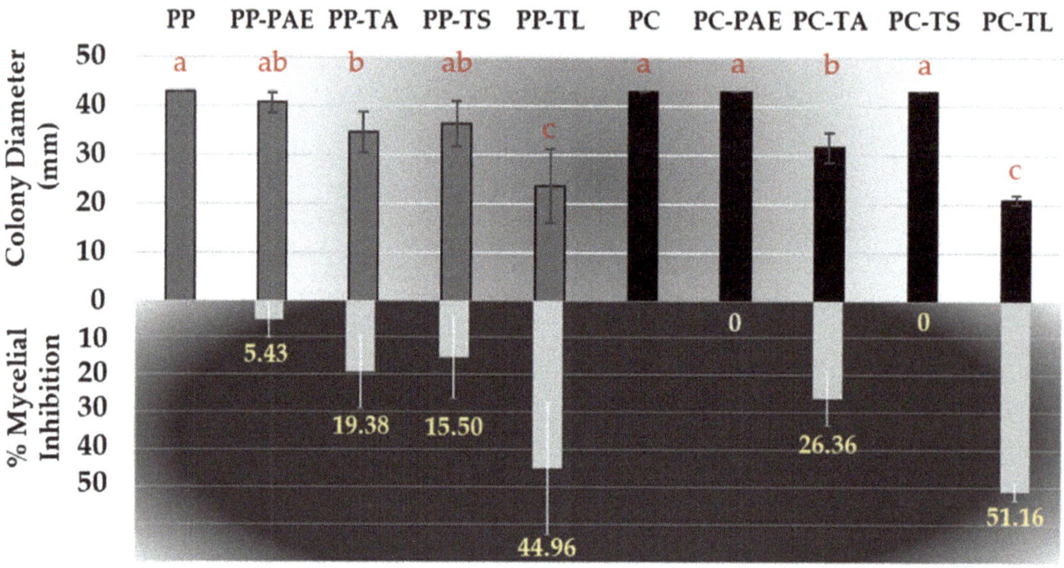

Figure 4. Colony diameter (mm) and mycelial inhibition (%) of *P. parasitica* (PP) and *P. capsici* (PC) by VOCs of *T. aggressivum* f. *europaeum* (TA), *P. variotii* (PAE), *T. longibrachiatum* (TL), and *T. saturnisporum* (TS). Mean values (±standard deviation) followed by different letters (line) indicate significant differences ($p < 0.05$) using the least significant difference (LSD) test.

In turn, all N-VOCs showed a slight growth inhibition of both phytopathogens (Table 2). Unexpectedly, the N-VOCs presented an inhibition range of PC and PP lower than 20 and 15%, respectively, as well as lower PAE and higher TL growth inhibition values.

3.3. Effects of TA, TS, TL, and PAE Isolates on the Severity of Phytophthora Blight in Pepper

At the end of the assay, the plants not treated with the pathogens (T0) were asymptomatic. Plants inoculated with *P. capsici* and *P. parasitica* showed a mean disease rating of 5 and 4.2, respectively, with 100% incidence in both cases.

The in-plant antagonistic effectiveness of the BCA test strains against *P. parasitica* was higher than that against *P. capsici*. In both cases, nevertheless, the percentage of plants without symptoms was higher in plants reinoculated with the BCAs. No plants died when treated with PAE (considering a disease rating of 4 and 5), and plants treated with TL and with no symptoms reached a maximum disease rating of 2 (Figure 5).

Based on the results outlined in Table 3, the treatments with TL, PAE, and TA substantially reduced the severity of the disease caused by *P. capsici*, with 76, 70, and 54%, respectively, over untreated control plants (100% of mortality). Similarly, all reinoculation treatments provided better results, with TL showing the strongest antagonistic effect, reaching 78% disease severity reduction.

Table 2. Mycelial growth of *P. parasitica* and *P. capsici* in PDA medium supplemented with 5, 10, and 15% microfiltered extract of TA, TL, TS, and PAE at different incubation times (7, 15, 21 and 30 days) in the dark and at 25 °C.

	Colony Diameter (cm)							
	Phytophthora capsici				*Phytophthora parasitica*			
	Incubation Time (days)							
Treatments	**7**	**15**	**21**	**30**	**7**	**15**	**21**	**30**
Control	2.7 ± 0 a*	2.68 ± 0.06 a	2.7 ± 0 a	2.7 ± 0 a	2.7 ± 0.00 a	2.69 ± 0.03 a	2.69 ± 0.08 a	2.7 ± 0.00 a
PAE 5%	2.65 ± 0.00 a	2.68 ± 0.06 a	2.66 ± 0.07 ab	2.68 ± 0.06 a	2.51 ± 0.25 cd	2.54 ± 0.22 bcd	2.54 ± 0.22 bcd	2.62 ± 0.11 b
PAE 10%	2.61 ± 0.10 ab	2.61 ± 0.18 ab	2.61 ± 0.18 abc	2.61 ± 0.18 bc	2.62 ± 0.09 b	2.62 ± 0.09 ab	2.60 ± 0.10 abc	2.59 ± 0.09 b
PAE 15%	2.66 ± 0.18 a	2.62 ± 0.15 ab	2.63 ± 0.15 abcd	2.63 ± 0.15 ab	2.58 ± 0.08 bc	2.57 ± 0.08 bc	2.52 ± 0.08 bcd	2.58 ± 0.08
TA 5%	2.5 ± 0.14 cd	2.43 ± 0.1 cd	2.53 ± 0.19 cde	2.60 ± 0.09 bc	2.58 ± 0.08 bc	2.45 ± 0.07 de	2.61 ± 0.04 ab	2.58 ± 0.05 bcd
TA 10%	2.51 ± 0.07 cd	2.46 ± 0.07 cd	2.62 ± 0.07 bcde	2.58 ± 0.06 bcd	2.51 ± 0.12 cd	2.43 ± 0.11 e	2.52 0.0 bcd	2.58 ± 0.06 bc
TA 15%	2.55 ± 0.09 bc	2.54 ± 0.05 bc	2.52 ± 0.04 de	2.56 ± 0.05 bcd	2.54 ± 0.05 bcd	2.56 ± 0.05 bcd	2.49 ± 0.06 cd	2.52 ± 0.06 bcd
TS 5%	2.54 ± 0.09 bc	2.51 ± 0.08 bcd	2.54 ± 0.16 cde	2.52 ± 0.07 d	2.7 ± 0.00 a	2.65 ± 0.07 ab	2.43 ± 0.06 d	2.48 ± 0.1 e
TS 10%	2.47 ± 0.07 cd	2.39 ± 0.16 de	2.45 ± 0.08 ef	2.43 ± 0.07 e	2.57 ± 0.05 bc	2.45 ± 0.13 de	2.46 ± 0.05 d	2.48 ± 0.06 e
TS 15%	2.34 ± 0.08 e	2.41 ± 0.09 de	2.48 ± 0.11 ef	2.43 ± 0.07 e	2.52 ± 0.07 cd	2.46 ± 0.19 cde	2.48 ± 0.11 d	2.45 ± 0.05 ef
TL 5%	2.45 ± 0.13 cd	2.58 ± 0.14 de	2.5 ± 0.12 de	2.55 ± 0.08 cd	2.57 ± 0.05 bc	2.54 ± 0.10 bcd	2.43 ± 0.06 d	2.51 ± 0.07 de
TL 10%	2.43 ± 0.08 de	2.31 ± 0.25 ef	2.46 ± 0.13 ef	2.43 ± 0.1 e	2.54 ± 0.05 bcd	2.28 ± 0.19 f	2.44 ± 0.09 d	2.48 ± 0.13 e
TL 15%	2.18 ± 0.13 f	2.22 ± 0.11 f	2.38 ± 0.14 f	2.43 ± 0.06 e	2.48 ± 0.08 d	2.42 ± 0.11 e	2.44 ± 0.1 d	2.4 ± 0.09 f
p	0.0000	0.0000	0.0000	0.0000	0.0000	0.0000	0.0000	0.0000

* Values in same column with different letters are significantly different according to one-way analysis of variance (ANOVA) followed by Tukey's test at the 0.05 alpha level of confidence. Green: favourable; Orange: no effect compared to control.

Table 3. *Phytophthora parasitica* and *P. capsici* disease severity in pepper inoculated with TA, TS, TL, or PAE (10^5 spores per plant) in two experiments in greenhouse conditions in which plants were inoculated with antagonist before the pathogen and before/after (reinoculated, R). All plants, except for controls (T0), were then inoculated with 5 mL of zoospore suspension (10^4 zoospores·mL^{-1}). Disease severity was assessed on a 1–5 scale, where 1 indicated free of infection (plants without symptoms) and 5 indicated dead plant.

	P. parasitica		*P. capsici*	
Treatment	**Severity**	**Plants without Symptoms (%)**	**Severity**	**Plants without Symptoms (%)**
T0	1.00 ± 0.00 c*	100%	1.00 ± 0.00 c	100%
TI	4.20 ± 1.13 a	0%	5.00 ± 0.00 a	0%
TA	1.40 ± 0.69 c	70%	2.30 ± 1.76 b	60%
TS	2.60 ± 2.06 b	60%	4.10 ± 1.37 a	5%
TL	1.80 ± 1.31 bc	70%	1.20 ± 0.42 c	80%
PAE	1.50 ± 1.08 c	80%	1.50 ± 0.84 bc	65%
TAR	1.20 ± 0.42 c	80%	1.50 ± 0.84 bc	75%
TSR	1.10 ± 0.31 c	90%	2.40 ± 1.57 b	50%
TLR	1.20 ± 0.42 c	80%	1.10 ± 0.31 c	90%
PAER	1.30 ± 0.67 c	80%	1.50 ± 1.26 bc	70%

* Values in same column with different letters are significantly different according to one-way analysis of variance (ANOVA) followed by Tukey's test at the 0.05 alpha level of confidence.

Figure 5. Effectiveness of BCAs in controlling disease caused by *Phytophthora* at end of assay (60 DATs): (**A**) distribution of plants in the greenhouse; (**B**) state of controls plants and plants treated with (**C**) *T. longibrachiatum* (TL), (**D**) *T. aggressivum* f. *europaeum* (TA), (**E**) *T. saturnisporum* (TS), and (**F**) *P. variotii* (PAE).

3.4. Effect of Antagonists on the Chlorophyll Content of Peppers

The chlorophyll content of plants infected with both pathogens and non-pathogens did not differ between treatments at 15 and 45 DATs. However, at 45 days, the plants inoculated with *P. capsici*, *P. parasitica*, and *Trichoderma longibrachiatum* showed a substantial increase in chlorophyll content, which reached 67 and 80% in both treatments (TL and TLR) for *P. capsici* and 56% for *P. parasitica* (TLR) in relation to the control (T0) (data not shown).

3.5. Molecular Responses of Pepper Leaves

Figure 6 shows the results from the analysis of the relative expression of the defence-related genes *CaBGLU*, *CaRGA1*, *CaBPR1*, *CaPTI1*, and *CaSAR8.2* at disease onset. This analysis was performed in leaves for all treatments with BCAs, comparing the results of these pepper plants with and without pathogen inoculation when they started showing symptoms (the samples were collected from plants with a disease severity rating of 2).

Considering the levels of each gene in relation to the levels of constitutively expressed *CaActin*, we observed that the expression levels of *CaBGLU* transcripts were not increased in plants treated with *P. parasitica* in any treatment tested in this study. However, inoculation with *P. capsici* produced the highest increase in the level of expression (two-fold), which decreased again after applying BCAs. Similarly, the *CaBGLU* gene was induced at low levels when applying BCAs only, except for TL, which increased the expression of this gene six-fold, and for TA and TAR, which doubled the expression of this gene.

The expression of the gene *CaRGA1* was moderately induced by BCAs, ranging from 1.68 to 4.69 times. These increased expression levels were maintained when incubating the plants with the pathogens. Strong induction of the *CaSAR8.2* gene also was found in pepper leaves treated with BCAs. For *P. parasitica*, the expression increased between 1.2 and four times, and the values were even higher upon reinoculation with BCAs. Expression was activated not only when inoculating with *P. capsici*, but when inoculating with BCAs. The application of BCAs alone showed a moderate level of expression activation in some cases, such as TL and TAR. The expression of the genes *CaBPR1* and *CaPI1* did not increase in any treatment.

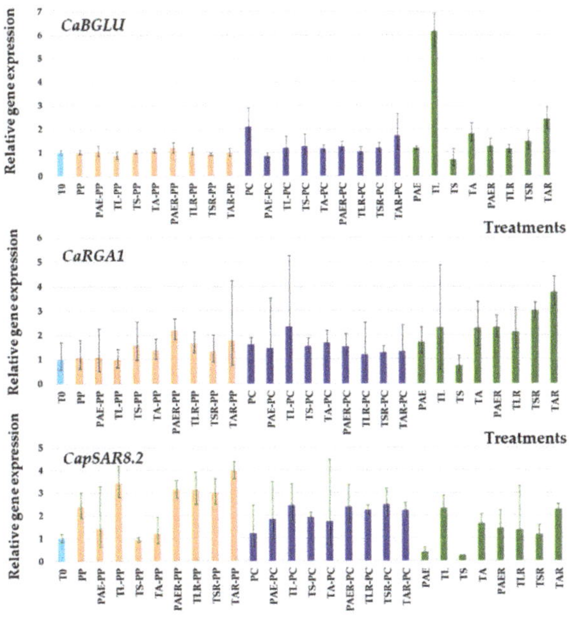

Figure 6. *Cont.*

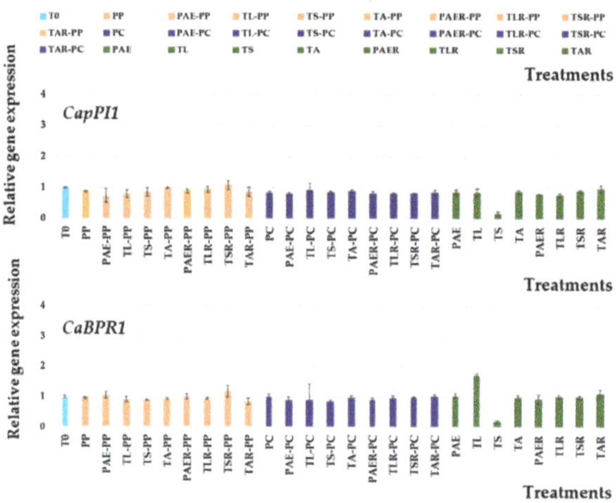

Figure 6. Analysis of the relative expression of defence-related genes *CaBGLU*, *CaRGA1*, *CaBPR1*, *CaPTI1*, and *CaSAR8.2*. Each bar represents relative gene expression for all conditions tested. Expressions of genes were normalised with respect to the *ACT* gene. Values were calculated following three replications, and standard deviations are shown.

4. Discussion

The importance of the biological activity of microorganisms close to plant roots has been highlighted in numerous studies on the biological control of oomycetes [137]. In soils rich in microorganisms, their competition for space and nutrients is intense, with a high production of numerous compounds and enzymes that limit the growth and development of plant pathogens, thus reducing the incidence of diseases. Moreover, many microorganisms stimulate plant growth or induce plant resistance to pathogens.

The control of diseases caused by oomycetes is particularly complex. Most of them produce effectors, which abolish or reduce plant defence responses against their attacks, and metabolites and enzymes, which degrade plant material, facilitating their penetration [138]. Numerous species of the genera *Trichoderma* and *Paecilomyces* have been reported to have fungicidal capacity against *Phytophthora* [72,139–147]. In the present study, the *P. capsici* and *P. parasitica* growth suppression effect of *T. aggressivum* f. *europaeum*, *T. longibrachiatum*, *T. saturnisporum*, and *P. variotii* was assessed in peppers.

In both in vitro and in vivo assays, the two plant pathogens showed differences in their relationship with BCAs. High antifungal activity (>80%) was observed in dual in vitro assays in PDA medium, with the plant pathogens reaching the maximum growth at 3 and 7 days of incubation for *Trichoderma* and *Paecilomyces* isolates, respectively. The three *Trichoderma* species completely overgrew the colony of the pathogen, showing hyperparasitism. Previous studies have shown the antifungal activity of these isolates against other plant pathogens, such as *Botrytis cinerea*, *Sclerotinia sclerotiorum*, and *Mycosphaerella melonis*, with a high efficacy [125,139]. These inhibition results are also very similar to those of Diánez et al. [72] when analysing *Trichoderma saturnisporum*. In turn, Ezziyyani et al. [146] found that *T. harzianum* provided inhibition values higher than 80% against *P. capsici*. Considering the variability of the protocols used in dual in vitro assays regarding the type and thickness of the solid culture medium in Petri dishes, the temperature, and the presence or absence of light, these results should be interpreted with caution because they are highly variable and often lack correlation between in vitro and in vivo conditions. Furthermore, the in vivo results depend on numerous factors, such as BCA dose, application time and method, and crop.

The *Trichoderma* species showed antifungal activity against *Phytophthora*. Nevertheless, the activity of N-VOCs and VOCs of *Paecilomyces* was very low or null. Volatile (VOCs) and non-volatile (N-VOCs) secondary metabolites of the *Trichoderma* and *Paecilomyces* species have different biological activities, such as biostimulation or biocontrol [147–151]. Li et al. [149] described 390 non-volatile components of 20 *Trichoderma* species, including *T. saturnisporum* and *T. longibrachiatum*, with antibacterial and antifungal capacity. In turn, Bae et al. [150] identified different non-volatile compounds produced by different *Trichoderma* species which showed the strongest inhibitory activities against *Phytophthora* isolates. Dai et al. [151] described 223 secondary metabolites and their biological activities isolated from different *Paecilomyces* species. Among them, only the compound farinomalein, isolated from *P. farinosus*, was a potent inhibitor of the plant pathogen *Phytophthora sojae*. Moreno et al. [152] did not assess high growth inhibition values of plant pathogens such as *F. solani* and *M. melonis* induced by N-VOCs and VOCs of *P. variotii*.

In our study, we found a high control of pepper plant diseases caused by *P. parasitica* and *P capsici*. The success of these results is derived mainly from inoculating BCAs in the seedbed phase. Consequently, when the plants were transplanted to the greenhouse, their roots were already colonised by BCAs and therefore "prepared" for a possible attack by phytopathogens, as clearly shown by the inability of reinoculation to significantly reduce disease severity, except for *T. saturnisporum*. In this case, reinoculation reduced disease severity by 57 and 70% for *P. parasitica* and *P. capsici*, respectively.

It has been reported that the addition of different species of *Trichoderma* in a plant's rhizosphere induces resistance due to the rise in the amounts of defensive metabolites as well as enzymes, which act as elicitors [153]. In contrast to this assay, most studies aimed at identifying genes related to plant defence mechanisms against different pathogens are usually performed in the first hours after contact with elicitors, whether they are plant pathogens or beneficial microorganisms. We observed an increase in the relative expression of some plant defence-related genes, such as *CaBGLU*, *CaRGA1* and *CaSAR8.2*, when applying BCAs, except in the TS treatment. This exception could be directly related to the high disease expression shown despite the application of *T. saturnisporum*, which requires plant reinoculation for improved control. However, this hypothesis was not confirmed when applying BCAs together with both phytopathogens. Jung and Hwang [154] showed that the accumulation of *CaBGLU* mRNA on the stems of peppers infected with *P. capsici* was greatly reduced between 48 and 96 h, possibly due to deterioration of the infected stems. They concluded that pepper basic b-1,3-glucanase may mediate a part of the defence responses to pathogen infections. Conversely, the induction of defence-related genes, such as *CaPR1* and *CaBGLU*, is essential for SAR in pepper plants [155,156]. Additionally, some researchers have reported that the level and onset of β-glucanase expression is often positively correlated with the degree of resistance to the pathogen [156]. Accordingly, Jung and Hwang [154] observed that *CaBGLU* mRNA increased in the first stage of infection to similar levels in both compatible and incompatible interactions with *P. capsici*, but at later times, the gene had higher expression in the incompatible interaction. In our study, the expression of these genes was not increased in plants inoculated with *P. parasitica*, even though plants co-inoculated with BCAs showed some degree of resistance against disease and did not die. For *P. capsici*, the relative expression of the gene *CaBGLU* in leaves was low, but no correlation with a defence response was found since the maximum disease severity was reached in all control plants without BCAs, which showed a 2-fold expression induction.

A high number of disease resistance genes are induced by *P. capsici* invasion, such as *CaRGA1* [157]. The study of the expression of RGA genes under pathogen attack would facilitate the determination of whether they play an active role in resistance or if they are merely linked to resistance genes [157]. Our results showed a low induction of the expression of this gene in all treatments, which reached higher values when BCAs were applied without the pathogen, except for TS. The application of *P. parasitica* did not increase the relative expression of this gene.

Silvar et al. [158] observed a strong and rapid induction of the *CaBPR1* gene in an incompatible interaction of pepper plants with *P. capsici*. Similarly, overexpressing this gene in tobacco plants increased tolerance to *P. nicotianae* and to the bacterial pathogens *Ralstonia solanacearum* and *Pseudomonas syringae* pv. *tabaci* [159]. This gene apparently plays a key role in the ability of resistant pepper cultivars to restrict pathogen colonisation, which is conversely weak in susceptible genotypes. In our study, the expression of this gene did not increase under the conditions of this assay, corroborating the findings of Sarowar et al. [159]. The expression of the gene *CaPTI1* did not change either, in contrast to the results reported by Jin et al. [98], who highlighted the high expression levels of the gene *CaPTI1* after inoculation with *P. capsici*, which were higher in stems than in leaves.

The *SAR8.2* gene is a gene that controls plant resistance to *P. nicotianae* [160]. Lee and Hook [135] suggested that *CaSAR8.2* functions as a molecular marker gene for various biotic and abiotic stresses in pepper plants. The relative expression results may be directly related to the resistance response observed for both plant pathogens tested in this study, further highlighting that BCA reinoculation decreases disease incidence and severity, and in turn increases the expression of the *CaSAR8.2* gene.

The expression of these genes does not seem to be linked to the ability to develop symptoms, except for the gene *CaSAR8.2*. The results of these plants must be compared with those of plants without any symptoms and at times near inoculation as well. In our study, we were unable to clearly identify the genes involved in improving plant resistance. Genes involved in the early response of plants with resistant genotypes, such as *CaBPR1*, showed no changes in expression. However, the relative expression levels of genes involved in SAR responses were increased. In any event, the high variability of the results found in different samples or replicates makes it difficult to interpret the results. Therefore, further studies are needed to clarify the role that these genes play in reducing disease severity.

Biological control is presented as an ecological and healthy alternative to chemical control. As commented above, numerous studies have shown the different mechanisms of action that microorganisms use to control the growth and multiplication of the plant pathogens and pests that affect crops. This scientific development contrasts with the reality in the field. The preventive nature of this type of control, possible changes in crop management, and new pest and disease problems resulting from climate change, make it difficult to broadly implement such solutions. The withdrawal of numerous commonly used phytosanitary active ingredients has forced production systems to search for and develop new biological control agents. In addition, farmers and technicians must change their mindset for biological control to work. The use of BCAs from the seedbed, which reduces the inoculum levels of the pathogen in crops, combined with the use of plant varieties with some degree of resistance to some diseases and reduced doses of fungicide could provide high levels of disease control.

5. Conclusions

We are the first to describe *P. parasitica* and *P. capsici* control using *T. aggressivum* f. *europaeum* and *Paecilomyces variotii*. In addition, a marine isolate, *T. longibrachiatum*, showed a high capacity to suppress disease expression. BCA reinoculation increased plant survival and the percentage of plants without symptoms. Similarly, applying beneficial microorganisms moderately activated genes involved in the defence responses in pepper plants.

Author Contributions: Conceptualisation, M.S. and F.D.; methodology, B.S.-M., J.A.G.-C. and B.E.G.; software, B.E.G. and A.M.-G.; validation, M.S., F.D., B.E.G., J.A.G.-C. and F.J.G.; formal analysis, V.H., B.E.G. and B.S.-M.; investigation, M.S., F.D., B.S.-M., A.M.-G. and B.E.G.; resources, M.S., F.D. and F.J.G.; data curation, M.S. and F.D.; writing—original draft preparation, M.S. and F.D.; writing—review and editing, B.E.G., V.H., A.M.-G., F.J.G. and B.S.-M.; visualisation, F.J.G.; supervision, J.A.G.-C. All authors have read and agreed to the published version of the manuscript.

Funding: This study received no external funding.

Institutional Review Board Statement: Not applicable.

Informed Consent Statement: Not applicable.

Data Availability Statement: Not applicable.

Acknowledgments: This study benefited from the input of project UAL-2020-AGR-A1949 and was supported by the I+D+I Projects Andalucia Feder Operational Program 2014–2020.

Conflicts of Interest: The authors declare no conflict of interest.

References

1. Beakes, G.; Sekimoto, S. The evolutionary phylogeny of Oomycetes–insights gained from studies of holocarpic parasites of algae and invertebrates. In *Oomycete Genetics and Genomics: Diversity, Interactions, and Research Tools*; Lamour, K., Kamoun, S., Eds.; Wiley: New York, NY, USA, 2009; pp. 1–24. [CrossRef]
2. Jayawardena, R.S.; Hyde, K.D.; Chen, Y.J.; Stadler, M.; Wang, Y. One stop shop IV: Taxonomic update with molecular phylogeny for important phytopathogenic genera. *Fungal Divers.* **2020**, *103*, 87–218. [CrossRef]
3. Mitchell, D.J. Relationships of inoculum levels of several soilborne species of *Phytophthora* and *Pythium* to infection of severalhosts. *Phytopathology* **1978**, *68*, 1754–1759. [CrossRef]
4. Bala, K.; Robideau, G.P.; Lévesque, C.A.; de Cock, A.; Abad, Z.; Lodhi, A.; ShahzadS, S.; Ghaffar, A.; Coffey, M. Persoonia—Molecular Phylogeny and Evolution of Fungi. *Fungal Planet* **2010**, *24*, 136–137. [CrossRef]
5. Ho, H.H. The taxonomy and biology of *Phytophthora* and *Pythium*. *J. Bacteriol. Mycol.* **2018**, *6*, 40–45. [CrossRef]
6. Tkaczyk, M. *Phytopythium*: Origin, differences and meaning in modern plant pathology. *Folia For. Pol. A* **2020**, *62*, 227–232. [CrossRef]
7. de Cock, A.W.A.M.; Lodhi, A.M.; Rintoul, T.L.; Bala, K.; Robideau, G.P.; Abad, G.Z.; Coffey, M.D.; Shahzad, S.; Lévesque, C.A. *Phytopythium*: Molecular phylogeny and systematics. *Persoonia* **2015**, *34*, 25–39. [CrossRef]
8. Brasier, C.M.; Cooke, D.E.; Duncan, J.M. Origin of a new *Phytophthora* pathogen through interspecific hybridization. *Proc. Natl. Acad. Sci. USA* **1999**, *96*, 5878–5883. [CrossRef] [PubMed]
9. Goheen, E.M.; Frankel, S.J. Phytophthoras in forests and natural ecosystems. In Proceedings of the Fourth Meeting of the International Union of Forest Research Organizations (IUFRO) Working Party S07.02.09, Monterey, CA, USA, 27 September 2021.
10. Martin, F.N.; Abad, Z.G.; Balci, Y.; Ivors, K. Identification and detection of *Phytophthora*: Reviewing our progress, identifying our needs. *Plant Dis.* **2012**, *96*, 1080–1103. [CrossRef]
11. Božik, M.; Mrázková, M.; Novotná, K.; Klouček, P.; Černý, K. MALDI-TOF MS as a method for rapid identification of *Phytophthora* de Bary, 1876. *Peer. J.* **2021**, *9*, e11662. [CrossRef]
12. Yang, X.; Tyler, B.M.; Hong, C. An expanded phylogeny for the genus *Phytophthora*. *IMA Fungus* **2017**, *8*, 355–384. [CrossRef]
13. Blair, J.E.; Coffey, M.D.; Park, S.Y.; Geiser, D.M.; Kang, S. A multi-locus phylogeny for *Phytophthora* utilizing markers derived from complete genome sequences. *Fungal Genet. Biol.* **2008**, *45*, 266–277. [CrossRef] [PubMed]
14. Goss, E.M.; Cardenas, M.E.; Myers, K. The plant pathogen *Phytophthora andina* emerged via hybridization of an unknown *Phytophthora* species and the Irish potato famine pathogen, *P. infestans*. *PLoS ONE* **2011**, *6*, e24543. [CrossRef] [PubMed]
15. Brasier, C.; Scanu, B.; Cooke, D.; Jung, T. *Phytophthora*: An ancient, historic, biologically and structurally cohesive and evolutionarily successful generic concept in need of preservation. *IMA Fungus* **2022**, *13*, 12. [CrossRef] [PubMed]
16. Yang, X.; Richardson, P.A.; Hong, C. *Phytophthora* × stagnum notho sp. Nov., a new hybrid from irrigation reservoirs at ornamental plant nurseries in Virginia. *PLoS ONE* **2014**, *9*, e103450. [CrossRef]
17. Brasier, C.M.; Kirk, S.A.; Delcan, J.; Cooke, D.E.L.; Jung, T.; Manin't Veld, W.A. *Phytophthora alni* sp. nov. and its variants: Designation of emerging heteroploid hybrid pathogens spreading on *Alnus* trees. *Mycol. Res.* **2004**, *108*, 1172–1184. [CrossRef] [PubMed]
18. Burgess, T.I. Molecular characterization of natural hybrids formed between five related indigenous clade 6 *Phytophthora* species. *PLoS ONE* **2015**, *10*, e0134225. [CrossRef]
19. Van Poucke, K.; Haegeman, A.; Goedefroit, T.; Ruttink, T.; Heungens, K. Unravelling hybridization in *Phytophthora* using phylogenomics and genome size estimation. *IMA Fungus* **2021**, *12*, 1–24. [CrossRef]
20. Rodríguez-Molina, M.C.; Morales-Rodríguez, M.C.; Palo Osorio, C.; Duarte Maya, M.S.; Picón-Toro, J. *Phytophthora nicotianae*, the causal agent of root and crown rot (Tristeza disease) of red pepper in La Vera region (Cáceres, Spain). *Span. J. Agri. Res.* **2010**, *8*, 770–774. [CrossRef]
21. Lacasa, C.M.; Martínez, V.; Hernández, A.; Serrano-Pérez, P.; Larregla, S. Survival reduction of *Phytophthora capsici* oospores and *P. nicotianae* chlamydospores with Brassica green manures combined with solarization. *Sci. Hortic.* **2015**, *197*, 607–618. [CrossRef]
22. Alcantara, T.P.; Bosland, P.W. An inexpensive disease screening technique for foliar blight of chile pepper seedlings. *HortScience* **1994**, *29*, 1182–1183. [CrossRef]
23. Lamour, K.H.; Stam, R.; Jupe, J. The oomycete broad-host-range pathogen *Phytophthora capsici*. *Mol. Plant Pathol.* **2012**, *13*, 329–337. [CrossRef] [PubMed]
24. Liu, X.; Zhou, Y.; Li, L. Infection of *Phytophthora capsici* on pepper-Models and affecting factors. *Front. Agri. China* **2008**, *2*, 66–71. [CrossRef]
25. Café-Filho, A.C.; Duniway, J.M. Effect of location of drip irrigation emitters and position of *Phytophthora capsici* infections in roots on *Phytophthora* root rot of pepper. *Phytopathology* **1996**, *86*, 1364–1369.

26. Hausbeck, M.K.; Lamour, K.H. *Phytophthora capsici* on vegetable crops: Research progress and management challenges. *Plant Dis.* **2004**, *88*, 1292–1303. [CrossRef]
27. European Commission. Regulation (EC) no. 1005/2009 of the European Parliament and of the council of 16 September 2009 on substances that deplete the ozone layer (recast). *Off. J. Eur. Union* **2009**, *286*, 1–30.
28. Ploetz, R.C.; Heine, G.; Haynes, J.L.; Watson, M. Investigating biological attributes that may contribute to *Phytophthora capsici* importance as a vegetable pathogen in Florida. *Ann. Appl. Biol.* **2002**, *140*, 61–67. [CrossRef]
29. Parra, G.; Ristaino, J.B. Resistance to mefenoxam and metalaxyl among field isolates of *Phytophthora capsici* causing *Phytophthora* blight of bell pepper. *Plant Dis.* **2001**, *85*, 1069–1075. [CrossRef]
30. Gisbert, C.; Sánchez-Torres, P.; Raigón, M.D.; Nuez, F. *Phytophthora capsici* resistance evaluation in pepper hybrids: Agronomic performance and fruit quality of pepper grafted plants. *J. Agric. Food Inf.* **2010**, *8*, 116–121. [CrossRef]
31. Ros, C.; Lacasa, C.M.; Martínez, V.; Bielza, P.; Lacasa, A. Response of pepper rootstocks to co-infection of Meloidogyne incognita and *Phytophthora* spp. *Eur. J. Hortic. Sci.* **2014**, *79*, 22–28.
32. López-Marín, J.; Galvez, A.; Porras, I.; Brotons-Martínez, J.M. Pepper grafting (*Capsicum annuum*): Benefits and profitability. *ITEA* **2016**, *112*, 127–146.
33. Guigón-López, C. Differential development of wilt and stem rot diseases in grafted bell pepper (*Capsicum annuum*, L.). *Eur. J. Plant Pathol.* **2019**, *154*, 347–357. [CrossRef]
34. Kousik, C.S.; Ikerd, J.L.; Hassell, R. Grafting onto resistant rootstocks for managing *Phytophthora* crown rot of peppers. *Acta Hort.* **2021**, *13*, 163–168. [CrossRef]
35. Guerrero, M.M.; Ros, C.; Guirao, P.; Martínez, M.A.; Martínez, M.C.; Barceló, N. Biofumigation plus solarisation efficacy for soil disinfestation in sweet pepper greenhouses in the Southeast of Spain. *Acta Hort.* **2005**, *698*, 293–297. [CrossRef]
36. Guerrero, M.M.; Ros, C.; Lacasa, C.; Martínez, V.; Lacasa, A.; Fernández, P. Effect of biosolarization using pellets of *Brassica carinata* on soilborne pathogens in protected pepper crops. *Acta Hort.* **2010**, *883*, 337–344. [CrossRef]
37. Butler, D.M.; Kokalis-Burelle, N.; Muramoto, J.; McCollum, T.G.; Rosskopf, E.N. Impact of anaerobic soil disinfestation combined with soil solarization on plant-parasitic nematodes and introduced inoculum of soilborne plant pathogens in raised-bed vegetable production. *Crop. Prot.* **2012**, *39*, 33–40. [CrossRef]
38. Núñez-Zofío, M.; Larregla, S.; Garbisu, C. Repeated biodisinfection controls the incidence of *Phytophthora* root and crown rot of pepper while improving soil quality. *Span. J. Agric. Res.* **2012**, *10*, 794–805. [CrossRef]
39. Larregla, S.; Núñez-Zofío, M.; Fernández-Molina, P.; Lacasa-Martinez, C.M.; Guerrero-Díaz, M.M. Reduction of *Phytophthora capsici* oospores viability by gases released during soil biosolarization of protected pepper crops in southeastern Spain. *Acta Hort.* **2014**, *1044*, 113–118. [CrossRef]
40. Rodríguez-Molina, M.C.; Serrano-Pérez, P.; Palo, C. Effect of biofumigation with brassica pellets combined with Brassicaceae cover crops and plastic cover on the survival and infectivity of inoculum of *Phytophthora nicotianae* Breda de Haan. *Pest Manag. Sci.* **2016**, *72*, 1295–1301. [CrossRef]
41. Gandariasbeitia, M.; Ojinaga, M.; Orbegozo, E.; Mendarte, S.; Larregla, S. Winter biodisinfestation with Brassica green manure is a promising management strategy for *Phytophthora capsici* control of protected pepper crops in humid temperate climate regions of northern Spain. *Span. J. Agric. Res.* **2019**, *17*, 1005. [CrossRef]
42. Larregla, S.; Gandariasbeitia, M.; Ojinaga, M.; Guerrero, M.D.M.; Ortiz-Barredo, A. Gases Released During Soil Biodisinfestation of Pepper Greenhouses Reduce Survival of *Phytophthora capsici* Oospores in Northern Spain. *Front. Sustain. Food Syst.* **2021**, *5*, 663–915. [CrossRef]
43. Guerrero, M.D.M.; Lacasa, C.M.; Martínez, V.; Monserrat, A.; Larregla, S. Low Temperature Biodisinfection Effectiveness for *Phytophthora capsici* Control of Protected Sweet Pepper Crops in the Southeast of Spain. *Front. Sustain. Food Syst.* **2021**, *5*, 659290. [CrossRef]
44. Babadoost, M.; Pavón, C. Survival of Oospores of *Phytophthora capsici* in Soil. *Plant Dis.* **2013**, *97*, 1478–1483. [CrossRef]
45. Cheng, H.; Zhang, D.; Ren, L.; Wang, Q.; Cao, A. Bio-activation of soil with beneficial microbes after soil fumigation reduces soil-borne pathogens and increases tomato yield. *Environ. Pollut.* **2021**, *283*, 117–160. [CrossRef] [PubMed]
46. Lorito, M.; Farkas, V.; Rebuffat, S.; Bodo, B.; Kubicek, C.P. Cell wall synthesis is a major target of mycoparasitic antagonism by *Trichoderma harzianum*. *J. Bacteriol.* **1996**, *178*, 6382–6385. [CrossRef] [PubMed]
47. Santos, M.; Diánez, F. Los antagonistas microbianos en el manejo de micosis de la parte aérea de la planta. In *Organismos Para el Control de Patógenos en los Cultivos Protegidos. Prácticas Culturales Para una Agricultura Sostenible*, 2nd ed.; Tello, J.C., Camacho, F., Eds.; Fundación Cajamar: Almería, España, 2010; pp. 523–528.
48. Wang, Q.; Ma, Y.; Wang, G.; Gu, Z.; Sun, D.; Andyz, X.; Chang, Z. Integration of biofumigation with antagonistic microorganism can control *Phytophthora* blight of pepper plants by regulating soil bacterial community structure. *Eur. J. Soil Biol.* **2014**, *61*, 58–67. [CrossRef]
49. Galletti, S.; Sala, E.; Leoni, O.; Burzi, P.L.; Cerato, C. *Trichoderma* spp. tolerance to *Brassica carinata* seed meal for a combined use in biofumigation. *Biol. Cont.* **2008**, *45*, 319–327. [CrossRef]
50. Joo, G.J. Production of an anti-fungal substance for biological control of *Phytophthora capsici* causing phytophthora blight in red-peppers by *Streptomyces halstedii*. *Biotechnol. Lett.* **2005**, *27*, 201–205. [CrossRef]
51. Nguyen, X.H.; Naing, K.W.; Lee, Y.S.; Kim, Y.H.; Moon, J.H.; Kim, K.Y. Antagonism of antifungal metabolites from *Streptomyces griseus* H7602 against *Phytophthora capsici*. *J. Basic Microbiol.* **2015**, *5*, 45–53. [CrossRef]

52. Nguyen, X.H.; Naing, K.W.; Lee, Y.S.; Tindwa, H.; Lee, G.H.; Jeong, B.K.; Kim, K.Y. Biocontrol potential of *Streptomyces griseus* H7602 against root rot disease (*Phytophthora capsici*) in pepper. *Plant Pathol. J.* **2012**, *28*, 282–289. [CrossRef]
53. Sakineh, A.; Ayme, S.; Akram, S.; Naser, S. *Streptomyces* strains modulate dynamics of soil bacterial communities and their efficacy in disease suppression caused by *Phytophthora capsici*. *Sci. Rep.* **2021**, *11*, 9317. [CrossRef]
54. Trinidad-Cruz, J.R.; Rincón-Enríquez, G.; Evangelista-Martínez, Z.; Quiñones-Aguilar, E.E. Biorational control of *Phytophthora capsici* in pepper plants using *Streptomyces* spp. *Rev. Chapingo Ser. Hort.* **2021**, *27*, 85–99. [CrossRef]
55. Chen, Y.Y.; Chen, P.C.; Tsay, T.T. The biocontrol efficacy and antibiotic activity of *Streptomyces plicatus* on the oomycete *Phytophthora capsici*. *Biol. Cont.* **2016**, *98*, 34–42. [CrossRef]
56. Yang, Y.; Zhang, S.W.; Li, K.T. Antagonistic activity and mechanism of an isolated *Streptomyces corchorusii* stain AUH-1 against phytopathogenic fungi. *World J. Microbiol. Biotechnol.* **2019**, *35*, 145. [CrossRef] [PubMed]
57. Abbasi, S.; Safaie, N.; Sadeghi, A.; Shamsbakhsh, M. Tissue-specific synergistic bio-priming of pepper by two *Streptomyces* species against *Phytophthora capsici*. *PLoS ONE* **2020**, *15*, 230–531. [CrossRef]
58. Sid, A.; Ezziyyani, M.; Egea-Gilabert, C.; Candela, M.E. Selecting bacterial strains for use in the biocontrol of diseases caused by *Phytophthora capsici* and *Alternaria alternata* in sweet pepper plants. *Biol. Plant* **2003**, *47*, 569–574. [CrossRef]
59. Jiang, Z.; Guo, Y.; Li, S.; Qi, H.; Guo, J. Evaluation of biocontrol efficiency of different *Bacillus* preparations and field application methods against *Phytophthora* blight of bell pepper. *Biol Cont.* **2006**, *36*, 216–223. [CrossRef]
60. Zhang, S.; White, T.L.; Martinez, M.C.; Mcinroy, J.A.; Kloepper, J.W.; Klassen, W. Evaluation of plant growth-promoting rhizobacteria for control of *Phytophthora* blight on squash under greenhouse conditions. *Biol Cont.* **2010**, *53*, 129–135. [CrossRef]
61. Akgül, D.S.; Mirik, M. Biocontrol of *Phytophthora capsici* on pepper plants by *Bacillus megaterium* strains. *J. Plant Pathol.* **2008**, *90*, 29–34. [CrossRef]
62. Lim, J.H.; Sang, D.K. Biocontrol of *Phytophthora* blight of red pepper caused by *Phytophthora capsici* using *Bacillus subtilis* AH18 and *B. licheniformis* K11 formulations. *J. Korean Soc. Appl. Biol. Chem.* **2010**, *53*, 766–773. [CrossRef]
63. Yau, J.A.; Diánez, F.; Marín, F.; Carretero, C.; Santos, M. Screening and evaluation of potential biocontrol fungi and bacteria foliar endophytes against *Phytophthora capsici* and *Phytophthora parasitica* on pepper. *J. Food Agric. Environ.* **2013**, *11*, 490–495.
64. Li, Y.; Feng, X.; Wang, X.; Zheng, L.; Liu, H. Inhibitory effects of *Bacillus licheniformis* BL06 on *Phytophthora capsici* in pepper by multiple modes of action. *Biol. Cont.* **2020**, *144*, 104–210. [CrossRef]
65. Bhusal, B.; Mmbaga, M.T. Biological control of *Phytophthora* blight and growth promotion in sweet pepper by Bacillus species. *Biol. Cont.* **2020**, *150*, 104–373. [CrossRef]
66. Naing, K.W.; Anees, M.; Nguyen, X.H.; Lee, Y.S.; Jeon, S.W.; Kim, S.J.; Kim, M.H.; Kim, K.Y. Biocontrol of late blight disease (*Phytophthora capsici*) of pepper and the plant growth promotion by *Paenibacillus ehimensis* KWN38. *J. Phytopathol.* **2014**, *162*, 367–376. [CrossRef]
67. Mei, X.L.; Zhao, Q.Y.; Tan, S.Y.; Shen, B.; Shen, Q.R. Screening, identification, and biocontrol effect of antagonistic bacteria against *Phytophthora capsici*. *Chin. J. Appl. Ecol.* **2010**, *21*, 2652–2658.
68. Gilardi, G.; Vasileiadou, A.; Garibaldi, A.; Gullino, M.L. Biocontrol agents and resistance inducers reduce *Phytophthora* crown rot (*Phytophthora capsici*) of sweet pepper in closed soilless culture. *Phytopathol. Mediterr.* **2021**, *60*, 149–163. [CrossRef]
69. Plett, J.M. Order of microbial succession affects rhizobia-mediated biocontrol efforts against *Phytophthora* root rot. *Microbiol. Res.* **2021**, *242*, 126–628. [CrossRef]
70. Okamoto, H.; Sato, M.; Sato, Z.; Isaka, M. Biocontrol of *Phytophthora capsici* by *Serratia marcescens* F-1-1 and analysis of biocontrol mechanisms using transposon-insertion mutants. *Ann. Phytopathol. Soc. Jpn.* **1998**, *64*, 287–293. [CrossRef]
71. Segarra, G.; Avilés, M.; Casanova, E.; Borrero, A.; Trillas, I. Effectiveness of biological control of *Phytophthora capsici* in pepper by *Trichoderma asperellum* strain T34. *Phytopathol. Mediterr.* **2013**, *52*, 77–83. [CrossRef]
72. Diánez, F.; Santos, M.; Carretero, F.; Marín, F. *Trichoderma saturnisporum*, a new biological control agent. *J. Sci. Food Agric.* **2016**, *96*, 1934–1944. [CrossRef]
73. Chemeltorit, P.P.; Mutaqin, K.H.; Widodo, W. Combining *Trichoderma hamatum* THSW13 and *Pseudomonas aeruginosa* BJ10–86: A synergistic chili pepper seed treatment for *Phytophthora capsici* infested soil. *Eur. J. Plant Pathol.* **2017**, *147*, 157–166. [CrossRef]
74. de Oliveira, T.A.S.; Blum, L.E.B.; Duarte, E.A.A.; Luz, E.D.M.N. Control of *Phytophthora palmivora* on postharvest papaya with *Trichoderma asperellum*, *T. virens*, *T. harzianum* and *T. longibrachiatum*. *Bioscience* **2018**, *34*, 1513–1521. [CrossRef]
75. Tomah, A.A.; Abd Alamer, I.S.; Li, B.; Zhang, J.Z. A new species of *Trichoderma* and gliotoxin role: A new observation in enhancing biocontrol potential of *T. virens* against *Phytophthora capsici* on chili pepper. *Biol. Cont.* **2020**, *145*, 104–261. [CrossRef]
76. Gilardi, G.; Pugliese, M.; Gullino, M.L.; Garibaldi, A. Effect of biocontrol agents and potassium phosphite against *Phytophthora* crown rot, caused by *Phytophthora capsici*, on zucchini in a closed soilless system. *Sci. Hortic.* **2020**, *265*, 109–207. [CrossRef]
77. La Spada, F.; Stracquadanio, C.; Riolo, M.; Pane, A.; Cacciola, S.O. Trichoderma counteracts the challenge of *Phytophthora nicotianae* cnfections on tomato by modulating plant defense mechanisms and the expression of crinkler, necrosis-inducing *Phytophthora* Protein 1, and cellulose-Binding Elicitor Lectin Pathogenic Effectors. *Front. Plant Sci.* **2020**, *11*, 539–583. [CrossRef]
78. Kang, S.W.; Kim, S.W. New antifungal activity of penicillic acid against *Phytophthora* species. *Biotechnol. Lett.* **2004**, *26*, 695–698. [CrossRef] [PubMed]
79. El-Sayed, A.S.A.; Ali, G.S. *Aspergillus flavipes* is a novel efficient biocontrol agent of *Phytophthora parasitica*. *Biol. Cont.* **2020**, *140*, 72–104. [CrossRef]

80. Wang, H.; Shi, J.; Luo, Z.; Chen, Q.; Wang, C. Isolation and identification of *Penicillium* strain QMYCS-2 and its effect on tobacco black shank Tobacco. *Sci. Technol. Stud.* **2020**, *49*, 2663. [CrossRef]
81. Mondol, M.A.M.; Farthouse, J.; Islam, M.T.; Schüffler, A.; Laatsch, H. Metabolites from the Endophytic Fungus *Curvularia* sp. M12 Act as Motility Inhibitors against *Phytophthora capsici* zoospores. *J. Nat. Prod.* **2017**, *80*, 347–355. [CrossRef]
82. Chen, J.T.; Su, H.J.; Huang, J.W. Isolation and identification of secondary metabolites of *Clitocybe nuda* responsible for inhibition of zoospore germination of *Phytophthora capsici*. *J. Agric. Food Chem.* **2012**, *60*, 7341–7344. [CrossRef]
83. Santos, M.; Diánez, F.; Moreno-Gavíra, A.; Sánchez-Montesinos, B.; Gea, F.J. *Cladobotryum mycophilum* as potential biocontrol agent. *Agronomy* **2019**, *9*, 891. [CrossRef]
84. Iqbal, M.; Jamshaid, M.; Zahid, M.A.; Vetukuri, R.R.; Stenberg, J.A. Biological control of strawberry crown rot, root rot and grey mould by the beneficial fungus *Aureobasidium pullulans*. *Bio. Cont.* **2021**, *66*, 535–545. [CrossRef]
85. Safitri, D.; Wiyono, S.; Soekarno, B.P.W. Mode of action of the endophytic yeast *Rhodotorula mucilaginosa* in controlling basal stem rot caused by *Phytopthora capsici*. *IOP Conf. Ser. Earth Environ. Sci.* **2021**, *667*, 12–50. [CrossRef]
86. Mercier, D.C. Manker Biocontrol of soil-borne diseases and plant growth enhancement in greenhouse soilless mix by the volatile producing fungus *Muscodor albus*. *Crop. Prot.* **2005**, *24*, 355–362. [CrossRef]
87. Pozo, M.J.; Cordier, C.; Dumas-Gaudot, E.; Barea, J.M.; Azcón-Aguilar, C. Localized versus systemic effect of arbuscular mycorrhizal fungi on defence responses to *Phytophthora* infection in tomato plants. *J. Exp. Bot.* **2002**, *53*, 525–534. [CrossRef]
88. Hu, J.; Hou, S.; Li, M.; Wu, F.; Lin, X. The better suppression of pepper *Phytophthora* blight by arbuscular mycorrhizal (AM) fungus than *Purpureocillium lilacinum* alone or combined with AM fungus. *J. Soils Sed.* **2020**, *20*, 792–800. [CrossRef]
89. González-Hernández, A.I.; Suárez-Fernández, M.B.; Pérez-Sánchez, R.; Gómez-Sánchez, M.Á.; Morales-Corts, M.R. Compost tea induces growth and resistance against Rhizoctonia solani and *Phytophthora capsici* in pepper. *Agronomy* **2021**, *11*, 781. [CrossRef]
90. Manasfi, Y.; Cannesan, M.A.; Riah, W.; Vicré, M.; Trinsoutrot-Gattin, I. Potential of combined biological control agents to cope with *Phytophthora parasitica*, a major pathogen of *Choisya ternata*. *Eur. J. Plant Pathol.* **2018**, *152*, 1011–1025. [CrossRef]
91. Marín, F.; Diánez, F.; Santos, M.; Navarro, M.J.; Yau, J.A. Control of *Phytophthora capsici* and *Phytophthora parasitica* on pepper (*Capsicum annuum* L.) with compost teas from different sources, and their effects on plant growth promotion. *Phytopathol. Med.* **2014**, *53*, 216–228. [CrossRef]
92. Diánez, F.; Santos, M.; Tello, J.C. Suppressive effects of grape marc compost on phytopathogenic oomycetes. *Arch. Phytopathol. Pflanzenschutz* **2007**, *40*, 1–18. [CrossRef]
93. Fontana, D.C.; de Paula, S.; Torres, A.G.; Schmidt, D.; Neto, D.D. Endophytic fungi: Biological control and induced resistance to phytopathogens and abiotic stresses. *Pathogens* **2021**, *10*, 570. [CrossRef]
94. Yyani, M.E.; Requena, C.; Egea-Gilabert, M.E. Candela biological control of *Phytophthora* root rot of pepper using *Trichoderma harzianum* and *Streptomyces rochei* in combination. *J. Phytopathol.* **2007**, *155*, 342–349. [CrossRef]
95. Jiang, H.; Zhang, L.; Zhang, J.Z.; Ojaghian, M.R.; Hyde, K.D. Antagonistic interaction between *Trichoderma asperellum* and *Phytophthora capsici* in vitro. *J. Zhejiang Univ. Sci. B* **2016**, *17*, 271–281. [CrossRef]
96. Trotta, A.; Varese, G.C.; Gnavi, E.; Fusconi, A.; Sampò, S.; Berta, G. Interactions between the soilborne root pathogen *Phytophthora nicotianae var.parasitica* and the arbuscular mycorrhizal fungus *Glomus mosseae* in tomato plants. *Plant Soil* **1996**, *185*, 199–209. [CrossRef]
97. Phoka, N.; Suwannarach, N.; Lumyong, S.; Ito, S.; Matsui, K.; Arikit, S.; Sunpapao, A. Role of volatiles from the endophytic fungus *Trichoderma asperelloides* PSU-P1 in biocontrol potential and in promoting the plant growth of *Arabidopsis thaliana*. *J. Fungi* **2020**, *6*, 341. [CrossRef]
98. Jin, J.H.; Zhang, H.X.; Tan, J.Y.; Yan, M.J.; Li, D.W.; Khan, A. A new ethylene-responsive factor *CaPTI1* gene of pepper (*Capsicum annuum* L.) involved in the regulation of defense response to *Phytophthora capsici*. *Front. Plant Sci.* **2015**, *6*, 1217. [CrossRef] [PubMed]
99. Lee, S.C.; Hwang, B.K. Induction of some defense-related genes and oxidative burst is required for the establishment of systemic acquired resistance in *Capsicum annuum*. *Planta* **2005**, *221*, 790–800. [CrossRef]
100. El-Sayed, A.S.A.; Akbar, A.; Iqrar, I. A glucanolytic *Pseudomonas* sp. associated with *Smilax bona-nox* L. displays strong activity against *Phytophthora parasitica*. *Microbiol. Res.* **2018**, *207*, 140–152. [CrossRef]
101. Shobha, M.S.; Mahadeva, M.S. Effect of Endophytic and plant growth promoting rhizobacteria against foot rot disease of *Piper nigrum* L. *Int. J. Environ. Agric. Biotechnol.* **2018**, *3*, 6. [CrossRef]
102. Spence, C.; Alff, E.; Johnson, C.; Ramos, C.; Donofrio, N.; Sundaresan, V.; Bais, H. Natural rice rhizospheric microbes suppress rice blast infections. *BMC Plant Biol.* **2014**, *14*, 130. [CrossRef]
103. Zhang, J.; Zhou, J.M. Plant immunity triggered by microbial molecular signatures. *Mol. Plant* **2010**, *3*, 783–793. [CrossRef]
104. Saijo, Y.; Loo, E.P.; Yasuda, S. Pattern recognition receptors and signaling in plant–microbe interactions. *Plant J.* **2018**, *93*, 592–613. [CrossRef] [PubMed]
105. Dong, X.; Ai, G.; Xia, C.; Pan, W.; Yin, Z.; Dou, D. Different requirement of immunity pathway components by oomycete effectors-induced cell death. *Phytopathol. Res.* **2022**, *4*, 4. [CrossRef]
106. Dou, D.L.; Zhou, J.M. Phytopathogen effectors subverting host immunity: Different foes, similar battleground. *Cell Host Microbe* **2012**, *12*, 484–495. [CrossRef]
107. Silvar, C.; Merino, F.; Díaz, J. Resistance in pepper plants induced by *Fusarium oxysporum f. sp. lycopersici* involves different defence-related genes. *Plant Biol.* **2009**, *11*, 68–74. [CrossRef]

108. Stamler, R.A.; Holguin, O.; Dungan, B.; Goldberg, N.; Randall, J.J. *BABA* and *Phytophthora nicotianae* induce resistance to *Phytophthora capsici* in chile pepper (*Capsicum annuum*). *PLoS ONE* **2015**, *10*, e0128327. [CrossRef] [PubMed]
109. Pieterse, C.M.J.; Zamioudis, C.B.R.L.; Weller, D.M.; Van Wees, S.C.M.; Bakker, P.A.H.M. Induced systemic resistance by beneficial microbes. *Ann. Rev. Phytopathol.* **2014**, *52*, 347–375. [CrossRef] [PubMed]
110. Das, K.; Roychoudhury, A. Reactive oxygen species (ROS) and response of antioxidants as ROS-scavengers during environmental stress in plants. *Front. Environ. Sci.* **2014**, *2*, 53. [CrossRef]
111. Savitha, M.J.; Sriram, S. Induced systemic resistance (ISR) in hot pepper against *Phytophthora capsici* infection triggered by cell wall oligosaccharide elicitors from *Trichoderma* species. *Indian J. Hortic.* **2017**, *74*, 233–239. [CrossRef]
112. Sriram, S.; Manasa, S.B.; Savitha, M.J. Potential use of elicitors from Trichoderma in induced systemic resistance for the management of *Phytophthora capsici* in red pepper. *J. Biol. Cont.* **2009**, *23*, 449–456. [CrossRef]
113. Umadevi, P.; Anandaraj, M. Proteomic analysis of the tripartite interaction between black pepper, *Trichoderma harzianum* and *Phytophthora capsici* provides insights into induced systemic resistance mediated by *Trichoderma* spp. *Eur. J. Plant Pathol.* **2019**, *154*, 607–620. [CrossRef]
114. Bellini, A.; Pugliese, M.; Guarnaccia, V.; Meloni, G.R.; Gullino, L.M. Calcium oxide, potassium phosphite and a *Trichoderma* enriched compost water suspension protect *Capsicum annuum* against *Phytophthora capsici* by priming the immune system. *Pest Manag. Sci.* **2021**, *77*, 3484–3490. [CrossRef] [PubMed]
115. Chowdappa, P.; Mohan Kumar, S.P.; Jyothi Lakshmi, M.; Upreti, K.K. Growth stimulation and induction of systemic resistance in tomato against early and late blight by *Bacillus subtilis* OTPB1 or *Trichoderma harzianum* OTPB3. *Biol. Cont.* **2013**, *65*, 109–117. [CrossRef]
116. Qiu, Y.; Yan, H.H.; Sun, S.M.; Wang, Y.Q.; Zhao, X.R.; Wang, H.Y. Use of *Bacillus velezensis* SDTB022 against tobacco black shank (TBS) and the biochemical mechanism involved. *Biol. Cont.* **2022**, *165*, 104–785. [CrossRef]
117. Yarullina, L.G.; Sorokan, A.V.; Tsvetkov, V.O.; Burkhanova, G.F.; Kalatskaja, J.N. Influence of the genus *Bacillus* bacteria on the content of H_2O_2 and the activity of hydrolases and their inhibitors in potato plants during *Phytophthora infestans* Mont. de Bary infection. *BIO Web Conf.* **2020**, *23*, 02010. [CrossRef]
118. Sorokan, A.; Benkovskaya, G.; Burkhanova, G.; Blagova, D.; Maksimov, I. Endophytic strain *Bacillus subtilis* 26DCryChS producing Cry1Ia toxin from *Bacillus thuringiensis* promotes multifaceted potato defense against *Phytophthora infestans* (Mont.) de Bary and pest *Leptinotarsa decemlineata* Say. *Plants* **2020**, *9*, 1115. [CrossRef]
119. Park, J.W.; Balaraju, K.; Kim, J.W.; Lee, S.W.; Park, K. Systemic resistance and growth promotion of chili pepper induced by an antibiotic producing *Bacillus vallismortis* strain BS07. *Biol. Cont.* **2013**, *65*, 246–257. [CrossRef]
120. Lee, B.D.; Dutta, S.; Ryu, H.; Suh, D.S.; Park, K. Induction of systemic resistance in panax ginseng against *Phytophthora cactorum* by native *Bacillus amyloliquefaciens* HK34. *J. Ginseng Res.* **2015**, *39*, 213–220. [CrossRef]
121. Kang, S.M.; Hamayun, M.; Waqas, M.; Shinwari, Z.K.; Lee, I.J. *Burkholderia* sp. KCTC 11096BP modulates pepper growth and resistance against *Phytophthora capsici*. *Pak. J. Bot.* **2016**, *48*, 1965–1970.
122. Sang, M.K.; Kim, J.G.; Kim, K.D. Biocontrol activity and induction of systemic resistance in pepper by compost water extracts against *Phytophthora capsici*. *Phytopathology* **2010**, *100*, 774–783. [CrossRef]
123. Ferrusquía-Jiménez, N.I.; Serrano-Jamaica, L.M.; Martínez-Camacho, J.E.; Sáenz de la O, D.; Villagomez-Aranda, A.L.; González-Chavira, M.M.; Guerrero-Aguilar, B.Z.; Torres-Pacheco, I.; Feregrino-Pérez, A.A.; Medina-Ramos, G.; et al. Extracellular self-DNA plays a role as a damage-associated molecular pattern (DAMP) delaying zoospore germination rate and inducing stress-related responses in *Phytophthora capsici*. *Plant Pathol.* **2022**, *71*, 1066–1075. [CrossRef]
124. Analisis de la Campaña Hortofruticola. Campaña 2020/21. Cajamar. Available online: https://publicacionescajamar.es/publicacionescajamar/public/pdf/series-tematicas/informes-coyuntura-analisis-de-campana/informe-71-campana-almeria-20-21.pdf (accessed on 10 March 2023).
125. Sánchez-Montesinos, B.; Diánez, F.; Moreno-Gavíra, A.; Gea, F.J.; Santos, M. Plant growth promotion and biocontrol of *Pythium ultimum* by saline tolerant *Trichoderma* isolates under salinity stress. *Int. J. Environ. Res. Public Health* **2019**, *16*, 2053. [CrossRef] [PubMed]
126. Sánchez-Montesinos, B.; Santos, M.; Moreno-Gavíra, A.; Marín-Rodulfo, T.; Gea, F.J.; Diánez, F. Biological control of fungal diseases by *Trichoderma aggressivum* f. *europaeum* and its compatibility with fungicides. *J. Fungi* **2021**, *7*, 598. [CrossRef] [PubMed]
127. Moreno-Gavíra, A.; Diánez, F.; Sánchez-Montesinos, B.; Santos, M. *Paecilomyces variotii* as a plant-growth promoter in horticulture. *Agronomy* **2020**, *10*, 597. [CrossRef]
128. Santos, M.; Dianez, F.; Gonzalez del Valle, M.; Tello, J.C. Grape marc compost: Microbial studies and suppression of soil-borne mycosis in vegetable seedlings. *World J. Microbiol. Biotechnol.* **2008**, *24*, 1493–1505. [CrossRef]
129. Diánez, F.; Boix, A.; de Cara, M.; Trillas, I.; Avilés, M.; Tello, J.C. Grape marc compost tea suppressiveness to plant pathogenic fungi: Role of siderophores. *Compost. Sci. Util.* **2006**, *14*, 48–53. [CrossRef]
130. Sánchez-Montesinos, B.; Diánez, F.; Moreno-Gavira, A.; Gea, F.J.; Santos, M. Role of *Trichoderma aggressivum* f. *europaeum* as plant-growth promoter in horticulture. *Agronomy* **2020**, *10*, 1004. [CrossRef]
131. Ling, Q.; Huang, W.; Jarvis, P. Use of a SPAD-502 meter to measure leaf chlorophyll concentration in *Arabidopsis thaliana*. *Photosyn. Res.* **2011**, *107*, 209–214. [CrossRef]
132. Livak, K.J.; Schmittgen, T.D. Analysis of Relative Gene Expression Data Using Real-Time Quantitative PCR and the $2^{-\Delta\Delta CT}$ Method. *Methods* **2001**, *25*, 402–408. [CrossRef]

133. Wang, J.E.; Li, D.W.; Zhang, Y.L.; Zhao, Q.; He, Y.M.; Gong, Z.H. Defence responses of pepper (*Capsicum annuum* L.) infected with incompatible and compatible strains of *Phytophthora capsici*. *Eur. J. Plant Pathol.* **2013**, *136*, 625–638. [CrossRef]
134. Gayoso, C.; Martinez de ilarduya, O.; Pomar, F.; Merino, F. Assessment of real-time PCR as a method for determining the presence of *Verticillium dahliae* in different *Solanaceae* cultivars. *Eur. J. Plant Pathol.* **2007**, *118*, 199–209. [CrossRef]
135. Lee, S.C.; Hwang, B.K. Identification of the pepper *SAR8.2* gene as a molecular marker for pathogen infection, abiotic elicitors and environmental stresses in *Capsicum annuum*. *Planta* **2003**, *216*, 387–396. [CrossRef] [PubMed]
136. Cheng, Y.; Pang, X.; Wan, H.; Ahammed, G.J.; Yu, J.; Yao, Z.; Ruan, M.; Ye, Q.; Li, Z.; Wang, R.; et al. Identification of optimal reference genes for normalization of qPCR analysis during pepper fruit development. *Front. Plant Sci.* **2017**, *8*, 1128. [CrossRef]
137. Volynchikova, E.; Kim, K.D. Biological control of oomycete soilborne diseases caused by *Phytophthora capsici*, *Phytophthora infestans*, and *Phytophthora nicotianae* in Solanaceous crops. *Mycobiology* **2022**, *50*, 269–293. [CrossRef]
138. Ren, Y.; Armstrong, M.; Qi, Y.; Mclellan, H.; Zhong, C.; Du, B. *Phytophthora infestans* RXLR effectors target parallel steps in an immune signal transduction pathway. *Plant Physiol.* **2019**, *180*, 2227–2239. [CrossRef]
139. Moreno-Gavira, A.; Huertas, V.; Diánez, F.; Sánchez-Montesinos, B.; Santos, M. *Paecilomyces* and its importance in the biological control of agricultural pests and diseases. *Plants* **2020**, *9*, 1746. [CrossRef]
140. Bononi, L.; Chiaramonte, J.B.; Pansa, C.C.; Moitinho, M.A.; Melo, I.S. Phosphorus-solubilizing *Trichoderma* spp. from Amazon soils improve soybean plant growth. *Sci. Rep.* **2020**, *10*, 2858. [CrossRef]
141. Khan, R.A.A.; Najeeb, S.; Hussain, S.; Xie, B.; Li, Y. Bioactive secondary metabolites from *Trichoderma* spp. against phytopathogenic fungi. *Microorganisms* **2020**, *8*, 817. [CrossRef] [PubMed]
142. Liu, Y.; He, P.; He, P.; Tian, Y.; He, Y. Potential biocontrol efficiency of *Trichoderma* species against oomycete pathogens. *Front. Microbiol.* **2022**, *13*, 974024. [CrossRef] [PubMed]
143. Ros, M.; Raut, I.; Santisima-Trinidad, A.B.; Pascual, J.A. Relationship of microbial communities and suppressiveness of *Trichoderma* fortified composts for pepper seedlings infected by *Phytophthora nicotianae*. *PLoS ONE* **2017**, *12*, e0174069. [CrossRef] [PubMed]
144. Choudhary, A.K.; Singh, N.; Singh, D. Evaluation of the bioformulation of potent native strains of *Trichoderma* spp. against the foot rot/gummosis of Kinnow mandarin. *Egypt J. Biol. Pest Cont.* **2021**, *31*, 90. [CrossRef]
145. Roberts, D.P.; Maul, J.E.; McKenna, L.F.; Emche, S.E.; Meyer, S.L.F.; Collins, R.T.; Bowers, J.H. Selection of genetically diverse *Trichoderma* spp. isolates for suppression of *Phytophthora capsici* on bell pepper. *Can. J. Microbiol.* **2010**, *56*, 864–873. [CrossRef] [PubMed]
146. Ezziyyani, M.; Pérez Sánchez, C.; Sid Ahmed, A.; Requena, M.E.; Candela, M.E. *Trichoderma harzianum* como biofungicida para el biocontrol de *Phytophthora capsici* en plantas de pimiento (*Capsicum annuum* L.). *Anal. Biol.* **2004**, *26*, 35–45.
147. Li, M.F.; Li, G.H.; Zhang, K.Q. Non-volatile metabolites from *Trichoderma* spp. *Metabolites* **2019**, *9*, 58. [CrossRef] [PubMed]
148. Lee, S.; Yap, M.; Behringer, G.; Hung, R.; Joan, W.B. Volatile organic compounds emitted by *Trichoderma* species mediate plant growth. *Fungal Biol. Biotechnol.* **2016**, *3*, 7. [CrossRef] [PubMed]
149. Li, X.Q.; Kuo, X.; Liu, X.M.; Zhang, P. A systematic review on secondary metabolites of *paecilomyces* species: Chemical diversity and biological activity. *Planta Med.* **2020**, *86*, 805–821. [CrossRef] [PubMed]
150. Bae, S.J.; Mohanta, T.K.; Chung, J.Y.; Ryu, M.; Park, G.; Shim, S.; Hong, S.B.; Seo, H.; Bae, D.W.; Bae, I.; et al. *Trichoderma* metabolites as biological control agents against *Phytophthora* pathogens. *Biol. Cont.* **2016**, *92*, 128–138. [CrossRef]
151. Dai, Z.B.; Wang, X.; Li, G.H. Secondary metabolites and their bioactivities produced by *Paecilomyces*. *Molecules* **2020**, *25*, 5077. [CrossRef]
152. Moreno-Gavíra, A.; Diánez, F.; Sánchez-Montesinos, B.; Santos, M. Biocontrol effects of *Paecilomyces variotii* against fungal plant diseases. *J. Fungi* **2021**, *7*, 415. [CrossRef]
153. Sood, M.; Kapoor, D.; Kumar, V.; Sheteiwy, M.S.; Ramakrishnan, M.; Landi, M. Trichoderma: The "secrets" of a multitalented biocontrol agent. *Plants* **2020**, *9*, 762. [CrossRef]
154. Jung, H.W.; Hwang, B.K. Pepper gene encoding a basic b-1,3-glucanase is differentially expressed in pepper tissues upon pathogen infection and ethephon or methyl jasmonate treatment. *Plant Sci.* **2000**, *156*, 23–34. [CrossRef]
155. Kim, Y.J.; Hwang, B.K. Pepper gene encoding a basic pathogenesis-related 1 protein is pathogen and ethylene inducible. *Physiol. Plantarum* **2000**, *108*, 51–60. [CrossRef]
156. Kang, D.S.; Min, K.J.; Kwak, A.M.; Lee, S.Y.; Kang, H.W. Defense response and suppression of *Phytophthora* blight disease of pepper by water extract from spent mushroom substrate of *Lentinula edodes*. *Plant Pathol. J.* **2017**, *33*, 264–275. [CrossRef] [PubMed]
157. Gutierrez, N.; Giménez, M.J.; Torres, A.M.; Atienza, S.G.; Avila, C.M.; Palomino, C. Up-regulation of resistance gene analogs (RGA) in chickpea in the early response to Fusarium wilt. *Euphytica* **2012**, *186*, 793–804. [CrossRef]
158. Silvar, C.; Merino, F.; Díaz, J. Differential activation of defense-related genes in susceptible and resistant pepper cultivars infected with *Phytophthora capsici*. *J. Plant Physiol.* **2008**, *165*, 1120–1124. [CrossRef]

159. Sarowar, S.; Kim, Y.J.; Kim, E.N. Overexpression of a pepper basic pathogenesis-related protein 1 gene in tobacco plants enhances resistance to heavy metal and pathogen stresses. *Plant Cell Rep.* **2005**, *24*, 216–224. [CrossRef]
160. Shi, R.; Jin, J.; Nifong, J.M.; Shew, D.; Lewis, R.S. Homoeologous chromosome exchange explains the creation of a QTL affecting soil-borne pathogen resistance in tobacco. *Plant Biotechnol. J.* **2022**, *20*, 47–58. [CrossRef]

Journal of *Fungi*

MDPI

Article

β-Glucan Enhances the Biocontrol Efficacy of Marine Yeast *Scheffersomyeces spartinae* W9 against *Botrytis cinerea* in Strawberries

Xueyan Chen [1], Yingying Wei [1,*], Xiurong Zou [1,2], Zichang Zhao [1], Shu Jiang [1], Yi Chen [1], Feng Xu [1] and Xingfeng Shao [1]

1 State Key Laboratory for Managing Biotic and Chemical Threats to the Quality and Safety of Agro-Products, Zhejiang-Malaysia Joint Research Laboratory for Agricultural Product Processing and Nutrition, College of Food and Pharmaceutical Sciences, Ningbo University, Ningbo 315800, China
2 Henry Fok School of Food Science and Engineering, Shaoguan University, Shaoguan 512005, China
* Correspondence: weiyingying@nbu.edu.cn

Abstract: The marine yeast *Scheffersomyeces spartinae* W9 is a promising biocontrol agent for gray mold caused by *Botrytis cinerea* in strawberries. Improving the biocontrol efficacy of *S. spartinae* W9 is necessary for its commercial application. In this study, different concentrations of β-glucan were added to the culture medium to evaluate its effect on the biocontrol efficacy of *S. spartinae* W9. The results showed that 0.1% β-glucan could increase the biocontrol effect of *S. spartinae* W9 against *B. cinerea* in strawberries and in vitro. We found that adding 0.1% β-glucan to the culture medium promoted the growth of *S. spartinae* W9 in wounds of strawberries, enhanced biofilm formation ability, and secreted more β-1,3-glucanase. In addition, 0.1% β-glucan increased the survival rate of *S. spartinae* W9 under oxidative, thermal, osmotic, and plasma membrane stressors. Transcriptome analysis revealed 188 differential expressed genes in *S. spartinae* W9 cultured with or without 0.1% β-glucan, including 120 upregulated and 68 downregulated genes. The upregulated genes were associated with stress response, cell wall formation, energy production, growth, and reproduction. Thus, culturing with 0.1% β-glucan is an effective way to improve the biocontrol ability of *S. spartinae* W9 against gray mold in strawberries.

Keywords: biological control; marine yeast; β-glucan; gray mold; strawberry

1. Introduction

Gray mold caused by *Botrytis cinerea* is the primary fungal disease of strawberry fruit during planting and postharvest storage [1]. Chemical fungicides have been used as the most effective method to control gray mold for decades. However, frequent use of chemical fungicides has increased the resistance of fungal pathogens and has had a negative influence on human health and environmental safety [2,3]. In recent years, researchers have been actively seeking alternatives to chemical fungicides. Biological control using microbial agents is considered to be a promising candidate in the management of postharvest diseases. Antagonistic yeasts are widely used to control postharvest diseases because of their biosafety, well-defined mechanism of action, and good commercial potential [4–6]. A variety of antagonistic yeasts have been reported to control gray mold decay of fruits, including *Pichia guilliermondii* [7], *Wickerhamomyces anomalus* [8], *Metschnikowia pulcherrima* [9], *Candida pseudolambica* [10], *Metarhizium anisopliae* [11], *Scheffersomyeces spartinae* [12], and so on. However, antagonistic yeasts have not been able to achieve the effect of traditional chemical fungicides in controlling postharvest diseases of fruit [13]. Therefore, it is important to improve the biocontrol efficacy of antagonistic yeasts.

Adding polysaccharides to the culture media may increase the biocontrol activity of antagonistic yeasts. β-glucan is a natural polysaccharide and the main component of yeast cell walls and has various biological activities [14]. β-glucan has been reported to improve various stress tolerances and biocontrol efficacy of *Cryptococcus laurentii* (*Papiliotrema laurentii*) [15,16] and enhance the biocontrol efficacy of *Cryptococcus podzolicus* (*Saitozyma podzolica*) [16] against postharvest decay of apples [17] and pears [18] by upregulating the genes related to cell wall synthesis and increasing its tolerance to oxidative stress. *S. spartinae* W9 is an antagonistic yeast isolated from the intertidal zone marine sediment by our laboratory and a promising biocontrol agent for gray mold in strawberries [12]. To promote the commercial application process, improving the biocontrol efficacy of *S. spartinae* W9 is urgent. Whether β-glucan can enhance the biocontrol efficacy of marine yeast *S. spartinae* W9 has yet to be studied.

Therefore, this study aimed to investigate the effect of β-glucan on the biocontrol efficacy of marine yeast *S. spartinae* W9 and reveal its possible mechanism. In this study, we investigated (1) the effect of different concentrations of β-glucan on antagonistic activities of *S. spartinae* W9 against *B. cinerea* in vitro and in strawberry fruit, (2) the effect of β-glucan on the colonization ability and biofilm forming ability of *S. spartinae* W9, (3) the effect of β-glucan on extracellular hydrolase activity of *S. spartinae* W9, (4) the effect of β-glucan on various forms of stress tolerance of *S. spartinae* W9, and (5) the effect of β-glucan on the transcriptome of *S. spartinae* W9.

2. Materials and Methods

2.1. Antagonistic Yeast and Fungal Pathogen

The antagonistic yeast *S. spartinae* W9 was preserved in our laboratory and had previously been isolated from the marine sediment in the South China Sea [12]. Consistent with the method used by Wang et al. (2018) and Zhao et al. (2020) [17,18], the activated yeast was suspended in NYDB (nutrient yeast dextrose broth medium) or NYDB supplemented with 0.1%, 0.5%, 1%, and 2% β-glucan (*w/v*), and incubated in an oscillating incubator at 28 °C for 24 h.

B. cinerea was isolated from infected strawberry fruit and preserved in our laboratory. The strain was maintained and propagated on PDA (potato dextrose agar medium). Spore suspensions were washed from the mycelium, cultured for 7 d, and adjusted to 1×10^5 spores mL/L using a hemocytometer.

2.2. Fruit

Commercially mature strawberries (*Fragaria ananassa* Duch. cv. Zhangji) were harvested from a greenhouse in the Meishan District, Ningbo, China. Strawberries with uniform color and size and absence of visual injury or lesion were selected for experiments. The strawberries were wiped with 75% ethanol (*v*/*v*) for disinfection and air-dried at 20 °C [19].

2.3. Biocontrol Efficacy of S. spartinae W9 Cultured with Different Concentrations of β-Glucan against B. cinerea in Strawberries

This experiment was conducted according to the method described by Fu et al. (2015) [15]. A uniform wound was made at the strawberry equator with a sterilized nail (3 mm deep × 3 mm wide), and each wound was inoculated with 10 μL of any one of the following solutions: (1) sterile water (control), (2) *S. spartinae* W9 cell suspension cultured in NYDB medium (1×10^8 cells/mL), or (3) *S. spartinae* W9 cell suspension cultured in NYDB medium with 0.1%, 0.5%, 1%, and 2% β-glucan (*w/v*) (1×10^8 cells/mL). After two hours, each wound was injected with 10 μL of *B. cinerea* spore suspension (1×10^5 spores/mL). Then all strawberries were stored in a thermostatic chamber at 20 °C with 90% relative humidity. The incidence of gray mold and the diameter of each lesion were observed every day. Every treatment was replicated three times, with 20 fruits selected at random in each replicate.

2.4. Antagonistic Activity of S. spartinae W9 Cultured with β-Glucan against B. cinerea In Vitro

For the mycelia growth test, 100 μL of *S. spartinae* W9 suspensions (cultured in NYDB and NYDB with 0.1%, 0.5%, 1%, and 2% β-glucan) at a concentration of 1×10^5 cells/mL was spread into the PDA medium. A total of 100 μL of sterile water was spread to the PDA and used as the control. Afterwards, a 9-mm diameter plug taken from the edge of the actively growing *B. cinerea* colony was inoculated into the center of each PDA plate. The plates were incubated at 28 °C for 5 d, and the mycelia diameters were measured daily. The experiment consisted of three replicates, with 5 plates per replicate.

For the spore germination test, cell suspensions of 1×10^9 cells/mL *S. spartinae* W9 (cultured in NYDB and NYDB with 0.1%, 0.5%, 1%, and 2% β-glucan) were prepared. Then 200 μL of yeast suspension was mixed with 200 μL of *B. cinerea* (1×10^6 spores/mL) spore suspension and added into a 1.6-μL PDB medium. An amount of 200 μL sterile water mixed with 200 μL *B. cinerea* spore suspension was used as the control. The spores were incubated for 6 h at 28 °C and 180 rpm on a shaker. The germination rate of *B. cinerea* spores were counted by light microscopy and hemocytometer. At least 100 spores were observed in each treatment, with 3 replicates per treatment.

2.5. Population of Yeasts in Strawberry Wounds

In accordance with the method of Zou et al. (2022) [10], strawberries were wounded as described in 2.3 and then inoculated with *S. spartinae* W9 or 0.1% β-glucan-treated *S. spartinae* W9. All strawberries were stored at 20 °C with 90% relative humidity. The amount of yeast at the strawberry fruit wounds was measured at 0 h, 12 h, 24 h, 48 h, and 72 h. The fruit wound tissue was removed with a sterile blade and homogenized in a mortar containing 10 mL of sterile water. A total of 100 μL of the homogenate diluted to the appropriate concentration was spread on NYDA and incubated at 28 °C for 48 h to count the number of colonies. The number of yeast colonies at each wound was expressed as $\log_{10}$CFU/wound. The experiment consisted of three replicates, with three strawberries per replicate.

2.6. The Effect of 0.1% β-Glucan on the Growth of S. spartinae W9

S. spartinae W9 cells (1×10^8 cells/mL) were cultured in NYDB or NYDB with 0.1% β-glucan at 28 °C. The growth of *S. spartinae* W9 was recorded by measuring the absorbance at 600 nm at 12 h, 24 h, 36 h, and 48 h. Three biological replicates were available for each treatment.

2.7. Determination of Biofilm Formation

The biofilm formation of *S. spartinae* W9 was assessed using the method of Qiu et al. (2022) [20] with some modifications. An amount of 100 μL of *S. spartinae* W9 suspension (1×10^7 cells/mL) was pipetted into 50 mL of YNB medium and cultured for about 12 h. Yeast precipitate was obtained by centrifugation at 8000 rpm, 4 °C, for 5 min and was washed twice to adjust the concentration of yeast to 1×10^8 cells/mL. Equal amounts of yeast suspension were injected into YNB or YNB containing 0.1% β-glucan and then were incubated at 28 °C, 75 rpm for 12, 18, 24, 36, and 48 h. For biofilm assay, the culture solutions were washed twice with phosphate buffer and treated with methanol for 15 min. Then, 200 μL of 0.4% crystalline violet was added for staining, and after 40 min, the excess dye was removed and washed with sterile water. Lastly, 200 μL of 33% glacial acetic acid was added and incubated for 30 min, and the supernatant was used to measure the absorbance at 590 nm.

2.8. Measurement of Extracellular Hydrolases

The ability of *S. spartinae* W9 to produce extracellular hydrolases, mainly including chitinase and β-1,3-glucanase (GLU), was determined by an extracellular hydrolase assay plate [21]. Chitinase and CLU can decompose the colloidal chitin and laminarin, respectively. If *S. spartinae* W9 secreted chitinase and CLU, the hyaline rings would appear in the plates.

According to the result of an extracellular hydrolase assay plate, we determined the GLU activity of *S. spartinae* W9. *S. spartinae* W9 cultured in NYDB or NYDB with 0.1% β-glucan were sampled at 12, 16, 20, 24, 36, and 48 h. The fermentation broth was centrifuged at 6000 rpm for 10 min, and the supernatant was filtered through a sterile membrane for enzyme assay. The GLU activity was determined using the 3, 5-dinitrosalicylic acid method [22], and the absorbance at 540 nm was measured to calculate GLU activity. One unit (U) of GLU was defined as the amount of the enzyme required to produce 1 nmol of glucose per minute, and the results were expressed as U/mL. Each group contained three independent replicates.

2.9. Stress Resistance Assays of S. spartinae W9

As described by Huang et al. (2021) [23], high temperature, NaCl, H_2O_2, and sodium dodecyl sulfate (SDS) were used to simulate stress. *S. spartinae* W9 were cultured in NYDB or NYDB with 0.1% β-glucan for 24 h, and the yeast suspensions (1×10^8 cells/mL) were prepared.

To determine the tolerance of *S. spartinae* W9 to high temperatures, the yeast suspensions were placed into a water bath at 40 °C, 45 °C, and 50 °C for 30 min. The heat-treated yeast suspensions were cooled at room temperature for 10 min and diluted to the appropriate concentration. Then, 0.1 mL of each yeast suspension was spread on a NYDA plate to calculate the survival rate.

S. spartinae W9 or β-glucan-treated *S. spartinae* W9 yeast suspensions were injected into NYDB (control) and a selective medium (NYDB-4% NaCl, NYDB-7% NaCl, NYDB-10% NaCl, NYDB-1.25mM H_2O_2, NYDB-2.5mM H_2O_2, NYDB-5mM H_2O_2, NYDB-10mM H_2O_2, NYDB-0.1‰ SDS, NYDB-0.5‰ SDS, NYDB-1‰ SDS). Yeasts were incubated at 28 °C and 180 rpm for 24 h and then diluted to the appropriate concentration to spread on NYDA plates. All plates were incubated at 28 °C for 48 h, and the numbers of yeast colony were recorded. The survival rates of *S. spartinae* W9 or β-glucan-treated *S. spartinae* W9 without stress treatment were used as respective controls. Each treatment contained three independent replicates.

2.10. Transcriptomic Analysis

RNA from *S. spartinae* W9 cultured in NYDB with or without 0.1% β-glucan was isolated using TRIzol (Thermofisher, MA, USA) according to the manufacturer's instructions. The mRNA with PolyA (polyadenylation) was specifically captured and fragmented with a magnesium ion interruption kit (NEBNext^R Magnesium RNA Fragmentation Module, cat. E6150S, New England Biolabs, MA, USA). The fragmented RNA was synthesized into cDNA. Then, double-stranded DNA were synthesized and repaired. The fragment sizes were screened and purified using magnetic beads, and the cDNA library was enriched with PCR. RNA-seq was performed by LC-Bio Technology CO., Ltd. Hangzhou, China. Genes with a fold change in expression level ≥ 1.2 and adjusted $p < 0.01$ were identified as differentially expressed genes (DEGs). The DEGs' data was analyzed for GO and KEGG enrichment analysis.

2.11. Real-Time Quantitative PCR

RNA extracted from *S. spartinae* W9 cultured in NYDB with or without 0.1% β-glucan were first reverse transcribed to cDNA. RT-qPCR was performed using ChamQ Universal SYBR qPCR Master Mix (Vazyme, Nanjing, China). Twelve DEGs were randomly selected for validation by qPCR, and *ACT1* was used as an internal reference gene. Relative gene expression was calculated according to the $2^{-\Delta\Delta Ct}$ method. Table 1 contains a list of the selected genes and the specific primers.

Table 1. Primers used for RT-qPCR analysis of *S. spartinae* genes.

Gene ID	Gene Name	Forward and Reverse Primers (5′ to 3′)
gene-KQ657_000363	*CYS3*	F: TGGGTGTTTTGGCAACCAAC R: ACCTCTGTGAGCCAACCAAG
gene-KQ657_002550	*GRE2_4*	F: TTTCACATTGCGTCTCCCGT R: GCTGAGGTGCCTTATCCGTT
gene-KQ657_002037	*ERG3*	F: CGGATGGTCTCTTCCACTCC R: CTGTTGCCCATATCGAGGGT
gene-KQ657_001934	*ACH1*	F: CGTTTCTACGCCAACTGGGA R: GGTGGAGTTAGCATGAGCGT
gene-KQ657_000095	*AOX2*	F: CATGCCGCACCTGTTTTAGT R: GGCTCTTTAACCGGTGGTGT
gene-KQ657_001274	*TDH1*	F: ACAAGGACTGGAGAGGTGGT R: TTACCGACAGCCTTAGCAGC
gene-KQ657_000537	*ECM4_1*	F: TTCGCTTCCGACCAAGAGAC R: TGACCACCAGCAGTGTGAAT
gene-KQ657_001686	*KRI1*	F: AGGCCGAGACCATTGAAGTC R: CTGGTGCTTCTTGGTCGCTA
gene-KQ657_001371	*IRS4*	F: GCTACGAAGCTTGTGGGAGT R: GCAGACAATCGAGCAGCAAC
gene-KQ657_001105	*RRP36*	F: ACCTGTTTCGGTAGTCAGGG R: CATCTTGCTCCGTTGCTTGG
gene-KQ657_000758	*ADH1_1*	F: ATGGGTTGCAGTCTCTGGTG R: CCTTCTCTTCGCCACCATCA
gene-KQ657_003270	*RNR2*	F: AGATGCCCTTCCAGTGTCTT R: AGGAGAAAGCATTGGCTGCT
gene-KQ657_004479	*ACT1*	F:CAGACCTGCTGACTTGGGTT R:AGAGGATGGGGCCAACAAAG

2.12. Statistical Analysis

SPSS statistics software version 20 was used to analyze the data. When the number of comparisons within the group was three or more, the data were compared with the mean by implementing Duncan's multiple analysis using ANOVA; when the number of comparisons within the group was two, the mean was compared using the independent samples *t*-test. Significance was assessed when $p < 0.05$. Data were expressed as mean ± standard deviation.

3. Results

3.1. Biocontrol Efficacy of S. spartinae W9 Cultured with β-Glucan against B. cinerea in Strawberries

As shown in Figure 1A, gray mold decay of strawberries in all groups was observed at the fourth day after inoculation, and *S. spartinae* W9 cultured with or without β-glucan reduced the disease incidence of strawberries. *S. spartinae* W9 cultured with 0.1% β-glucan significantly reduced the disease incidence when compared to yeast without β-glucan after the fourth day ($p < 0.05$). At the end of storage, the disease incidence was 96.7% in the control group, 63.3% in the *S. spartinae* W9 group, and only 40% in the 0.1% β-glucan-treated *S. spartinae* W9 group (Figure 1A,B). Adding 0.1% β-glucan into NYDB effectively improved the biocontrol efficacy of *S. spartinae* W9.

3.2. In Vitro Inhibition of S. spartinae W9 Cultured with β-Glucan against B. cinerea

As shown in Figure 2A, β-glucan enhanced the antagonistic effect of *S. spartinae* W9 against *B. cinerea* in vitro. On the fourth day, the mycelia of *B. cinerea* in the control group covered almost the whole plate, while *B. cinerea* mycelia in the other groups were significantly inhibited ($p < 0.05$). Compared to the *S. spartinae* W9 cultured without β-glucan, 0.1% and 1% β-glucan were effective in increasing the inhibition of *B. cinerea* mycelial growth, for which the diameter of *B. cinerea* plaques in the *S. spartinae* W9 treatment group was 23.4 mm, while the diameters of the 0.1% and 1% β-glucan-induced *S. spartinae* W9 treated groups were 18.8 mm and 18 mm, respectively.

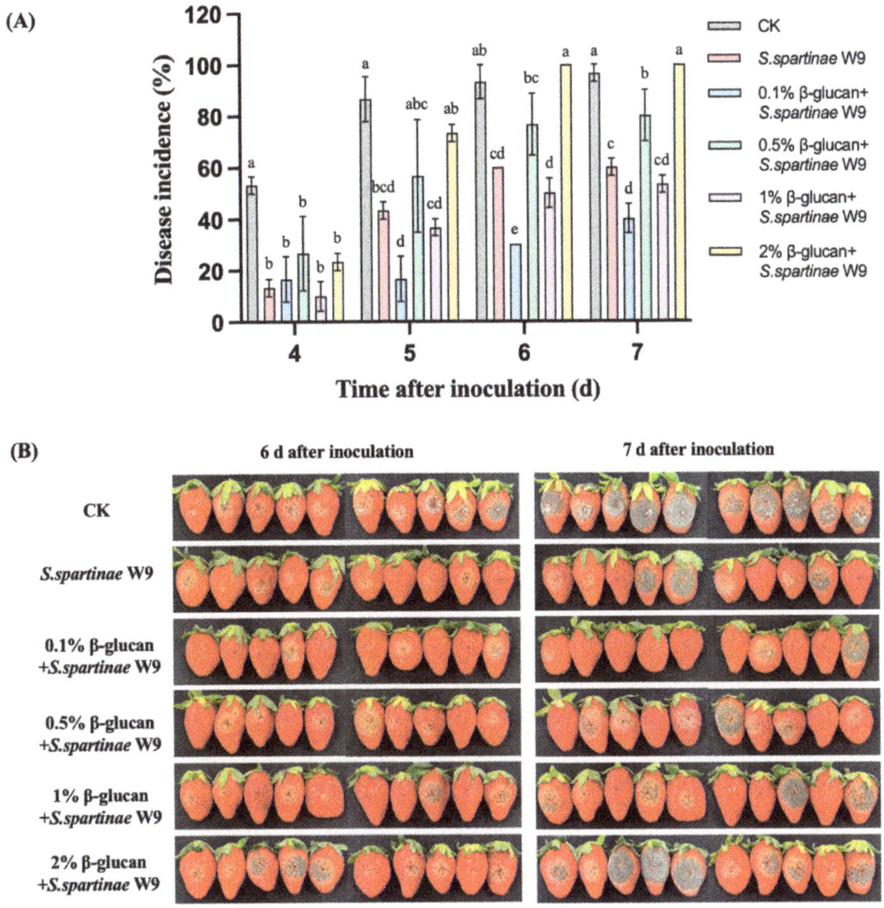

Figure 1. Biocontrol efficacy of *S. spartinae* W9 cultured with different concentrations of β-glucan against *B. cinerea* in strawberries. (**A**) Disease incidence of gray mold in strawberries after the treatment with *S. spartinae* W9 cultured with different concentrations (0, 0.1%, 0.5%, 1%, and 2%) of β-glucan during storage. (**B**) Phenomenon of gray mold in strawberries in each group at day 6 and day 7 after inoculation with *B. cinerea*. Different lowercase letters indicate significant differences between different treatment groups according to Duncan's test ($p < 0.05$).

Figure 2B shows that *S. spartinae* W9 cultured with or without β-glucan significantly inhibited spore germination of *B. cinerea* ($p < 0.05$). *S. spartinae* W9 cultured with different concentrations of glucan showed different inhibitory effects on spore germination of *B. cinerea*. The germination rate of *B. cinerea* spores in 0.1% β-glucan-treated *S. spartinae* W9 was 9.5%, which was significantly lower than the 17.1% in the *S. spartinae* W9 group ($p < 0.05$). Only 0.1% β-glucan increased the ability of *S. spartinae* W9 to inhibit spore germination of *B. cinerea*.

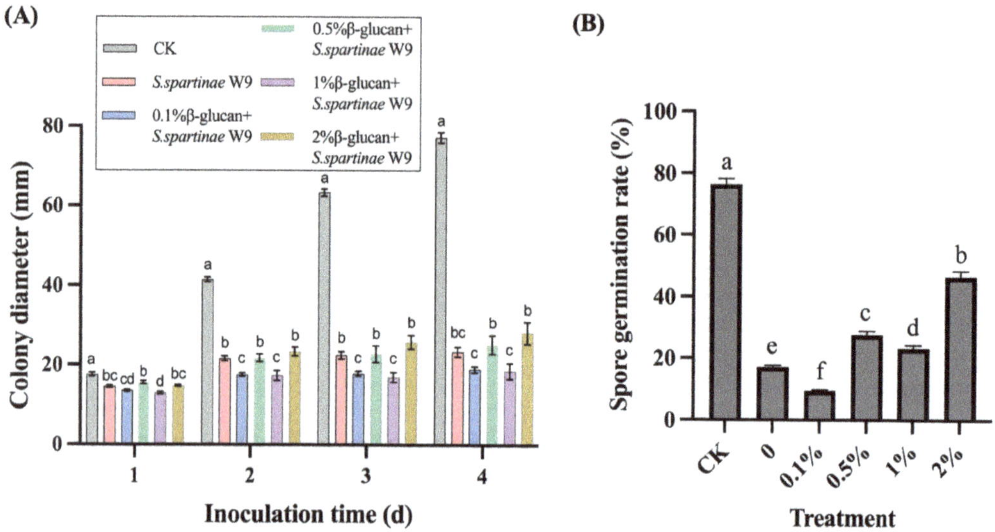

Figure 2. In vitro inhibition of *S. spartinae* W9 cultured with different concentrations of β-glucan against *B. cinerea*. (**A**) Mycelia diameter of *B. cinere* after treatment with *S. spartinae* W9 cultured with different concentrations (0, 0.1%, 0.5%, 1%, and 2%) of β-glucan. (**B**) Spore germination rate of *B. cinerea* after treatment with *S. spartinae* W9 cultured with different concentrations (0, 0.1%, 0.5%, 1%, and 2%) of β-glucan. Different lowercase letters indicated significant differences between different treatment groups according to Duncan's test ($p < 0.05$).

3.3. Effects of 0.1% β-Glucan on the Growth of S. spartinae W9 In Vitro

S. spartinae W9 yeasts were cultured in NYDB and NYDB containing 0.1% β-glucan at a concentration of 0.1% for 48 h, respectively. The OD_{600} of *S. spartinae* W9 in both media showed no significant difference ($p > 0.05$; Table 2). The results indicated that 0.1% β-glucan had no significant effect on the growth of *S. spartinae* W9 in vitro.

Table 2. Growth of *S. spartinae* W9 in NYDB or NYDB containing 0.1% β-glucan.

Culture Conditions	OD_{600}			
	12 h	24 h	36 h	48 h
NYDB	1.25 ± 0.011 a	2.22 ± 0.006 a	2.42 ± 0.005 a	2.51 ± 0.001 a
NYDB with 0.1%β-glucan	1.27 ± 0.009 a	2.23 ± 0.014 a	2.41 ± 0.007 a	2.51 ± 0.018 a

Same letter in each column means no significant difference according to the independent samples *t*-test ($p > 0.05$).

3.4. Effects of 0.1% β-Glucan on Colonization Ability in Strawberries and Biofilm Forming Ability of S. spartinae W9

S. spartinae W9 cultured with or without β-glucan was able to grow rapidly in the wounds of strawberries, and its population levels reached the highest value at 24 h (Figure 3A). The population of the β-glucan-induced *S. spartinae* W9 increased from 6.1 $\log_{10}$ CFU wound^{-1} to 7.7 $\log_{10}$ CFU wound^{-1} at 24 h, which was significantly higher than that of *S. spartinae* W9 cultured in NYDB ($p < 0.05$). The growth rate of the β-glucan-treated *S. spartinae* W9 within 24 h was higher than *S. spartinae* W9 cultured in NYDB.

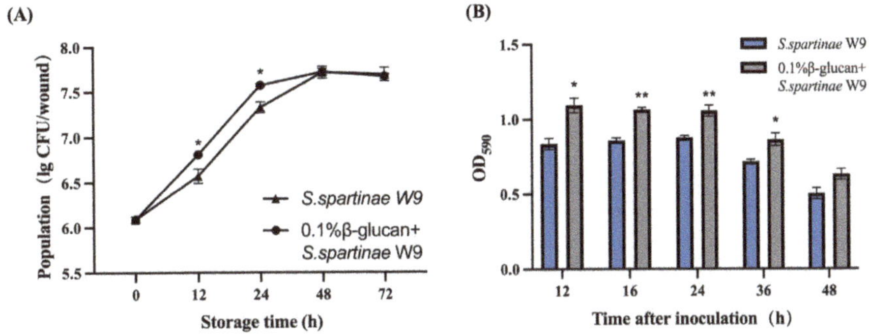

Figure 3. Effects of 0.1% β-glucan on colonization ability in strawberries and biofilm forming ability of *S. spartinae* W9. (**A**) Population dynamics of *S. spartinae* W9 cultured in NYDB or NYDB with 0.1% β-glucan in wounds of strawberries at 20 °C. (**B**) Biofilm forming ability of *S. spartinae* W9 cultured with or without 0.1% β-glucan stained with crystal violet. * indicate a significant difference between *S. spartinae* W9 cultured with or without 0.1% β-glucan identified by the *t*-test (*, $p < 0.05$; **, $p < 0.01$).

Crystal violet staining was used to further evaluate the biofilm formation capacity of *S. spartinae* W9. During the 24 h of yeast culture, the biofilm formation capacity remained stable in each group (Figure 3B). The OD_{590} of *S. spartinae* W9 cultured with β-glucan was consistently higher than that of *S. spartinae* W9 cultured without β-glucan. This indicated that β-glucan promoted the formation of biofilm for *S. spartinae* W9 and enhanced its biofilm formation capacity.

3.5. Effects of 0.1% β-Glucan on Extracellular Hydrolase of S. spartinae W9

The extracellular hydrolase assay plates showed that *S. spartinae* W9 could secrete GLU but not chitinase (Figure 4A). The 0.1% β-glucan could not induce *S. spartinae* W9 to secrete chitinase. The extracellular GLU activities of *S. spartinae* W9 cultured in NYDB with 0.1% β-glucan were significantly higher than those of *S. spartinae* W9 cultured in NYDB during the culture period (Figure 4B), indicating that β-glucan could induce *S. spartinae* W9 to secrete more GLU.

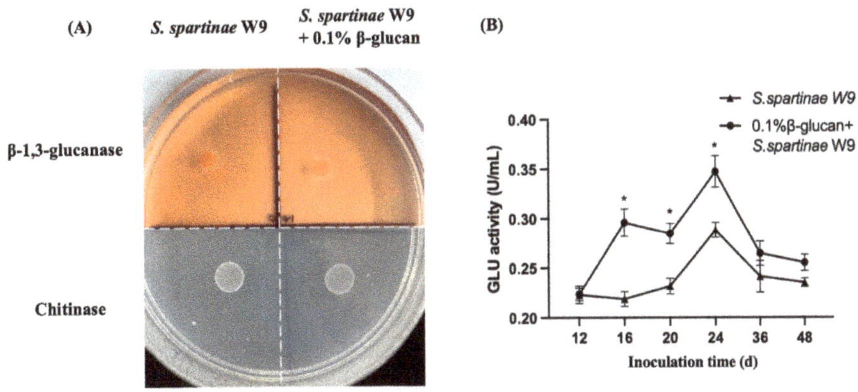

Figure 4. Effects of 0.1% β-glucan on extracellular hydrolase activity of *S. spartinae* W9. (**A**) Results of the extracellular hydrolase assay plates of *S. spartinae* W9 cultured with or without 0.1% β-glucan. (**B**) GLU activity of *S. spartinae* W9 cultured in NYDB or NYDB containing 0.1% β-glucan. * indicate a significant difference between *S. spartinae* W9 cultured with or without 0.1% β-glucan identified by the *t*-test ($p < 0.05$).

3.6. 0.1% β-Glucan Increased the Survival rate of S. spartinae W9 under Different Stresses

S. spartinae W9 cells harvested from NYDB with 0.1% β-glucan showed higher survival rates under oxidative, thermal, osmotic, and plasma membrane stresses, compared with the yeast cells harvested from NYDB (Figure 5). The survival rate of *S. spartinae* W9 under 10 mM H_2O_2, 10% NaCl, 1% SDS, and 50 °C were 66.2%, 68.2% 67.7%, and 76.3%, respectively, while those of β-glucan-treated *S. spartinae* W9 were 87.4%, 71.0%, 70.6%, and 81.63%, respectively (Figure 5). The 0.1% β-glucan increased the survival rate of *S. spartinae* W9 under the 10 mM H_2O_2 environment by 32% compared to the control and significantly improved its tolerance to oxidative stress.

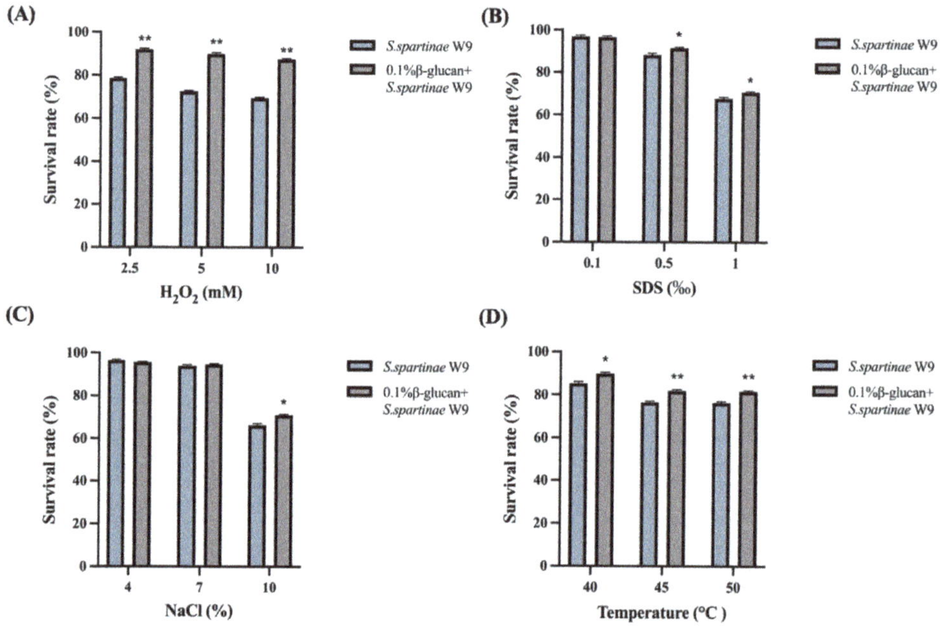

Figure 5. Effects of 0.1% β-glucan on the survival rate of *S. spartinae* W9 under different stresses. (**A**) The survival rate of *S. spartinae* W9 under oxidative stress caused by H_2O_2. (**B**) The survival rate of *S. spartinae* W9 under plasma membrane stress caused by SDS. (**C**) The survival rate of *S. spartinae* W9 under hyperosmotic stress caused by NaCl. (**D**) The survival rate of *S. spartinae* W9 under high temperature stressor. Asterisks indicate a significant difference between *S. spartinae* W9 cultured with or without 0.1% β-glucan identified by the *t*-test (*, $p < 0.05$; **, $p < 0.01$).

3.7. RNA-seq Analysis and RT-qPCR Validation

A total of 40.87 G clean reads were obtained from *S. spartinae* W9 cultured in NYDB with or without β-glucan. The Q30 levels were both higher than 70%, indicating that the base quality was up to standard. GC contents were both above 40%, and the percentage of valid reads exceeded 95%, which indicated that the data were reliable. By comparing the gene expression levels between *S. spartinae* W9 cultured in NYDB with and without β-glucan, 188 genes were identified as DEGs, including 120 upregulated genes and 68 downregulated genes.

A total of 172 DEGs were able to be annotated in the GO database and were enriched into three major categories (Figure 6A). In the biological process category, the DEGs were involved in translation, the biological process, the oxidation-reduction process, and so on. The highly enriched DEGs of the cellular components were involved in cell organelles and cell membranes. The subclass with the most enriched DEGs by molecular function was the structural constituent of ribosome (34 DEGs) subclass, followed by RNA binding (12 DEGs).

The DEGs were classified into five categories by KEGG enrichment analysis (Figure 6B). Among them, the pathway with the most DEGs was translation, involving 40 genes. The second pathway was the carbohydrate metabolism pathway, with 13 DEGs.

(A)

(B)

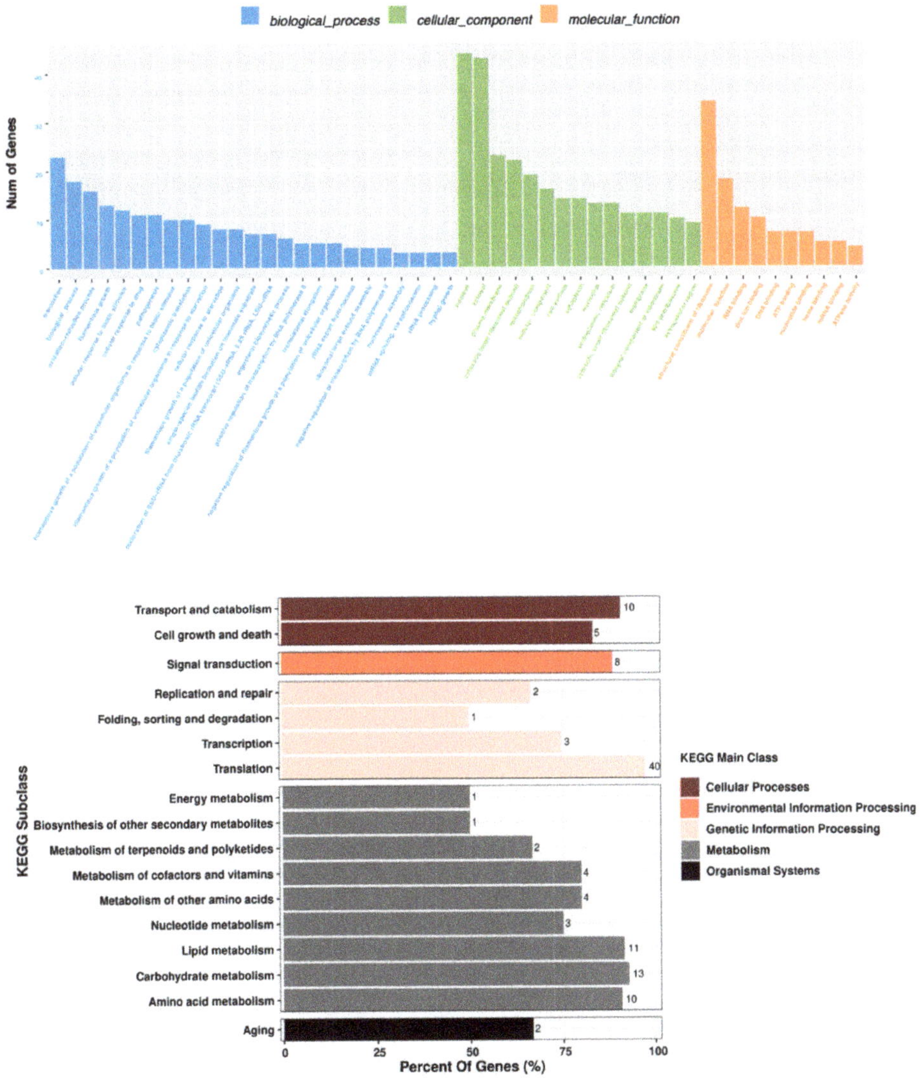

Figure 6. GO (**A**) and KEGG (**B**) enrichment analysis of DEGs in *S. spartinae* W9 cultured with and without 0.1% β-glucan. The numbers behind the bars stand for the number of genes involved in the metabolism.

RT-qPCR was performed on 12 genes to validate the results of RNA-seq. The transcript levels of all selected genes determined by RT-qPCR were consistent with those obtained using RNA-seq (Figure 7).

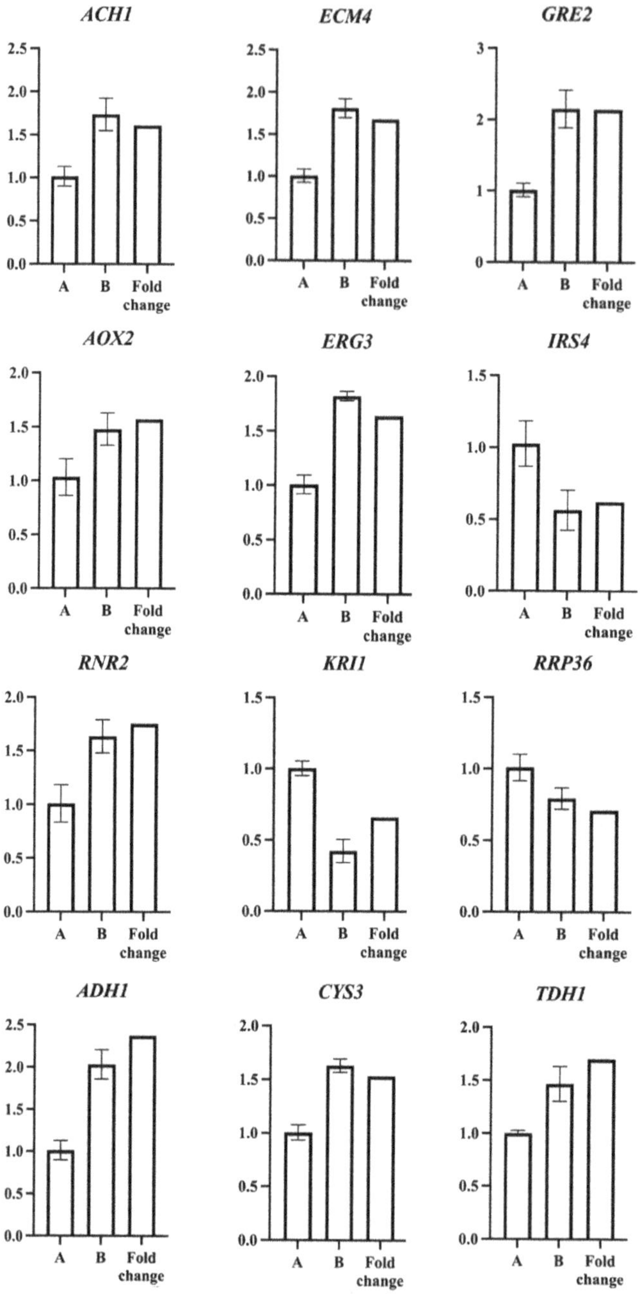

Figure 7. Results of verification of differently expressed genes by RT-qPCR. A: Relative gene expression level of *S. spartinae* W9 cultured in NYDB. B: Relative gene expression level of *S. spartinae* W9 cultured in NYDB with 0.1% β-glucan. Fold Change: Transcriptome sequencing results. Bars represent standard errors.

4. Discussion

A bottleneck in the development of antagonistic yeasts has been the limited biocontrol effect in practical application. Therefore, it is very important to enhance the biocontrol efficacy of antagonistic yeasts. It has been reported that β-glucan could improve the biocontrol efficacy of *C. laurentii* and *C. podzolicus* against postharvest blue mold of pears [24]. In the present study, our results also confirmed that β-glucan was an effective elicitor to improve the biocontrol ability of *S. spartinae* W9 on gray mold of strawberry fruit. The 0.1% β-glucan enhanced the biocontrol efficacy and antagonistic activity of *S. spartinae* W9 against *B. cinerea* (Figures 1 and 2). Despite the differences in PDA medium and in fruit [25], the increased antagonistic effects of *S. spartinae* W9 by β-glucan were consistent. Therefore, we selected 0.1% β-glucan to culture *S. spartinae* W9 and reveal the involved mechanism.

The key factor affecting the biocontrol activity of antagonistic yeasts has been proposed to be the rate of colonization [26]. Yeasts rapidly consume limited nutrients, thereby inhibiting spore germination of fungal pathogens [27]. This study showed that the addition of β-glucan to the medium had no effect on the growth and reproduction of *S. spartinae* W9 in vitro, while β-glucan could promote the colonization of *S. spartinae* W9 in strawberry wounds within 24 h of storage (Figure 3A). Wang et al. (2018) also found that β-glucan increased the growth ability of *C. podzolicus* in apple wounds [28]. Di Francesco et al. (2017) found that the colonization ability of yeast on the wounds has a certain correlation with the biofilm-forming ability [28]. High-density yeast forms exopolysaccharide capsules, which can more firmly adhere to fruit wounds and prevent pathogen infection [29,30]. Our results showed that the addition of 0.1% β-glucan to the culture medium effectively improved the biofilm formation capacity of *S. spartinae* W9 (Figure 3B). β-glucan improved the colonization ability in strawberries and biofilm forming ability of *S. spartinae* W9, which inhibited the growth of *B. cinerea*, thus reducing the incidence of gray mold in strawberries. Our findings are similar to the previous reports, which showed that some exogenous nutrients could induce antagonistic yeasts to enhance the biocontrol efficacy by improving the growth ability at fruit wounds and biofilm formation ability [31,32].

The ability of antagonistic yeast to resist pathogens also depends on the secretion of extracellular hydrolase. GLU can hydrolyze glucan, which is the major component of fungal cell walls, leading to cell malformation and cytoplasmic leakage, and is one of the possible mechanisms of biocontrol of antagonistic yeasts [33]. GLU is an inducible hydrolase enzyme, and the presence of β-glucan could induce the secretion of GLU from antagonistic yeasts. Our results showed that *S. spartinae* W9 could secret GLU, and 0.1% β-glucan induced the higher extracellular GLU activities (Figure 4). This suggests that 0.1% β-glucan could induce *S. spartinae* W9 to secrete more GLU, which might be an important factor for the enhanced biocontrol efficacy of *S. spartinae* W9. This finding is consistent with the report of Zhao et al. (2020), which showed that higher GLU activity of *C. podzolicus* is an important factor for the increased biocontrol effect of *C. podzolicus* cultured with β-glucan [17,18].

In practical application environments such as orchards and postharvest storage, various abiotic stressors could decrease the biocontrol efficacy of antagonistic yeasts [34]. Thus, improving the ecological adaptability and resistance of antagonistic yeasts is an effective way to ensure their biocontrol efficacy. SDS is an organic compound that penetrates cell membranes, causes damage to the plasma membrane, and limits cell growth [35,36]. Oxidative adversity was simulated using hydrogen peroxide because a fruit wound is an environment with high levels of reactive oxygen species [37], which cause oxidative damage to the cell membrane component membrane lipids, thereby destroying the integrity of the cell [38]. NaCl is a common osmotic stressor, and high NaCl has the effect of inhibiting cell growth [39]. The present study showed that the growth of *S. spartinae* W9 was significantly inhibited under these stressors, while *S. spartinae* W9 cultured with 0.1% β-glucan could enhance tolerance under these stressors (Figure 5). Fu et al. (2015) also found that 0.5%-β-glucan-induced *C. laurentii* can survive better under heat and oxidative stress [15].

These findings suggest that β-glucan could improve the stress tolerance of antagonistic yeasts and help to improve their biocontrol efficacy in practical applications.

The transcriptome analysis was conducted to explore the molecular mechanism of β-glucan, enhancing the biocontrol efficacy of *S. spartinae* W9. The results showed that β-glucan caused 188 DEGs, and the DEGs were related to cell wall formation and cell integrity, energy production, growth, and reproduction (Figure 6). Some upregulated DEGs are associated with stress response, including *GRE2*, *CYS3*, *AOX2*, and *ERG3*. GRE2 has been reported to be induced in yeast cells in response to oxidative stress [40] and osmotic stress [41]. Similarly, elevated AOX2 implies an acceleration of oxidative processes in mitochondria and limitation of ROS production [42]. The upregulation of the *ERG3* gene due to β-glucan was beneficial for maintaining the stability of yeast cell plasma membranes and improving the resistance of yeast to oxidative stress [43]. *CYS3* can play a protective role for yeast in a state of oxidative stress and promote the efficiency of yeast cell proliferation [44–46]. The upregulation of *GRE2*, *CYS3*, *AOX2*, and *ERG3* could explain the increased oxidative stress tolerance of *S. spartinae* W9 cultured with 0.1% β-glucan.

The yeast cell wall serves a variety of biological functions, including protecting against environmental stresses. It also affects the permeability of the cell wall when it is attacked by substances that disrupt cell integrity [47]. The genes *TDH1* and *ECM4* have been shown to be involved in cell wall structure [48,49]. Compared to the control, the higher expression of *TDH1* and *ECM4* in *S. spartinae* W9 cultured with 0.1% β-glucan promoted cell wall biosynthesis and biocontrol efficacy.

Intracellular energy is heavily depleted when yeast cells are under abiotic stress [50]. ACH1 provides acetyl coenzyme A (acetyl-CoA) to mitochondria by transferring the CoA group to acetate, which may accelerate metabolism in the TCA cycle and glycolysis [51]. The gene *ADH1* encodes alcohol dehydrogenase, which is the primary enzyme for the reoxidation of NADH in yeast. Zhong et al. (2021) showed that inactivation of ADH1 caused the accumulation of NADH, resulting in the imbalance of NADH/NAD and growth retardation [52]. The upregulation of *ACH1* and *ADH1* in *S. spartinae* W9 cells cultured with 0.1% β-glucan led to higher colonization ability in strawberry wounds and improved biocontrol efficacy. In addition, upregulation of the *RNR2* gene required for DNA repair may be helpful for maintaining cellular integrity [53]. In brief, β-glucan increased the energy metabolism and cellular integrity of *S. spartinae* W9 cells.

The transcriptome analysis demonstrated that the 0.1 % β-glucan-induced *S. spartinae* W9 achieved better proliferation and enhanced biological activity by altering cellular processes, energy metabolism, and stress responses. Similarly, Zhao et al. (2020) used transcriptomic analysis to reveal that β-glucan accelerated the synthesis of cell walls and energy of *C. podzolicus*, enhanced its antioxidant ability, and thus improved the biological control efficacy [18].

5. Conclusions

Adding 0.1% β-glucan into the culture medium significantly enhanced the biocontrol efficacy of *S. spartinae* W9 against *B. cinerea* in strawberries. β-glucan improved the colonization ability of yeast in strawberries, the biofilm forming ability, and the stress tolerance of *S. spartinae* W9. β-glucan also upregulated the genes involved in stress responses, cell wall formation, energy metabolism, growth, and reproduction of *S. spartinae* W9.

Author Contributions: Conceptualization, X.C. and Y.W.; methodology, X.C., X.Z. and Z.Z.; software, X.C. and S.J.; validation, X.C., X.Z. and Z.Z.; formal analysis, Y.C. and F.X.; investigation, X.C.; data curation, X.C. and Y.W.; writing—original draft preparation, X.C. and Y.W.; writing—review and editing, X.C., Y.W. and X.S; visualization, X.C. and Y.C.; supervision, Y.W. and X.S.; project administration, Y.W.; funding acquisition, Y.W., X.Z. and Z.Z. All authors have read and agreed to the published version of the manuscript.

Funding: This research was funded by the National Natural Science Foundation of China (No. 32202122), Ningbo Public Welfare Technology Project (No. 2022S143 and No. 2021S064), and the Research Innovation Foundation for Graduate Students of Ningbo University (IF2022173).

Data Availability Statement: The data presented in this study are included in the article. Further inquiries can be directed to the corresponding author.

Conflicts of Interest: The authors declare no conflict of interest.

References

1. Haile, Z.M.; Guzman, N.D.; Grace, E.; Moretto, M.; Sonego, P.; Engelen, K.; Zoli, L.; Moser, C.; Baraldi, E. Transcriptome Profiles of Strawberry (Fragaria vesca) Fruit Interacting With Botrytis cinerea at Different Ripening Stages. *Front. Plant Sci.* **2019**, *10*, 1131. [CrossRef] [PubMed]
2. Janisiewicz, W.J.; Korsten, L. Biological control of postharvest diseases of fruits. *Annu. Rev. Phytopathol.* **2002**, *40*, 411–441. [CrossRef]
3. Morales, H.; Marín, S.; Ramos, A.J.; Sanchis, V. Influence of post-harvest technologies applied during cold storage of apples in Penicillium expansum growth and patulin accumulation: A review. *Food Control* **2010**, *21*, 953–962. [CrossRef]
4. Sharma, R.R.; Singh, D.; Singh, R. Biological control of postharvest diseases of fruits and vegetables by microbial antagonists: A review. *Biol. Control* **2009**, *50*, 205–221. [CrossRef]
5. Zhang, X.; Li, B.; Zhang, Z.; Chen, Y.; Tian, S. Antagonistic Yeasts: A Promising Alternative to Chemical Fungicides for Controlling Postharvest Decay of Fruit. *J. Fungi* **2020**, *6*, 158. [CrossRef] [PubMed]
6. Muccilli, S.; Restuccia, C. Bioprotective Role of Yeasts. *Microorganisms* **2015**, *3*, 588–611. [CrossRef] [PubMed]
7. Sadeghi, R.; Aminian, H.; Remize, F.; Sheikh, M.; Ebrahimi, L. Integrated control of blue and gray molds of apples with antagonistic yeasts combined with carbon dioxide or ozone. *J. Plant Pathol.* **2021**, *103*, 943–953. [CrossRef]
8. Lanhuang, B.; Yang, Q.; Godana, E.A.; Zhang, H. Efficacy of the Yeast *Wickerhamomyces anomalus* in Biocontrol of Gray Mold Decay of Tomatoes and Study of the Mechanisms Involved. *Foods* **2022**, *11*, 720. [CrossRef]
9. Millan, A.F.-S.; Gamir, J.; Farran, I.; Larraya, L.; Veramendi, J. Identification of new antifungal metabolites produced by the yeast Metschnikowia pulcherrima involved in the biocontrol of postharvest plant pathogenic fungi. *Postharvest Biol. Technol.* **2022**, *192*, 111995. [CrossRef]
10. Zou, X.; Wei, Y.; Jiang, S.; Cao, Z.; Xu, F.; Wang, H.; Zhan, P.; Shao, X. Volatile organic compounds and rapid proliferation of Candida pseudolambica W16 are modes of action against gray mold in peach fruit. *Postharvest Biol. Technol.* **2021**, *183*, 111751. [CrossRef]
11. Guigón-López, C.; Holguín-Ibarra, P.D.; Torres-Zapien, J.H.; Cruz, I.G.; Villapando, I.; Salas-Salazar, N.A. Metarhizium anisopliae reduces conidial germination and mycelium growth of the apple gray mold Botrytis cinerea. *Biol. Control* **2021**, *160*, 104660. [CrossRef]
12. Zou, X.; Wei, Y.; Dai, K.; Xu, F.; Wang, H.; Shao, X. Yeasts from intertidal zone marine sediment demonstrate antagonistic activities against Botrytis cinerea in vitro and in strawberry fruit. *Biol. Control* **2021**, *158*, 104612. [CrossRef]
13. Roca-Couso, R.; Flores-Félix, J.D.; Rivas, R. Mechanisms of Action of Microbial Biocontrol Agents against *Botrytis cinerea*. *J. Fungi* **2021**, *7*, 1045. [CrossRef] [PubMed]
14. Dai, H.; Han, X.Q.; Gong, F.Y.; Dong, H.; Tu, P.F.; Gao, X.M. Structure elucidation and immunological function analysis of a novel beta-glucan from the fruit bodies of Polyporus umbellatus (Pers.) Fries. *Glycobiology* **2012**, *22*, 1673–1683. [CrossRef]
15. Fu, D.; Zeng, L.; Zheng, X.; Yu, T. Effect of β-glucan on stress tolerances and biocontrol efficacy of Cryptococcus laurentii against Penicillium expansum in pear fruit. *Biocontrol* **2015**, *60*, 669–679. [CrossRef]
16. Liu, X.Z.; Wang, G.M.; Göker, M.; Groenewald, M.; Kachalkin, A.V.; Lumbsch, H.T.; Millanes, A.M.; Wedin, M.; Yurkov, A.M.; Bai, F.Y.; et al. Towards an integrated phylogenetic classification of the Tremellomycetes. *Stud. Mycol.* **2015**, *81*, 85–147. [CrossRef]
17. Wang, Y.; Li, Y.; Xu, W.; Zheng, X.; Zhang, X.; Abdelhai, M.H.; Zhao, L.; Li, H.; Diao, J.; Zhang, H. Exploring the effect of β-glucan on the biocontrol activity of Cryptococcus podzolicus against postharvest decay of apples and the possible mechanisms involved. *Biol. Control* **2018**, *121*, 14–22. [CrossRef]
18. Zhao, L.; Wang, Y.; Wang, Y.; Li, B.; Gu, X.; Zhang, X.; Boateng, N.A.S.; Zhang, H. Effect of β-glucan on the biocontrol efficacy of Cryptococcus podzolicus against postharvest decay of pears and the possible mechanisms involved. *Postharvest Biol. Technol.* **2019**, *160*, 111057. [CrossRef]
19. Hassan, H.; Mohamed, M.; Yusoff, S.; Hata, E.; Tajidin, N. Selecting Antagonistic Yeast for Postharvest Biocontrol of *Colletotrichum gloeosporioides* in Papaya Fruit and Possible Mechanisms Involved. *Agronomy* **2021**, *11*, 760. [CrossRef]
20. Qiu, J.-E.; Zhao, L.; Jiang, S.; Godana, E.A.; Zhang, X.; Zhang, H. Efficacy of Meyerozyma caribbica in the biocontrol of blue mold in kiwifruit and mechanisms involved. *Biol. Control* **2022**, *173*, 105000. [CrossRef]
21. Agirman, B.; Erten, H. Biocontrol ability and action mechanisms of *Aureobasidium pullulans* GE17 and *Meyerozyma guilliermondii* KL3 against *Penicillium digitatum* DSM2750 and *Penicillium expansum* DSM62841 causing postharvest diseases. *Yeast* **2020**, *37*, 437–448. [CrossRef] [PubMed]
22. Zhou, Y.; Zhang, L.; Zeng, K. Efficacy of Pichia membranaefaciens combined with chitosan against Colletotrichum gloeosporioides in citrus fruits and possible modes of action. *Biol. Control* **2016**, *96*, 39–47. [CrossRef]

23. Huang, Y.; Fan, Z.; Cai, Y.; Jin, L.; Yu, T. The influence of N-acetylglucosamine: Inducing Rhodosporidium paludigenum to enhance the inhibition of Penicillium expansum on pears. *Postharvest Biol. Technol.* **2021**, *176*, 111486. [CrossRef]
24. Zeng, L.; Yu, C.; Fu, D.; Lu, H.; Zhu, R.; Lu, L.; Zheng, X.; Yu, T. Improvement in the effectiveness of Cryptococcus laurentii to control postharvest blue mold of pear by its culture in β-glucan amended nutrient broth. *Postharvest Biol. Technol.* **2015**, *104*, 26–32. [CrossRef]
25. Millan, A.F.-S.; Larraya, L.; Farran, I.; Ancin, M.; Veramendi, J. Successful biocontrol of major postharvest and soil-borne plant pathogenic fungi by antagonistic yeasts. *Biol. Control* **2021**, *160*, 104683. [CrossRef]
26. Yu, T.; Yu, C.; Chen, F.; Sheng, K.; Zhou, T.; Zunun, M.; Abudu, O.; Yang, S.; Zheng, X. Integrated control of blue mold in pear fruit by combined application of chitosan, a biocontrol yeast and calcium chloride. *Postharvest Biol. Technol.* **2012**, *69*, 49–53. [CrossRef]
27. Rivas-Garcia, T.; Murillo-Amador, B.; Nieto-Garibay, A.; Rincon-Enriquez, G.; Chiquito-Contreras, R.G.; Hernandez-Montiel, L.G. Enhanced biocontrol of fruit rot on muskmelon by combination treatment with marine Debaryomyces hansenii and Stenotrophomonas rhizophila and their potential modes of action. *Postharvest Biol. Technol.* **2019**, *151*, 61–67. [CrossRef]
28. Di Francesco, A.; Ugolini, L.; D'Aquino, S.; Pagnotta, E.; Mari, M. Biocontrol of Monilinia laxa by Aureobasidium pullulans strains: Insights on competition for nutrients and space. *Int. J. Food Microbiol.* **2017**, *248*, 32–38. [CrossRef]
29. Klein, M.N.; Kupper, K.C. Biofilm production by Aureobasidium pullulans improves biocontrol against sour rot in citrus. *Food Microbiol.* **2018**, *69*, 1–10. [CrossRef]
30. Dukare, A.S.; Paul, S.; Nambi, V.E.; Gupta, R.K.; Singh, R.; Sharma, K.; Vishwakarma, R.K. Exploitation of microbial antagonists for the control of postharvest diseases of fruits: A review. *Crit. Rev. Food Sci. Nutr.* **2018**, *59*, 1498–1513. [CrossRef]
31. Wang, S.; Zhang, H.; Qi, T.; Deng, L.; Yi, L.; Zeng, K. Influence of arginine on the biocontrol efficiency of Metschnikowia citriensis against Geotrichum citri-aurantii causing sour rot of postharvest citrus fruit. *Food Microbiol.* **2021**, *101*, 103888. [CrossRef] [PubMed]
32. Zhang, H.; Ge, L.; Chen, K.; Zhao, L.; Zhang, X. Enhanced Biocontrol Activity of Rhodotorula mucilaginosa Cultured in Media Containing Chitosan against Postharvest Diseases in Strawberries: Possible Mechanisms Underlying the Effect. *J. Agric. Food Chem.* **2014**, *62*, 4214–4224. [CrossRef]
33. Masih, E.I.; Paul, B. Secretion of β-1,3-Glucanases by the Yeast Pichia membranifaciens and Its Possible Role in the Biocontrol of Botrytis cinerea Causing Grey Mold Disease of the Grapevine. *Curr. Microbiol.* **2002**, *44*, 391–395. [CrossRef]
34. Sui, Y.; Wisniewski, M.; Droby, S.; Liu, J. Responses of Yeast Biocontrol Agents to Environmental Stress. *Appl. Environ. Microbiol.* **2015**, *81*, 2968–2975. [CrossRef]
35. Schroeder, L.; Ikui, A.E. Tryptophan confers resistance to SDS-associated cell membrane stress in Saccharomyces cerevisiae. *PLoS ONE* **2019**, *14*, e0199484. [CrossRef] [PubMed]
36. Kono, K.; Al-Zain, A.; Schroeder, L.; Nakanishi, M.; Ikui, A.E. Plasma membrane/cell wall perturbation activates a novel cell cycle checkpoint during G1 in *Saccharomyces cerevisiae*. *Proc. Natl. Acad. Sci. USA* **2016**, *113*, 6910–6915. [CrossRef] [PubMed]
37. Nie, X.; Zhang, C.; Jiang, C.; Zhang, R.; Guo, F.; Fan, X. Trehalose increases the oxidative stress tolerance and biocontrol efficacy of Candida oleophila in the microenvironment of pear wounds. *Biol. Control* **2019**, *132*, 23–28. [CrossRef]
38. Reverter-Branchat, G.; Cabiscol, E.; Tamarit, J.; Ros, J. Oxidative damage to specific proteins in replicative and chronological-aged Saccharomyces cerevisiae: Common targets and prevention by calorie restriction. *J. Biol. Chem.* **2004**, *279*, 31983–31989. [CrossRef]
39. Wang, Y.; Luo, Y.; Sui, Y.; Xie, Z.; Liu, Y.; Jiang, M.; Liu, J. Exposure of Candida oleophila to sublethal salt stress induces an antioxidant response and improves biocontrol efficacy. *Biol. Control* **2018**, *127*, 109–115. [CrossRef]
40. Goud, B.S.; Kim, J.H.; Ulaganathan, K. Identification of Genes Associated with Stress Tolerance of High Ethanol–Producing Saccharomyces cerevisiae Strain, NCIM3186, by Differential Gene Expression Analysis. *BioEnergy Res.* **2022**, *15*, 1459–1471. [CrossRef]
41. Perez-Martinez, M.E.; Benet, M.; Alepuz, P.; Tordera, V. Nut1/Hos1 and Sas2/Rpd3 control the H3 acetylation of two different sets of osmotic stress-induced genes. *Epigenetics* **2020**, *15*, 251–271. [CrossRef] [PubMed]
42. She, X.; Zhang, P.; Gao, Y.; Zhang, L.; Wang, Q.; Chen, H.; Calderone, R.; Liu, W.; Li, D. A mitochondrial proteomics view of complex I deficiency in Candida albicans. *Mitochondrion* **2018**, *38*, 48–57. [CrossRef] [PubMed]
43. Horstmann, C.; Campbell, C.; Kim, D.S.; Kim, K. Transcriptome profile with 20 nm silver nanoparticles in yeast. *FEMS Yeast Res.* **2019**, *19*, foz003. [CrossRef] [PubMed]
44. Matiach, A.; Schröder-Köhne, S. Yeast cys3 and gsh1 mutant cells display overlapping but non-identical symptoms of oxidative stress with regard to subcellular protein localization and CDP-DAG metabolism. *Mol. Genet. Genom.* **2001**, *266*, 481–496. [CrossRef]
45. Nguyen, P.-T.; Toh-E, A.; Nguyen, N.-H.; Imanishi-Shimizu, Y.; Watanabe, A.; Kamei, K.; Shimizu, K. Identification and characterization of a sulfite reductase gene and new insights regarding the sulfur-containing amino acid metabolism in the basidiomycetous yeast Cryptococcus neoformans. *Curr. Genet.* **2020**, *67*, 115–128. [CrossRef] [PubMed]
46. Gu, Z.; Sun, Y.; Wu, F.; Wu, X. Mechanism of Growth Regulation of Yeast Involving Hydrogen Sulfide From S-Propargyl-Cysteine Catalyzed by Cystathionine-γ-Lyase. *Front. Microbiol.* **2021**, *12*, 679563. [CrossRef]
47. Klis, F.M.; Mol, P.; Hellingwerf, K.; Brul, S. Dynamics of cell wall structure in Saccharomyces cerevisiae. *FEMS Microbiol. Rev.* **2002**, *26*, 239–256. [CrossRef]

48. Delgado, M.L.; O'Connor, J.E.; Azorín, I.; Renau-Piqueras, J.; Gil, M.L.; Gozalbo, D. The glyceraldehyde-3-phosphate dehydrogenase polypeptides encoded by the Saccharomyces cerevisiae TDH1, TDH2 and TDH3 genes are also cell wall proteins. *Microbiology* **2001**, *147*, 411–417. [CrossRef]
49. Terzioğlu, E.; Alkım, C.; Arslan, M.; Balaban, B.G.; Holyavkin, C.; Kısakesen, H.I.; Topaloğlu, A.; Şahin, Y.; Işık, S.G.; Akman, S.; et al. Genomic, transcriptomic and physiological analyses of silver-resistant *Saccharomyces cerevisiae* obtained by evolutionary engineering. *Yeast* **2020**, *37*, 413–426. [CrossRef]
50. Dai, Y.; Wang, Z.; Leng, J.; Wang, Q.; Liu, J. Heat stress alters the transcriptome of Debaryomyces hansenii and reduces its biocontrol activity against postharvest gray mold on kiwifruit. *Postharvest Biol. Technol.* **2021**, *178*, 111541. [CrossRef]
51. van Rossum, H.M.; Kozak, B.U.; Niemeijer, M.S.; Duine, H.J.; Luttik, M.A.H.; Boer, V.M.; Kötter, P.; Daran, J.-M.G.; van Maris, A.J.A.; Pronk, J.T. Alternative reactions at the interface of glycolysis and citric acid cycle in *Saccharomyces cerevisiae*. *FEMS Yeast Res.* **2016**, *16*, fow017. [CrossRef] [PubMed]
52. Zhong, W.; Yang, M.; Hao, X.; Sharshar, M.M.; Wang, Q.; Xing, J. Improvement of D-Lactic acid productivity by introducing Escherichia coli acetyl-CoA synthesis pathway in engineered Saccharomyces cerevi. *J. Chem. Technol. Biotechnol.* **2021**, *96*, 2509–2519. [CrossRef]
53. Thakre, P.K.; Sv, A.; Golla, U.R.; Chauhan, S.; Tomar, R.S. Previously uncharacterized amino acid residues in histone H3 and H4 mutants with roles in DNA damage repair response and cellular aging. *FEBS J.* **2018**, *286*, 1154–1173. [CrossRef] [PubMed]

Journal of
Fungi

MDPI

Article

Adding Metal Ions to the *Bacillus mojavensis* D50 Promotes Biofilm Formation and Improves Ability of Biocontrol

Lining Zheng [1], Xuehu Gu [1], Liangpeng Sun [1], Meiqi Dong [1], Ao Gao [1], Zhe Han [1], Hongyu Pan [2] and Hao Zhang [1,*]

[1] College of Plant Protection, Jilin Agricultural University, Changchu 130118, China; 13074304872@163.com (L.Z.)
[2] College of Plant Sciences, Jilin University, Changchun 130062, China
* Correspondence: zhanghao100@jlau.edu.cn

Abstract: *Bacillus mojavensis* D50, a biocontrol strain, is used to prevent and treat the fungal plant pathogen *Botrytis cinerea*. *Bacillus mojavensis* D50's biofilms can affect its colonization; thus, the effects of different metal ions and culture conditions on biofilm formation were determined in this study. The results of medium optimization showed that Ca^{2+} had the best ability to promote biofilm formation. The optimal medium composition for the formation of biofilms contained tryptone (10 g/L), $CaCl_2$ (5.14 g/L), and yeast extract (5.0 g/L), and the optimal fermentation conditions included pH 7, a temperature of 31.4 °C, and a culture time of 51.8 h. We found that the antifungal activity and abilities to form biofilms and colonize roots were improved after optimization. In addition, the levels of expression of the genes *luxS*, *SinR*, *FlhA*, and *tasA* were up-regulated by 37.56-, 2.87-, 12.46-, and 6.22-fold, respectively. The soil enzymatic activities which related biocontrol-related enzymes were the highest when the soil was treated by strain D50 after optimization. In vivo biocontrol assays indicated that the biocontrol effect of strain D50 after optimization was improved.

Keywords: optimization; biofilm formation; *Bacillus mojavensis* D50; biocontrol effects; tomato gray mold

1. Introduction

Tomato (*Solanum lycopersicon* L.), an important vegetable crop, is widely planted all over the world [1]. However, the occurrence of diseases can lead to a reduction in yields. Particularly important pathogens of tomato include the fungus *Botrytis cinerea*, which causes gray mold [2], and the bacterium *Ralstonia solanacearum*, which causes bacterial wilt [3]. Therefore, immense efforts and vast amounts of money are being invested to control these pathogens [4].

Currently, the primary way to control pathogens is the use of chemical pesticides [5]. They are inexpensive and highly efficient. However, the long-term use of chemical pesticides will lead to environmental pollution, pesticide resistance, and negative effects on human health [6,7]. Therefore, it is urgent to find an environmentally friendly approach to protect tomato plants from pathogens. Biological control is an approach to control plant diseases, which introduces another organism to control the pathogen rather than chemical pesticides. This approach is frequently highly effective, and it is used on many plants. The modes of action of biocontrol agents include space or nutrient competition, antibiotics production and inducible resistances. There are many studies about the use of biological control, such as a study which suggests that *Burkholderia cenocepacia* ETR-B22 could product volatile organic compounds and suppress the tomato gray mold [8]. *Bacillus velezensis* HY19 could also produce salicylic acid and numerous antifungal substances to inhibit the growth of tomato gray mold [9]. *Wickerhamomyces anomalus* could inhibit the growth of gray mold by competition for nutrients and space, and induce the host tissue's disease resistance and producing of antifungal metabolites [10]. *Trichoderma harzianum* inhibited the tomato gray mold using competition for space [11].

Microbial biofilms are a microbial community that is attached to biological or abiotic surfaces. They can be composed of one or more microbial species [12]. Biofilm formation can improve the biocontrol effects on pathogens. The formation of biofilm can help the bacteria to rapidly ingest nutrition and become more competitive. In addition, it can also facilitate colonization of the plant surface, which can protect the plant from pathogens [13]. Many studies have reported that the amount of metal ions and culture conditions can influence the formation of biofilm. Yang et al. [14] reported that the morphology of the *B. subtilis* 1JN2 biofilm was flattened and the expression of biofilm related genes was down-regulated under high concentrations of Cd^{2+}, while the addition of Mg^{2+} increased the biocontrol effect of *Burkholderia pyrrocinia* JK-SH007 [15]. The three *Bacillus* spp. strains which could form biofilm were isolated to protect maize from pathogens, the temperature, culture medium, and culture time, which could influence the biofilm formation and thus further impact the biocontrol effect of these strains [16]. The application of phenylalanine could promote the biofilm formation of *Meyerozyma caribbica* and improved its biocontrol efficacy [17]. These studies have indicated that the biofilm formation of strains can influence their biocontrol effects.

The *luxS*/AI-2 quorum sensing (QS) system is an important factor affecting biofilm formation [18]. Many studies have reported that the gene of *luxS* influences the formation of biofilms, such as *Shewanella xiamenensis* [19] and *Lactobacillus reuteri* [20]. In addition, the QS system can also regulate the probiotic activities of lactic acid bacteria [21]. The transcriptional repressor *sinR* is a master regulator of biofilm formation and the production of the secreted, amyloid-like protein component of the matrix (*TasA*), thus blocking biofilm formation [22]. The gene of *flhA*, encoded flagellin protein *FlhA*, can influence the formation of biofilm. Minamino et al. [23] have reported that the gene of *flhA* can change the growth of flagella, further influencing biofilm formation. The *tasA* is a major gene which encodes the protein involved in antimicrobial activities, spore coat assembly, and germination. It is also found in the stationary phase, sporulating cultures, and the biofilm matrix [24].

B. mojavensis D50, which was isolated from tomato rhizosphere soil, has been reported to have a substantial effect on *B. cinerea*. Moreover, its fermentation supernatant also had great antifungal effect on tomato gray mold [5]. However, there have been few studies on the promotion of biofilm formation and biocontrol effects by the manipulation of concentrations of metal ions and culture conditions. In this study, to further determine the effect of biofilm formation on pathogen inhibition, the goals were as follows: (1) to screen the optimal concentration of metal ions and culture conditions for biofilm formation; (2) to determine the effect of optimal metal ion and culture conditions on the levels of expression of the genes related to biofilm formation; (3) to assess the effect of optimal metal ion and culture conditions of *B. mojavensis* D50 on soil enzyme activities; and (4) to assess the effect of optimal metal ion and culture conditions of *B. mojavensis* D50 on reducing tomato grey mold.

2. Materials and Methods

2.1. Strains, Medium, Culture Conditions and Plant Materials

Bacillus mojavensis D50, isolated from tomato rhizosphere soil, was used to prevent tomato from *Botrytis cinerea*. *B. mojavensis* D50 and *B. cinerea* were kindly provided by the Laboratory of Pesticide Bioassay, Plant Protection College at Jilin Agricultural University (Jilin, China). The LB broth that was used to culture *B. mojavensis* D50 contained 5 g yeast powder, 10 g tryptone, 10 g NaCl, and 1000 mL distilled water. The potato dextrose agar (PDA) medium that was used to culture *B. cinerea* contained 20 g glucose, 20 g agar, 200 g potato, and 1000 mL distilled water.

The initial culture conditions that were used to optimize the formation of biofilm were 30 °C, a culture time of 48 h, and pH of 7. The initial medium that was used to optimize the formation of biofilm formation was LB broth (excluding metal ions) to exclude the influence of metal ions on biofilm formation: 5 g/L yeast powder, 10 g/L tryptone, and 1000 mL water. The tomato plants were planted as described by Zheng et al. [25], the seeds were sterilized in 0.5% sodium hypochlorite solution for 1 min and rinsed with sterile distilled water three times, then the seeds were incubated at 28 °C for 4 days for germination. One seedling was planted in each plastic pot, and the tomato plants were grown in a greenhouse at 25–30 °C with a 14 h-light/10 h-dark cycle and 70% relative humidity with regular irrigation.

2.2. Preparation of the *B. mojavensis* D50 Suspension

To prepare the *B. mojavensis* D50 suspension, a single colony was inoculated into a 250-mL Erlenmeyer flask that contained 100 mL of LB broth, and it was shaken at 150 rpm and cultivated at 30 °C for 12 h. The suspension was centrifuged at 7000 rpm for 10 min, and the strains were washed three times with PBS (pH 7.3). The suspension was adjusted to 1.0×10^8 cfu/mL [25].

2.3. Semiquantitative Evaluation of the *B. mojavensis* D50 Biofilm

The crystal violet method was used to determine the amount of biofilm formed [26]. Briefly, a suspension of *B. mojavensis* D50 was prepared as described in Section 2.2. A volume of 5 μL/well was added to the wells in a 96-well plate and mixed with the relevant medium (95 μL/well). The plate was left stationary for cultivation. After that, the cell culture was removed and dyed with 1% crystal violet (CV, 100 μL/well) for 15 min. The plate was washed three times with distilled water to remove excess dye. The biofilm was extracted with absolute ethanol (200 μL/well), and the amount of biofilm formation was determined at 570 nm [27].

2.4. Screening of the Optimal Metal Ions and Conditions for Biofilm Formation

2.4.1. Determination of the Effect of Metal Ions on Biofilm Formation by a Single Factor Experiment

To determine the effect of metal ions on biofilm formation, the "one-factor-at-a-time" method was also used in this experiment. Briefly, the $CaCl_2$, $MnCl_2$, $FeCl_3 \cdot 6H_2O$, $MgSO_4 \cdot 7H_2O$ and NaCl were added into initial medium. The concentrations of $CaCl_2$, $MnCl_2$, $FeCl_3 \cdot 6H_2O$, $MgSO_4 \cdot 7H_2O$ were adjusted to 0 g/L, 5 g/L, 10 g/L, 15 g/L, 20 g/L, and 25 g/L, respectively, and the concentrations of NaCl were adjusted to 0 g/L 20 g/L, 25 g/L, 30 g/L, 35 g/L, and 40 g/L, respectively. The *B. mojavensis* D50 suspension (5 μL) was inoculated into the medium (which contained different metal ions) and cultured at 30 °C for 48 h. The amount of biofilm formation was determined as described in Section 2.3 [15]. The experiments were replicated three times and repeated three times.

2.4.2. Determination of the Effect of Culture Conditions on Biofilm Formation by a Single Factor Experiment

The "one-factor-at-a-time" method was also used in this experiment. Briefly, the culture conditions were established as follows: the pH of initial medium was adjusted to 5, 6, 7, 8, 9, and 10. The culture time was adjusted to 12 h, 24 h, 36 h, 48 h, 60 h, and 72 h. The culture temperature was incubated to 25 °C, 30 °C, 35 °C, 40 °C, and 45 °C. The amount of biofilm formed was determined as described in Section 2.3. The experiments were replicated three times and repeated three times.

2.4.3. Determination of the Optimal Amount of Metal Ions and Culture Conditions for Biofilm Formation Using a Box-Behnken Design

The results of the single factor experiment indicated the three factors that had the best effect on biofilm formation, and they were selected for coding. OD_{570} was used as the response value. A Box-Behnken central combined experiment was used to design the response surface test. The codes and levels of each factor are shown in Table 1. The Box-Behnken design (BBD) was used to conduct 17 tests, including five tests to estimate the error (Table 2). The concentration of metal ions and culture conditions were prepared based on the BBD. The ability to form biofilm was measured as described in Section 2.3. Finally, the prediction of metal ions and culture conditions were verified under the same fermentation conditions as the BBD. Based on the response surface methodology (RSM) results, the final metal ions and culture conditions were determined and verified experimentally [28,29]. The experiments were replicated three times and repeated three times.

Table 1. Box-Behnken design to optimize the biofilm formation medium and conditions using three components at three levels.

Design Variable (Factors)	Unit	Code	Real Values of the Coded Levels		
			−1	0	+1
$CaCl_2$	g/L	A	0	5	10
Temperature	°C	B	25	30	35
Time	h	C	36	48	60

Note: "+1" represents high level, "−1" represents low level.

Table 2. Box Behnken design scheme containing 17 experimental runs.

Runs	A	B	C	OD_{570}
1	0	−1	−1	2.546
2	0	0	0	3.224
3	0	1	−1	2.535
4	−1	0	−1	2.632
5	0	−1	1	2.604
6	0	0	0	3.226
7	−1	−1	0	2.609
8	−1	1	0	2.704
9	0	0	0	3.228
10	1	1	0	2.698
11	0	1	1	2.712
12	1	0	1	2.712
13	1	−1	0	2.688
14	0	0	0	3.221
15	0	0	0	3.03
16	−1	0	1	2.682
17	1	0	−1	2.632

Note: "+1" represents high level, "−1" represents low level.

2.5. Effect of Metal Ions and Culture Conditions on the Biocontrol Characteristics of *B. mojavensis* D50 against *B. cinerea* In Vitro

To determine the effect of metal ions and culture conditions on the characteristics of biocontrol, the flat-stand method was used to measure the antifungal activity in this experiment [30]. As described by Fu et al. [15], the ability to form biofilm and root colonization by *B. mojavensis* D50 before and after optimization were also measured. The experiments were replicated three times and repeated three times.

2.6. *B. mojavensis* D50 Biofilm-Related Analysis of Differentially Expressed Genes

To determine the effect of optimized metal ions and culture conditions on the differential expression of genes related to biofilm formation, the *B. mojavensis* D50 suspension was inoculated into a 250-mL Erlenmeyer flask that contained 100 mL of optimized culture medium. The culture was then cultivated at 31.4 °C and 150 rpm for 12 h and 51.8 h. The key genes related to biofilm formation, *luxS*, *SinR*, *FlhA*, and *tasA*, were analyzed. The primers were designed using Primer 6 software (Table S1), and the total RNA was extracted using a Trazol up Plus RNA Kit (TransGen Biotech., Beijing, China). The cDNA was generated according to the manufacturer's instructions for the TransGen all-in-one first-strand cDNA synthesis SuperMix (TransGen Biotech.). The *luxS*, *SinR*, *FlhA*, and *tasA* genes were measured in this study. Quantitative real-time PCR (qRT-PCR) was performed using PerfectStart Green qPCR SuperMix on a Roche Light Cycler® 96 (Roche Diagnostics Corporation, Indianapolis, IN, USA). The qRT-PCR program used was as follows: 95 °C for 2 min, followed by 32 cycles of 95 °C for 30 s, 60 °C for 30 s, 72 °C for 30 s and 78 °C for 11 s. The melting curve cycle program was as follows: 95 °C for 10 s, 60 °C for 1 min, and 97 °C for 1 s. The relative levels of expression of the genes were determined using the $2^{-\Delta\Delta CT}$ method, and the *16S rDNA* gene was used as a reference gene. Data are reported on a logarithmic scale as a relative gene expression ratio (RQ) after calibration on values obtained at *B. mojavensis* D50 that was cultured in basic medium [31]. The experiments were replicated three times and repeated three times.

2.7. Effect of Metal Ions and Culture Conditions on the Biocontrol Characteristics of B. mojavensis D50 against B. cinerea In Vivo

As described by Zheng et al., the tomato plant and spore suspension of *B. cinerea* were prepared [25]. The suspension of *B. mojavensis* D50 before or after optimization was prepared as described in Section 2.2. The 6-leaf-stages of tomato seedlings were used in this experiment. The suspension of *B. mojavensis* D50 before or after optimization (1.0×10^8 cfu/mL, 15 mL/pot) was irrigated into the roots of tomato plants as the treatment, whereas the sterile water (15 mL/pot) was irrigated into the roots as the control. After 7 days, the spores (5×10^4 spores/mL, 8 mL/pot) were sprayed on the tomato seedlings. The tomato seedlings were incubated in a growth chamber at 26 ± 0.5 °C and covered with plastic bags for 3 days to maintain high humidity. The disease incidence was determined 7 days after inoculation with *B. cinerea* [32,33]. Each treatment contained 16 pots with 1 seedling per pot. The experiments were replicated three times and repeated three times. The biochemical growth parameters of tomato plant were measured as described by Zhang et al. [31].

2.8. Effect of B. mojavensis D50 on Soil Enzyme Activities before and after Optimization

To determine the effect of optimized metal ions and culture conditions of *B. mojavensis* D50 on soil enzyme activities, the activities of invertase, catalase, urease, and dehydrogenase were measured. The treatments were the same as those described in Section 2.7. The soil samples were collected at 2 h and 1, 3, 5, 7, and 15 days. Invertase activity was measured by the 3,5-dinitrosalicylic acid method. Briefly, 2 g of soil sample was mixed with 15 mL of 8% sucrose solution, 5 mL of PBS (pH 5.5), and 0.2 mL drops of toluene, and it was cultured at 37 °C for 24 h. A volume of 1 mL of the supernatant was mixed with 5 mL of 3,5-dinitrosalicylic acid and 5 mL of water, and the invertase activity was determined at 508 nm [34]. To measure the catalase activity, 2 g of soil sample was mixed with 40 mL water and 5 mL of 0.5% H_2O_2 and incubated at 30 °C for 20 min. A volume of 5 mL of H_2SO_4 (1.5 mol/L) was added to terminate the reaction. Finally, the supernatant was titrated with $KMnO_4$ (20 mmol/L) [35]. Urease activity was determined by the Berthelot method. Briefly, 5 g of soil was mixed with 1 mL of PBS (pH 7.7, 20 mmol/L) and then incubated at 37 °C for 2 h. It was extracted with KCl (2 mmol/L), and the absorbance at 690 nm was measured using a visible spectrophotometer [36]. The dehydrogenase activity was determined by colorimetry. Briefly, 2 g of soil was mixed with 2 mL of tetrazolium chloride (TTC) and 2 mL water and then cultured at 37 °C for 24 h. The solution was extracted with 5 mL of alcohol. Its absorbance was measured at 485 nm using a visible spectrophotometer. The experiments were replicated three times and repeated three times.

2.9. Statistical Analysis

The data were subjected to analyses of variance (ANOVA) using SPSS 26.0 (IBM, Inc., Armonk, NY, USA). The means were separated by a least significant difference (LSD) test at the level $p < 0.05$. Respective significant differences were denoted using different letters (a, b, c, etc.).

3. Results

3.1. Screening of the Optimal Metal Ions and Culture Conditions for Biofilm Formation

The "one-by-one factor" experiments were used to select the metal ions and conditions. We found that the Ca^{2+} and Mg^{2+} could promote the formation of biofilm and the Na^{+} had less of an effect on the biofilm formation, while Mn^{2+}, Fe^{3+}, and Zn^{2+} inhibited it (Figure 1). When the concentrations of Ca^{2+} and Mg^{2+} were 5 g/L and 10 g/L, respectively, the biofilm formation was maximum: the biofilm formation increased by 24.7% and 21.1%, respectively. We found that the factors that were the most effective at stimulating biofilm formation were $CaCl_2$ (5 g/L), pH 7, and a culture time of 48 h at 30 °C (Figures 1 and 2). According to the principle of Box-Behnken design, the optimal concentration or conditions were taken to the climbing test as the central point of the response surface test factor level. A response surface analysis (three factors and three levels) was used to determine the optimal level of the primary factors that affected the biofilm formation of *B. mojavensis* D50. The design factors and levels of the Box-Behnken response surface are shown in Table 1, and the design results are shown in Table 2. A regression analysis of the results in Table 3 was used to obtain the regression equation of $Y = 3.19 + 0.0129A + 0.0253B + 0.0456C - 0.0213AB + 0.0075AC + 0.0297BC - 0.2229A^2 - 0.2882B^2 - 0.2984C^2$. The correlation coefficient of the model was $R^2 = 0.9710$, and the corrected determination coefficient was $R^2_{Adj} = 0.9336$. The fitting model could clarify the change of 93.36% of the response value, indicating that there was a strong fitting degree between the experimental data and that predicted. There was little experimental error, which indicated that it could be used to predict and analyze the factors that affected the biofilm formation.

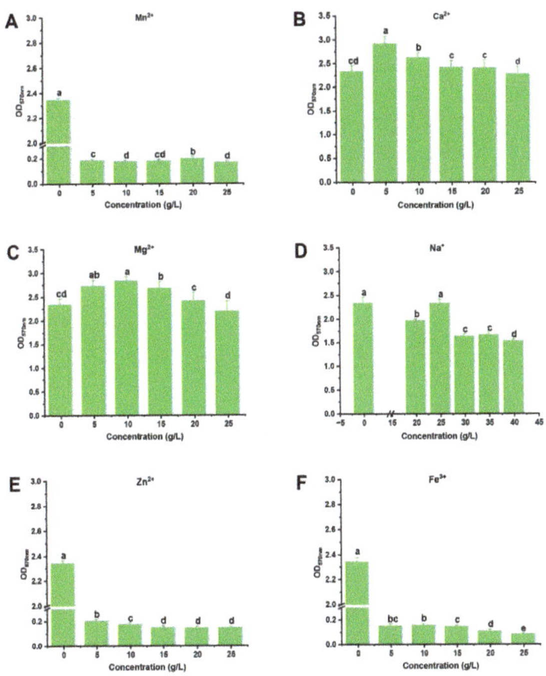

Figure 1. Effect of different concentrations of metal ions on *B. mojavensis* D50 biofilm formation: Mn^{2+} (**A**), Ca^{2+} (**B**), Mg^{2+} (**C**), Na^{+} (**D**), Zn^{2+} (**E**), and Fe^{3+} (**F**). The strain D50 (5 μL) was inoculated into mediums (containing different metal ions) and cultured at 30 °C for 48 h. Bars followed by the same letter are significantly different at $p < 0.05$ using a least significant difference (LSD) test. Error bars indicate the SD of three experiments. SD, standard deviation.

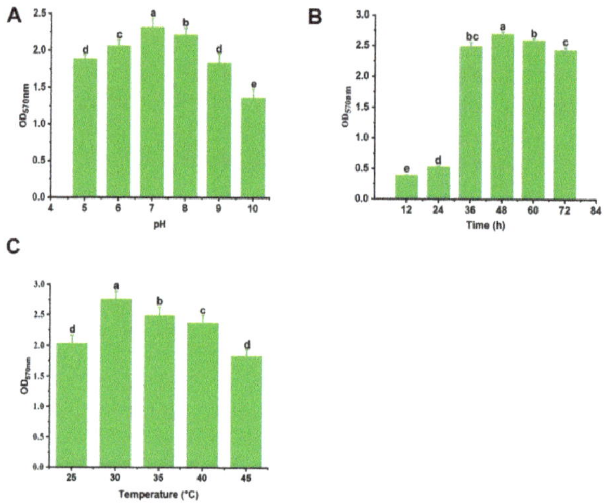

Figure 2. Effect of different concentrations of culture conditions on *B. mojavensis* D50 biofilm formation: pH (**A**), time (**B**), and temperature (**C**). The strain D50 (5 μL) was inoculated into mediums (without metal ions) and cultured for different amounts of time and at different temperatures. To determine the effect of pH on biofilm formation of strain D50, the strain D50 (5 μL) was inoculated into mediums (of different pH) and cultured at 30 °C for 48 h. Bars followed by the same letter are significantly different at $p < 0.05$ using a least significant difference (LSD) test. Error bars indicate the SD of three experiments. SD, standard deviation.

Table 3. ANOVA for response surface reduced quadratic model.

Source	Sum of Squares	df	Mean Square	*F*-Value	*p*-Value	
Model	1.07	9	0.1189	26.01	0.0001	Significant
A-$CaCl_2$	0.0013	1	0.0013	0.2902	0.6068	
B-Temperature	0.0051	1	0.0051	1.12	0.03258	
C-Time	0.0167	1	0.0167	3.64	0.0979	
AB	0.0018	1	0.0018	0.3953	0.5495	
AC	0.0002	1	0.0002	0.0492	0.8307	
BC	0.0035	1	0.0035	0.7748	0.4079	
A^2	0.2092	1	0.2092	45.78	0.0003	
B^2	0.3496	1	0.3496	76.51	<0.0001	
C^2	0.3749	1	0.3749	82.05	<0.0001	
Residual	0.0320	7	0.0046			
Lack of Fit	0.0016	3	0.0005	0.0710	0.9724	Not significant
Pure Error	0.0304	4	0.0076			
Cor Total	1.10	16				

Note: ANOVA, analysis of variance.

The variance analysis of response surface quadratic model is shown in Table 3. The response surface quadratic model was highly significant ($p < 0.0001$), and the lake of fit (lake of fit = 0.0710 > 0.05) was not significant. There were no abnormal points in the data, and there was a high fitting degree of the regression equation. The influence degree of each factor of F value on the response value, C > B > A, indicated that the influence intensity of each factor on the biofilm formation was as follows: time > temperature > $CaCl_2$. The response surface curve and contour map were drawn based on the results of the equation simulation (Figure 3). All three surface graphs were convex, and all the contour lines were oval. These results indicated that there was a stable maximum value of the response. As shown in Figure 3 and Table 3, the simulation factors A^2, B^2, and C^2 were significant, indicating that magnesium sulfate had a significant impact on the biofilm formation. In contrast, A, B, and C were not significant, indicating that $CaCl_2$, temperature, and time had no significant impact on the biofilm formation, and the impact of various influencing factors on biofilm formation was not linear. AB, AC, and BC were not significant, indicating that the interaction between various factors was not significant. The parity plot of biofilm formation shows the distribution of predicted yield and actual yield of biofilm formation under different conditions, as seen in Figure 4. The corresponding predicted data of the model was almost consistent with that of the actual data, indicating that the polynomial model was highly accurate and universal. Thus, it was reasonable to use the model to analyze the corresponding trend.

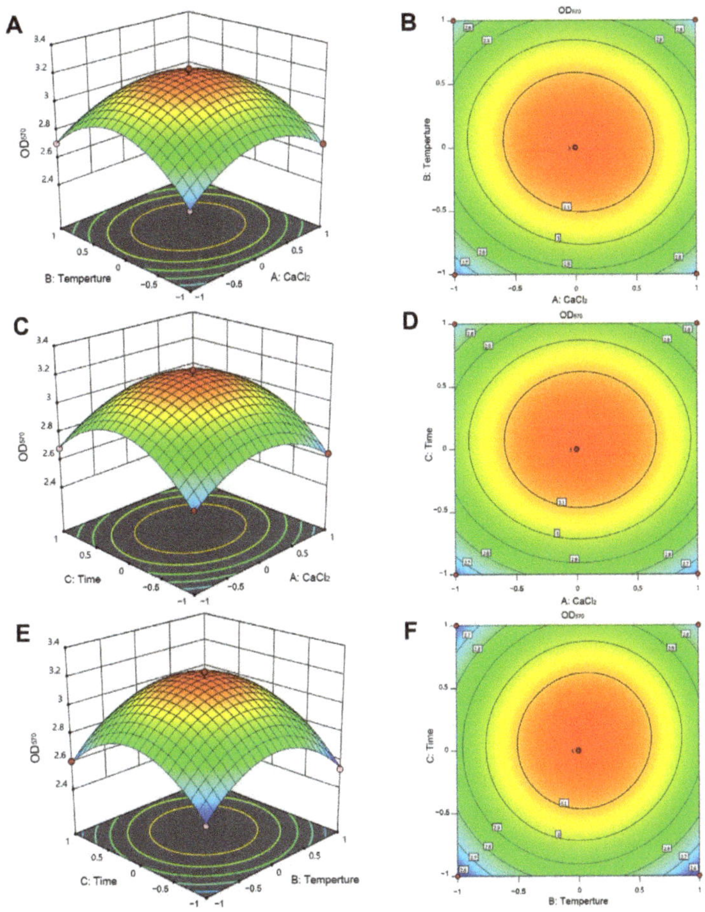

Figure 3. Three-dimensional response and two-dimensional surface contour for biofilm formation amount as evaluation indicators. The interactions between solution concentration of $CaCl_2$ and temperature (**A**,**B**); solution concentration of $CaCl_2$ and time (**C**,**D**); and temperature and time (**E**,**F**). The change of color from blue to red in the graph indicates a change in extraction quality from less to more, and the faster the change, the greater the slope, which has a more significant impact on the experimental results.

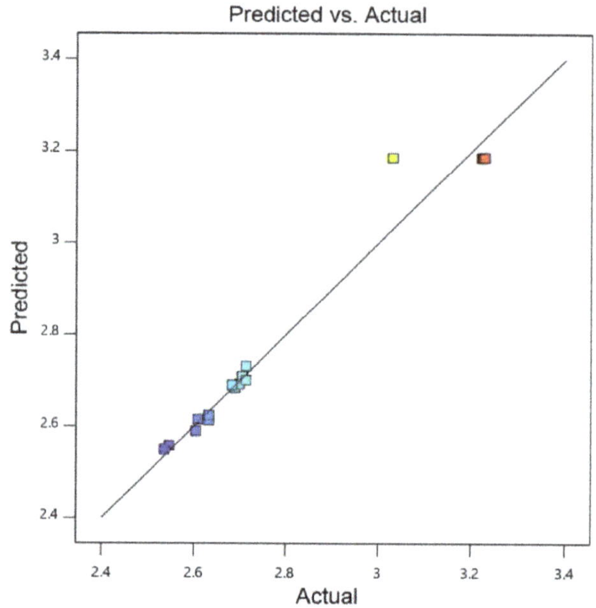

Figure 4. Parity plot of correlation between the actual value and predicted value of biofilm formation. The change of color from blue to red in the graph indicates a change in extraction quality from less to more, and the faster the change, the greater the slope, which has a more significant impact on the experimental results.

3.2. Determination of Antagonistic Abilities and Colonization Capacity Differences

Antagonistic abilities and colonization capacity are important targets to optimize for high biocontrol ability. Therefore, these abilities were measured. As shown in Figure 5A, the biofilm of *B. mojavensis* D50 (after optimization) was more viscous and denser than those in *B. mojavensis* D50 (before optimization). The biofilm yields of *B. mojavensis* D50 (after optimization) were more than those in *B. mojavensis* D50 (before optimization). The antagonistic abilities were improved after optimization, and the morphology of *B. cinerea* obviously changed (Figure 5B). After optimization, the colonization capacity of *B. mojavensis* D50 increased in the root of the tomato plant (Figure 5C). These results could indicate that the biocontrol ability of *B. mojavensis* D50 increased after optimization.

3.3. *B. mojavensis* D50 Biofilm-Related Gene Differential Expression Analysis

To determine the differential expression of biofilm-related genes (before or after optimization), the levels of expression of the genes *luxS*, *SinR*, *FlhA*, and *tasA* were measured. As shown in Figure 6, the level of expression of the *luxS* gene was upregulated by 27.87- and 37.56-fold when the culture times were 24 h and 51.8 h, respectively (Figure 6A). The level of expression of the *SinR* gene after optimization was higher than that before optimization at the different culture times (1.69- and 2.87-fold) (Figure 6B). Moreover, the level of expression of the *FlhA* gene was upregulated by 7.47- and 12.46-fold when the culture times were 24 h and 51.8 h, respectively (Figure 6C). The level of expression of the *tasA* gene after optimization was higher than that before optimization at the different culture times (4.68- and 6.22-fold, respectively) (Figure 6D). These results indicated that the levels of expression of biofilm-related genes were upregulated after optimization, and the ability to form biofilm improved owing to the increased expression of the genes involved in the production of biofilm.

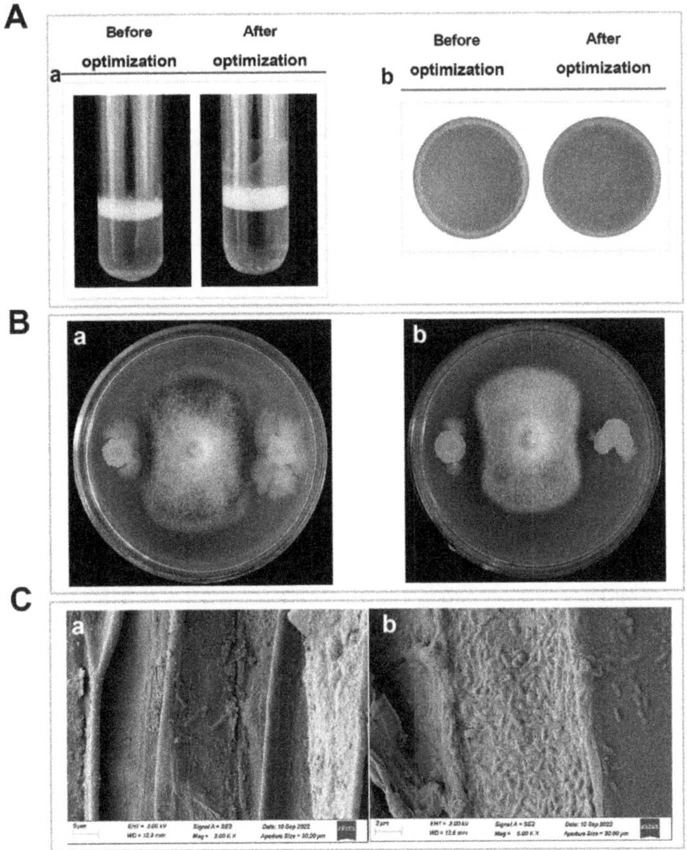

Figure 5. Effect of *B. mojavensis* D50 on biofilm formation, antifungal activity, and colonization ability on the plant root. (**A**) The biofilm formation of D50 ((a): before optimization, and (b): after optimization). The strain D50 (5 μL) was inoculated into initial or optimum medium and cultured at 30 °C or 31.4 °C for 48 h or 51.8 h. (**B**) The antifungal activity of D50 ((a): before optimization, (b): after optimization). The strain D50 (5 μL) which was cultured into initial or optimum medium was inoculated into PDA medium and cultured at 26 °C for 7 days to determine the biocontrol effect. (**C**) The colonization ability of D50 ((a): before optimization, (b): after optimization). The strain D50 (1.0×10^8 cfu/mL, 20 mL), before or after optimization, was poured on the roots of tomatoes, and the colonization ability was determined after 2 days.

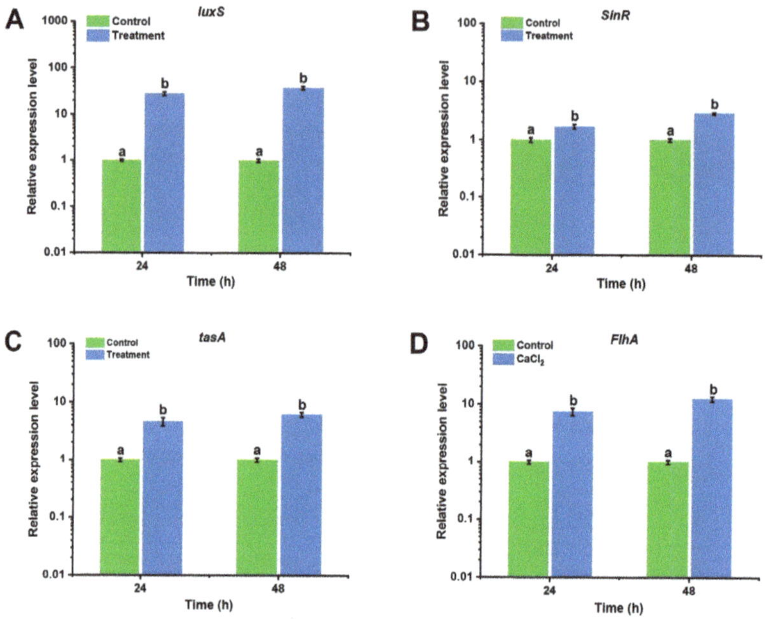

Figure 6. The levels of expression of the genes *luxS*, *SinR*, *FlhA*, and *tasA* related to biofilm formation were detected using qRT-PCR: (**A**) *luxS*, (**B**) *SinR*, (**C**) *tasA*, (**D**) *FlhA*. *B. mojavensis* D50 (1.0×10^8 cfu/mL, 1 mL) was inoculated into initial medium (100 mL) and cultured at 30 °C for 48 h as the control, while *B. mojavensis* D50 (1.0×10^8 cfu/mL, 1 mL) was inoculated into optimization medium (100 mL) and cultured at 31.4 °C for 51.8 h as the treatment. Bars followed by the same letter are significantly different at $p < 0.05$ using a least significant difference (LSD) test. Error bars indicate SD of three experiments. qRT-PCR, quantitative PCR; SD, standard deviation.

3.4. *Effect of Metal Ions and Culture Conditions on the Biocontrol Characteristics of B. mojavensis D50 against B. cinerea In Vivo*

Furthermore, *B. mojavensis* D50 (before and after optimization) was used in this experiment to assess its antifungal ability improvement. Three days after inoculation with *B. cinerea*, small spotted lesions appeared on the tomato. As time went on, lesions became larger and caused extensive yellowing of leaves in the group BS. In the group BSA and BSB, small spotted lesions appeared after five days' inoculation with *B. cinerea*. The lesions did not expand with time. After the inoculation with *B. cinerea* for seven days, large areas of tomato plants in the group BS turned yellow, while a small portion of the plant had diseased spots in the group BSB and BSA. The disease severity index was 85.6 in the group BS (*B. cinerea* + sterile water). In the group BSB (*B. cinerea* + strain D50 before optimization), the severity of disease and disease reduction was 58.4 and 39.7, respectively, while in the group BSA (*B. cinerea* + strain D50 after optimization), the severity of disease and disease reduction was 45.3 and 46.2, respectively (Table 4).

Table 4. Antifungal spectra of *B. mojavensis* D50 (before and after optimization) against *Botrytis cinerea* in vivo. Sterile water as control; *B. mojavensis* D50 before or after optimization as treatment.

Treatment	Inoculants of Treatment	Disease Severity Index (%)	Disease Reduction (%)
BS	*B. cinerea* + Sterile water	85.6	-
BSB	*B. cinerea* + strain D50 (before optimization)	58.4	39.7
BSA	*B. cinerea* + strain D50 (after optimization)	45.3	46.2

Data with sample size n = 16 plants per treatment.

As shown in Table 5, various growth and biochemical parameters of the tomato seedlings were measured. The fresh weight, shoot length, and chlorophyll content in the tomato seedlings differed significantly among BS (*B. cinerea* + sterile water), BSB (*B. cinerea* + strain D50 before optimization), and BSA (*B. cinerea* + strain D50 after optimization) ($p < 0.05$). The chlorophyll content, shoot length, and fresh weight in BSA were higher than those in BS and BSB. Moreover, the contents of chlorophyll a, chlorophyll b, and total chlorophyll in BSA were significantly higher than BS (24.2%, 22.4%, and 23.8%, respectively) and BSB (19.3%, 9.8%, and 12.8%, respectively). The total phenolic content was enhanced by 52.5% and 10.3% more than BS and BSB, respectively. The correlated parameters (chlorophyll a, chlorophyll b, total chlorophyll, fresh weight, shoot length, root length, total phenolic content, and total soluble protein) in BS and BSB were lower than that in BSA. This effect was associated with decreases in fresh weight, shoot length, and root length. The results indicated that the biofilm formation improvement could improve the antifungal effects in vivo.

Table 5. Growth and biochemical parameters (fresh weight, root length, shoot length, chlorophyll a, chlo-rophyll b, total chlorophyll, total phenolic content, and total soluble protein) of tomato plants after strain D50 (before and after optimization) treatments. Bars followed by the same letter are significantly different at $p < 0.05$ by LSD test, error bars indicate ± SD of triplicates.

Treatments	Fresh Weight (g)	Root Length (cm)	Shoot Length (cm)	Chl a (mg/mL)	Chl b (mg/mL)	Total Chlorophyll (mg/mL)	Total Phenolic Content (mg/100 g)	Total Soluble Protein (mg/100 g)
BS	1.45 ± 0.05 c	6.15 ± 0.13 c	17.82 ± 0.32 c	11.99 ± 0.08 c	3.75 ± 0.08 c	15.74 ± 0.15 c	163.38 ± 0.28 c	301.76 ± 1.42 c
BSB	2.06 ± 0.07 b	7.06 ± 0.12 b	20.23 ± 0.12 b	12.48 ± 0.13 b	4.18 ± 0.09 b	17.27 ± 0.08 b	225.82 ± 0.68 b	345.2 ± 2.68 b
BSA	2.73 ± 0.03 a	8.26 ± 0.24 a	24.36 ± 0.18 a	14.89 ± 0.06 a	4.59 ± 0.05 a	19.48 ± 0.12 a	249.17 ± 0.36 a	390.8 ± 3.87 a

Note: BS (*Botrytis cinerea* + sterile water), BSB (*B. cinerea* + strain *B. mojavensis* D50 before optimization) and BSA (*B. cinerea* + strain *B. mojavensis* D50 after optimization).

3.5. *Effect of* B. mojavensis *D50 (before and after Optimization) on Soil Enzyme Activities*

To determine the effect of *B. mojavensis* D50 (before or after optimization) on soil enzyme activities (related to biocontrol), the activities of invertase, catalase, urease, and dehydrogenase were measured. As shown in Figure 7, the invertase, catalase, and dehydrogenase activities increased from 2 h to 15 days (Figure 7A,C,D). The urease activity in BSB (*B. cinerea* + strain D50 before optimization) and BSA (*B. cinerea* + strain D50 after optimization) group increased from 2 h to 5 days, while it decreased from 5 days to 15 days. In addition, the urease activity in the BS (*B. cinerea* + sterile water) group decreased from 2 h to 3 days, while it decreased from 3 days to 15 days (Figure 7B). The soil enzyme activities in the T3 group were the highest among these groups (Figure 7) ($p < 0.05$). These results indicated that *B. mojavensis* D50 (after optimization) improved the related biocontrol soil enzyme activities and promoted the growth of tomato plants.

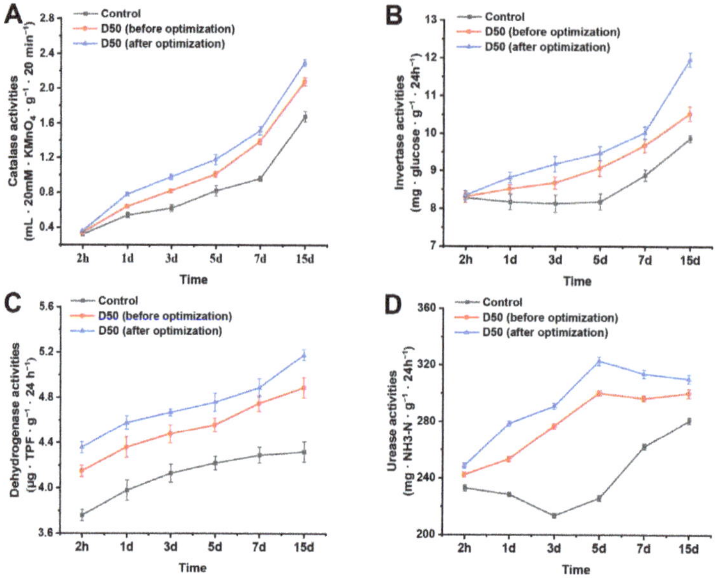

Figure 7. Soil enzyme activities in various treatments (sterile water as control; *B. mojavensis* D50 before or after optimization as treatment): (**A**) catalase, (**B**) invertase, (**C**) dehydrogenase, and (**D**) urease. The soil was collected at 2 h, 1 day, 3 days, 5 days, 7 days, and 15 days after the inoculation with strain D50 (before and after optimization). Error bars indicate SD of three experiments. SD, standard deviation.

4. Discussion

Most Gram-positive bacteria can form biofilms, but this process is difficult to observe under culture conditions [15]. The formation of biofilm facilitates its colonization in plants and increases its biocontrol ability [37]. In this study, the metal ion and culture conditions were optimized by the one-by-one method and response surface methodology. As shown in Figures 1 and 2, Ca^{2+} promoted biofilm formation, and the temperature and culture time were also important factors that influenced the biofilm formation. The optimized culture conditions and culture medium were determined using response surface methodology. Many studies have reported that the optimal culture medium and conditions can influence biofilm formation; for instance, the biofilm formation of *B. subtilis* 1JN2 decreased when it was cultured under a high concentration of Cd^{2+} [14]. The culture medium and culture time could also impact the biofilm formation of *Bacillus* spp. [16]. Glycerol, as a carbon source, and Ca^{2+} and Mg^{2+} have been reported to have a substantial effect on biofilm formation [15]. These studies indicated that optimizing the culture conditions and culture medium can improve the biofilm formation, and their results were consistent with ours.

Biofilm is an aggregated form of a growing microorganism. Biofilm formation can influence strain colonization, affecting its biocontrol effects [37]. To ensure the effect of optimizing the culture medium and culture conditions on biofilm formation, the antifungal abilities and colonization capacity were measured. We found that these biocontrol characteristics were improved (Figure 5). Many studies have reported that the biocontrol effects could improve by enhancing the biofilm formation [38]. Our results closely agreed with the findings in these studies.

The quorum sensing (QS) system, which regulates biofilm formation, is one of the important factors that regulates biofilm formation in many bacteria [39]. To determine the mechanism of biofilm formation, the levels of expression of the genes *luxS*, *SinR*, *FlhA*, and *tasA* involved in its formation were measured. We found that the level of expression of these genes increased after optimization. These results showed that the optimal culture medium and conditions could upregulate the level of expression of the biofilm-related genes to promote biofilm formation (Figure 6). The optimal culture medium and conditions of *C. albicans* could promote biofilm formation and growth [40]. These results were consistent with ours.

To determine the effect of *B. mojavensis* D50 in vivo, the disease incidence and severity, fresh weight, shoot length, and chlorophyll content of the tomato seedlings were measured. We found that *B. mojavensis* D50 (after optimization) effectively protected the tomato plants from *B. cinerea*. In addition, the growth and biochemical parameters of the tomato seedlings treated with *B. mojavensis* D50 (after optimization) were the highest among those in BS and BSB group ($p < 0.05$) (Table 5). There are some studies on the effect of biocontrol strain on pathogens, such as the ability of *B. subtilis* Y2 to inhibit the growth of *Alternaria brassicicola* in pear fruit [41]. The antagonistic activity of *B. velezensis* CMRP 4489 could inhibit the growth of pathogens by the formation of biofilm [42].

Soil enzyme activity is an important physical and chemical index of soil. It is generally believed that a higher activity of soil enzymes in the soil environment indicates that the soil is more fertile [43]. The results of this study showed that the activities of urease, invertase, catalase, and dehydrogenase in the soil treated with *B. mojavensis* D50 (after optimization) increased in the tomato rhizosphere (Figure 7). Simultaneously, they could promote the growth of root length, root weight, and fresh weight of tomato seedlings, indicating that the growth promotion of *B. mojavensis* D50 could be related to its ability to improve the activities of soil enzymes, which improves the fertility of soil.

5. Conclusions

In this study, the effects of different metal ions and culture conditions on biofilm formation were determined. After optimization, the characteristics of biocontrol and the ability of colonization were improved, and the levels of expression of the biofilm-related genes *luxS*, *SinR*, *FlhA*, and *tasA* were up-regulated. After optimization, *B. mojavensis* D50 could increase the activity of soil enzymes related to biological control, and the biocontrol effect was also improved in vivo. Further study is needed with a focus on searching for how the metal ions influence the biofilm formation of *B. mojavensis* D50, and the effect of key genes which are related to the biofilm formation and biocontrol of *B. cinerea*.

Supplementary Materials: The following supporting information can be downloaded at: https://www.mdpi.com/article/10.3390/jof9050526/s1, Table S1: Primers used for quantitative real time PCR (qRT-PCR).

Author Contributions: L.Z.: Conceptualization, Methodology, Validation, Formal analysis, Investigation, Resources, Data curation, Writing original draft. X.G. and L.S.: Methodology, Validation, Formal analysis, Investigation, Resources, Data curation. M.D. and A.G.: Methodology, Validation, Formal analysis, Investigation, Resources. Z.H.: Methodology, Validation, Formal analysis, Resources, Writing—review & editing. H.P.: Validation, Formal analysis, Writing—review & editing, Project administration, Funding acquisition. H.Z.: Methodology, Validation, Formal analysis, Investigation, Resources, Writing—review & editing, Supervision, Project administration, Funding acquisition. All authors have read and agreed to the published version of the manuscript.

Funding: This work was financially supported by grants from the Inter-governmental International Cooperation Special Project of National Key & D Program of China [grant number 2019YFE0114200]; and Key R&D projects of Jilin Provincial Science and Technology Development Plan [grant number 20230203014SF].

Institutional Review Board Statement: Not applicable.

Informed Consent Statement: Not applicable.

Data Availability Statement: Not applicable.

Acknowledgments: The authors would like to thank Xiao Y. and Wang S. who assisted in collection of some experimental samples.

Conflicts of Interest: The authors declare no conflict of interest.

References

1. Bu, S.; Munir, S.; He, P.; Li, Y.; Wu, Y.; Li, X.; Kong, B.; He, P.; He, Y. *Bacillus subtilis* L1-21 as a biocontrol agent for postharvest gray mold of tomato caused by *Botrytis cinerea*. *Biol. Control* **2021**, *157*, 104568. [CrossRef]
2. Hong, J.K.; Sook Jo, Y.; Jeong, D.H.; Woo, S.M.; Park, J.Y.; Yoon, D.J.; Lee, Y.H.; Choi, S.H.; Park, C.J. Vapours from plant essential oils to manage tomato grey mould caused by *Botrytis cinerea*. *Fungal Biol.* **2023**, *127*, 985–996. [CrossRef]
3. Suresh, P.; Shanmugaiah, V.; Rajakrishnan, R.; Muthusamy, K.; Ramamoorthy, V. *Pseudomonas fluorescens* VSMKU3054 mediated induced systemic resistance in tomato against *Ralstonia solanacearum*. *Physiol. Mol. Plant Pathol.* **2022**, *199*, 101836. [CrossRef]
4. Guo, J.; Xu, Y.; Liang, S.; Zhou, Z.; Zhang, C.; Li, K.; Peng, X.; Qin, S.; Xing, K. Antifungal activity of volatile compounds from *Bacillus tequilensis* XK29 against *Botrytis cinerea* causing gray mold on cherry tomatoes. *Postharvest Biol. Technol.* **2023**, *198*, 112239. [CrossRef]
5. Zheng, L.; Gu, X.; Xiao, Y.; Wang, S.; Liu, L.; Pan, H.; Zhang, H. Antifungal activity of *Bacillus mojavensis* D50 against *Botrytis cinerea* causing postharvest gray mold of tomato. *Sci. Hortic.* **2023**, *312*, 111841. [CrossRef]
6. Mali, H.; Shah, C.; Raghunandan, B.H.; Prajapati, A.S.; Patel, D.H.; Trivedi, U.; Subramanian, R.B. Organophosphate pesticides an emerging environmental contaminant: Pollution, toxicity, bioremediation progress, and remaining challenges. *J. Environ. Sci.* **2023**, *127*, 234–250. [CrossRef]
7. Yang, X.; Wang, Y.; Jiang, H.; Song, R.; Liu, Y.; Guo, H.; Meng, D. Antimicrobial peptide CB-M exhibits direct antifungal activity against *Botrytis cinerea* and induces disease resistance to gray mold in cherry tomato fruit. *Postharvest Biol. Technol.* **2023**, *196*, 112184. [CrossRef]
8. Chen, J.; Wei, X.; Lu, X.; Ming, R.; Huang, D.; Yao, Y.; Li, L.; Huang, R. *Burkholderia cenocepacia* ETR-B22 volatile organic compounds suppress postharvest grey mould infection and maintain aroma quality of tomato fruit. *LWT* **2022**, *165*, 113715. [CrossRef]
9. Li, S.; Xiao, Q.; Yang, H.; Huang, J.; Li, Y. Characterization of a new *Bacillus velezensis* as a powerful biocontrol agent against tomato gray mold. *Pestic. Biochem. Physiol.* **2022**, *187*, 105199. [CrossRef]
10. Zhao, Q.; Shi, Y.; Xu, C.; Jiang, Z.; Liu, J.; Sui, Y.; Zhang, H. Control of postharvest blue and gray mold in kiwifruit by *Wickerhamomyces anomalus* and its mechanism of antifungal activity. *Postharvest Biol. Technol.* **2023**, *201*, 112345. [CrossRef]
11. Geng, L.; Fu, Y.; Peng, X.; Yang, Z.; Zhang, M.; Song, Z.; Guo, N.; Chen, S.; Chen, J.; Bai, B.; et al. Biocontrol potential of *Trichoderma harzianum* against *Botrytis cinerea* in tomato plants. *Biol. Control.* **2022**, *174*, 105019. [CrossRef]
12. Wang, B.; Wang, C.; Hu, Y. Sorption behavior of Pb(II) onto polyvinyl chloride microplastics affects the formation and ecological functions of microbial biofilms. *Sci. Total Environ.* **2022**, *832*, 155026. [CrossRef] [PubMed]
13. Guo, Q.; Shi, M.; Chen, L.; Zhou, J.; Zhang, L.; Li, Y.; Xue, Q.; Lai, H. The biocontrol agent *Streptomyces pactum* increases *Pseudomonas koreensis* populations in the rhizosphere by enhancing chemotaxis and biofilm formation. *Soil Biol. Biochem.* **2020**, *144*, 107755. [CrossRef]
14. Yang, W.; Yan, H.; Dong, G.; Li, Z.; Jiang, C.; Gu, D.; Niu, D.; Zhou, D.; Luo, Y. Comparative transcriptomics reveal different genetic adaptations of biofilm formation in *Bacillus subtilis* isolate 1JN2 in response to Cd^{2+} treatment. *Front. Microbiol.* **2022**, *13*, 1002482. [CrossRef]
15. Fu, H.; Chen, F.; Liu, W.; Kong, W.; Wang, C.; Fang, X.; Ye, J. Adding nutrients to the biocontrol strain JK-SH007 promotes biofilm formation and improves resistance to stress. *AMB Express* **2020**, *10*, 32. [CrossRef]
16. Fessia, A.; Sartori, M.; Garcia, D.; Fernandez, L.; Ponzio, R.; Barros, G.; Nesci, A. In vitro studies of biofilm-forming *Bacillus* strains, biocontrol agents isolated from the maize phyllosphere. *Biofilm* **2022**, *4*, 100097. [CrossRef]
17. Deng, Q.; Lei, X.; Zhang, H.; Deng, L.; Yi, L.; Zeng, K. Phenylalanine Promotes Biofilm Formation of *Meyerozyma caribbica* to Improve Biocontrol Efficacy against Jujube Black Spot Rot. *J. Fungi* **2022**, *8*, 1313. [CrossRef]
18. Zhang, Y.; Gu, Y.; Zheng, Y.; Wang, Y.; Nie, L.; Qiao, R.; He, Y. Deletion of *luxS* gene mediated by λRed gene recombination technology reduces biofilm formation and stress resistance of *Lactobacillus fermentum*. *Food Biosci.* **2022**, *49*, 101892. [CrossRef]
19. Liu, J.; Liu, K.; Zhao, Z.; Wang, Z.; Wang, F.; Xin, Y.; Qu, J.; Song, F.; Li, Z. The *LuxS*/AI-2 Quorum-Sensing System Regulates the Algicidal Activity of *Shewanella xiamenensis* Lzh-2. *Front. Microbiol.* **2021**, *12*, 814929. [CrossRef]
20. Meng, F.; Zhang, F.; Chen, Q.; Yang, M.; Yang, Y.; Li, X.; Gu, W.; Yu, J. Virtual screening and in vitro experimental verification of *LuxS* inhibitors from natural products for *Lactobacillus reuteri*. *Biomed. Pharmacother.* **2022**, *147*, 112521. [CrossRef]
21. Meng, F.; Zhao, M.; Lu, Z. The *LuxS*/AI-2 system regulates the probiotic activities of lactic acid bacteria. *Trends Food Sci. Technol.* **2022**, *127*, 272–279. [CrossRef]
22. Colledge, V.L.; Fogg, M.J.; Levdikov, V.M.; Leech, A.; Dodson, E.J.; Wilkinson, A.J. Structure and organisation of *SinR*, the master regulator of biofilm formation in *Bacillus subtilis*. *J. Mol. Biol.* **2011**, *411*, 597–613. [CrossRef]
23. Minamino, T.; Kinoshita, M.; Inoue, Y.; Kitao, A.; Namba, K. Conserved GYXLI Motif of *FlhA* Is Involved in Dynamic Domain Motions of *FlhA* Required for Flagellar Protein Export. *Microbiol. Spectr.* **2022**, *10*, e0111022. [CrossRef] [PubMed]
24. Verma, N.; Srivastava, S.; Malik, R.; Goyal, P.; Pandey, J. Inhibition and disintegration of *Bacillus subtilis* biofilm with small molecule inhibitors identified through virtual screening for targeting $TasA_{(28\text{-}261)}$, the major protein component of ECM. *J. Biomol. Struct. Dyn.* **2022**, *41*, 2431–2447. [CrossRef] [PubMed]

25. Zheng, L.; Zhang, J.; Wu, X.; Gu, X.; Wang, S.; Zhang, H. A novel biocontrol strain *Pantoea jilinensis* D25 for effective biocontrol of tomato gray mold (causative agent *Botrytis cinerea*). *Biol. Control* **2021**, *164*, 104766. [CrossRef]
26. O'Toole, G.A. Microtiter dish biofilm formation assay. *J. Vis. Exp.* **2011**, *47*, 2437. [CrossRef]
27. Camara-Almiron, J.; Navarro, Y.; Diaz-Martinez, L.; Magno-Perez-Bryan, M.C.; Molina-Santiago, C.; Pearson, J.R.; de Vicente, A.; Perez-Garcia, A.; Romero, D. Dual functionality of the amyloid protein *TasA* in *Bacillus* physiology and fitness on the phylloplane. *Nat. Commun.* **2020**, *11*, 1859. [CrossRef]
28. El-Naggar, N.E.; Saber WI, A.; Zweil, A.M.; Bashir, S.I. An innovative green synthesis approach of chitosan nanoparticles and their inhibitory activity against phytopathogenic *Botrytis cinerea* on strawberry leaves. *Sci. Rep.* **2022**, *12*, 3515. [CrossRef]
29. Gao, N.; Zhang, J.; Pan, Z.; Zhao, X.; Ma, X.; Zhang, H. Biodegradation of Atrazine by Mixed Bacteria of *Klebsiella variicola* Strain FH-1 and *Arthrobacter* sp. NJ-1. *Bull. Environ. Contam. Toxicol.* **2020**, *105*, 481–489. [CrossRef]
30. Xi, X.; Fan, J.; Yang, X.; Liang, Y.; Zhao, X.; Wu, Y. Evaluation of the anti-oomycete bioactivity of rhizosphere soil-borne isolates and the biocontrol of soybean root rot caused by *Phytophthora sojae*. *Biol. Control* **2022**, *166*, 104818. [CrossRef]
31. Zhang, M.; Zhang, C.; Zhang, S.; Yu, H.; Pan, H.; Zhang, H. *Klebsiella jilinsis* 2N3 promotes maize growth and induces resistance to northern corn leaf blight. *Biol. Control* **2021**, *156*, 104554. [CrossRef]
32. Wang, Y.; Yu, T.; Xia, J.; Yu, D.; Wang, J.; Zheng, X. Biocontrol of postharvest gray mold of cherry tomatoes with the marine yeast *Rhodosporidium paludigenum*. *Biol. Control* **2010**, *53*, 178–182. [CrossRef]
33. Raynaldo, F.A.; Dhanasekaran, S.; Ngea GL, N.; Yang, Q.; Zhang, X.; Zhang, H. Investigating the biocontrol potentiality of *Wickerhamomyces anomalus* against postharvest gray mold decay in cherry tomatoes. *Sci. Hortic.* **2021**, *285*, 110137. [CrossRef]
34. Chen, N.; Li, X.; Shi, H.; Hu, Q.; Zhang, Y.; Leng, X. Effect of biodegradable film mulching on crop yield, soil microbial and enzymatic activities, and optimal levels of irrigation and nitrogen fertilizer for the *Zea mays* crops in arid region. *Sci. Total Environ.* **2021**, *776*, 145970. [CrossRef]
35. Ghiloufi, W.; Seo, J.; Kim, J.; Chaieb, M.; Kang, H. Effects of Biological Soil Crusts on Enzyme Activities and Microbial Community in Soils of an Arid Ecosystem. *Microb. Ecol.* **2019**, *77*, 201–216. [CrossRef] [PubMed]
36. Kandeler, E.; Gerber, H. Short-term assay of soil urease activity using colorimetric determination of ammonium. *Biol. Fertil. Soils* **1988**, *6*, 68–72. [CrossRef]
37. Zhang, P.; Xin, H.; van der Lee, T. Tree pathogens *Armillaria solidipes* influence the biocontrol activity of *Bacillus velezensis* BY6. *Biol. Control* **2023**, *179*, 105176. [CrossRef]
38. Gao, T.; Wang, X.; Qin, Y.; Ren, Z.; Zhao, X. Watermelon Root Exudates Enhance Root Colonization of *Bacillus amyloliquefaciens* TR2. *Curr. Microbiol.* **2023**, *80*, 110. [CrossRef]
39. Mauritzen, J.J.; Sondberg, E.; Kalatzis, P.G.; Roager, L.; Gram, L.; Svenningsen, S.L.; Middelboe, M. Strain-specific quorum-sensing responses determine virulence properties in *Vibrio anguillarum*. *Environ. Microbiol.* **2023**. [CrossRef]
40. Sadanandan, B.; Vaniyamparambath, V.; Lokesh, K.N.; Shetty, K.; Joglekar, A.P.; Ashrit, P.; Hemanth, B. Candida albicans biofilm formation and growth optimization for functional studies using response surface methodology. *J. Appl. Microbiol.* **2022**, *132*, 3277–3292. [CrossRef]
41. Wang, X.; Xie, S.; Mu, X.; Guan, B.; Hu, Y.; Ni, Y. Investigating the resistance responses to *Alternaria brassicicola* in 'Korla' fragrant pear fruit induced by a biocontrol strain *Bacillus subtilis* Y2. *Postharvest Biol. Technol.* **2023**, *199*, 112293. [CrossRef]
42. Baptista, J.P.; Teixeira, G.M.; de Jesus ML, A.; Berte, R.; Higashi, A.; Mosela, M.; da Silva, D.V.; de Oliveira, J.P.; Sanches, D.S.; Brancher, J.D.; et al. Antifungal activity and genomic characterization of the biocontrol agent *Bacillus velezensis* CMRP 4489. *Sci. Rep.* **2022**, *12*, 17401. [CrossRef] [PubMed]
43. Zhu, C.; Zhang, C.; Zhang, M.; Wu, Y.; Zhang, Z.; Zhang, H. Degradation characteristics and soil remediation of thifensulfuron-methyl by immobilized *Serratia marcecens* N80 beads. *Environ. Technol. Innov.* **2021**, *24*, 102059. [CrossRef]

Journal of
Fungi

MDPI

Article

Erg4 Is Involved in Ergosterol Biosynthesis, Conidiation and Stress Response in *Penicillium expansum*

Zhanhong Han [1,†], Yuanyuan Zong [1,†], Xuemei Zhang [1], Di Gong [1], Bin Wang [1], Dov Prusky [1,2], Edward Sionov [3], Huali Xue [4,*] and Yang Bi [1,*]

[1] College of Food Science and Engineering, Gansu Agricultural University, Lanzhou 730070, China; hanzhanhong009@163.com (Z.H.); zongyy@gsau.edu.cn (Y.Z.); gongdi531@163.com (D.G.); wangbin_1519@163.com (B.W.); dovprusk@volcani.agri.gov.il (D.P.)
[2] Department of Postharvest Science of Fresh Produce, Agricultural Research Organization, Volcani Center, Rishon LeZion 50250, Israel
[3] Department of Food Science, Agricultural Research Organization, Volcani Center, Rishon LeZion 50250, Israel; edwardsio@volcani.agri.gov.il
[4] College of Science, Gansu Agricultural University, Lanzhou 730070, China
* Correspondence: xuehuali77@sina.com (H.X.); biyang@gsau.edu.cn (Y.B.)
† These authors have contributed equally to the work.

Abstract: *erg4* is a key gene for ergosterol biosynthesis in filamentous fungi, but its function in *Penicillium expansum* remains unknown. Our results showed that *P. expansum* contains three *erg4* genes, including *erg4A*, *erg4B* and *erg4C*. The expression levels of the three genes showed differences in the wild-type (WT) strain, and the expression level of *erg4B* was the highest, followed by *erg4C*. Deletion of *erg4A*, *erg4B* or *erg4C* in the WT strain revealed functional redundancy between them. Compared to the WT strain, *erg4A*, *erg4B* or *erg4C* knockout mutants reduced ergosterol levels, with *erg4B* deletion having the greatest effect. Furthermore, deletion of the three genes reduced sporulation of the strain, and Δ*erg4B* and Δ*erg4C* mutants showed defective spore morphology. In addition, Δ*erg4B* and Δ*erg4C* mutants were found to be more sensitive to cell wall integrity and oxidative stress. However, deletion of *erg4A*, *erg4B* or *erg4C* had no significant effect on colony diameter, spore germination rate, conidiophore structure of *P. expansum* or pathogenicity to apple fruit. Taken together, *erg4A*, *erg4B* and *erg4C* have redundant functions and are all involved in ergosterol synthesis and sporulation in *P. expansum*. In addition, *erg4B* and *erg4C* contribute to spore morphogenesis, cell wall integrity and response to oxidative stress in *P. expansum*.

Keywords: *Penicillium expansum*; *erg4s*; ergosterol; growth and development; pathogenicity

1. Introduction

Penicillium expansum is an important postharvest pathogenic fungus that causes blue mold in several temperate fruits. During colonization, the pathogen produces patulin and citrinin in the fruit, posing a potential threat to consumer health [1]. Ergosterol is a fungal-specific sterol found in the plasma membrane of fungi [2]. Ergosterol plays an important role in maintaining the integrity and fluidity of cell membranes and is involved in the activity of membrane proteins, transduction of signaling molecules and various biological processes [3]. A total of 20 enzymes are involved in the synthesis of ergosterol in *Saccharomyces cerevisiae* [2]. Among them, the sterol C-24 reductase, encoded by *erg4*, catalyzes the conversion of ergosta-5,7,22,24-tetraenol to ergosterol in the final step of ergosterol biosynthesis [4]. In *S. cerevisiae*, deletion of *erg4* completely blocked ergosterol biosynthesis, resulting in the accumulation of ergosta-5,7,22,24(28)-tetraenol, a precursor compound of ergosterol biosynthesis [5]. In contrast, overexpression of *erg4* promoted ergosterol production in *S. cerevisiae* [6]. Furthermore, *erg4* deletion inhibited ergosterol biosynthesis in *Xanthophyllomyces dendrorhous* [7]. The *erg4* deletion mutant of *Fusarium graminearum* showed a decrease in mycelial growth and conidiation and produced abnormal conidia as well as lower levels of DON production. Furthermore, *erg4* deletion increased the sensitivity of *F. graminearum* to osmotic and oxidative stress, but inhibited ergosterol biosynthesis and virulence against wheat heads and tomato fruits [8]. In *Aspergillus fumigatus*, *erg4A* or *erg4B* deletion had no significant effect on ergosterol synthesis and mycelial growth of the fungus. However, the mutant with both *erg4A* and *erg4B* knockout showed impaired colony growth, complete blockage of ergosterol synthesis and severe conidiation defects, but had no effect on virulence in mice [9]. Although the *erg4* gene family has been reported to regulate growth, development and pathogenicity of *S. cerevisiae* and filamentous fungi, how the *erg4* gene family affects growth, development, ergosterol biosynthesis and pathogenicity of *P. expansum* has not been reported. Our previous results showed that *P. expansum* has three homologous *erg4* genes, *erg4A*, *erg4B* and *erg4C* [10]. Transmembrane domain analysis of the corresponding encoded proteins revealed that Erg4A and Erg4C proteins contain nine transmembrane structures, while Erg4B protein contains seven transmembrane structures. In addition, subcellular localization results showed that Erg4A, Erg4B and Erg4C proteins were all localized to the endoplasmic reticulum membrane [10]. Therefore, the objectives of this study were to (1) analyze the sequence characteristics and phylogenetic relationships of Erg4A, Erg4B and Erg4C proteins; (2) construct *erg4A*, *erg4B* and *erg4C* deletion mutants and their corresponding complementation strains; (3) determine the transcription levels of *erg4A*, *erg4B* and *erg4C* and the ergosterol content in WT and mutant strains; (4) elucidate the role of the three genes in colony growth, sporulation, spore germination rate, conidiophore development and spore morphology of *P. expansum*; (5) compare the sensitivity of Δ*erg4A*, Δ*erg4B* and Δ*erg4C* mutants to osmotic stress, cell wall integrity and oxidative stress; (6) observe the pathogenicity of the three knockout mutants on apple fruit.

2. Materials and Methods

2.1. Fungal Strains, Culture Conditions and Fruit

The WT strain of *P. expansum* T01 was kindly provided by Prof. Shiping Tian, the Institute of Botany, Chinese Academy of Sciences. The WT strain and mutants used in this study were cultured on PDA medium for 7 days. Spore suspensions of each strain were collected with 10 mL of sterile water (containing 0.05% Tween-20) and then filtered through four layers of sterile gauze. A total of 100 μL of spore suspension containing 1×10^6 spores/mL of each strain was inoculated into 200 mL of Czapek Yeast Extract (CY) liquid medium (containing 3 g $NaNO_3$, 1 g $K_2HPO_4 \cdot 3H_2O$, 0.5 g KCl, 0.5 g $MgSO_4 \cdot 7H_2O$, 0.01 g $FeSO_4 \cdot 7H_2O$, 30 g sucrose, 5 g yeast extract and 1000 mL of distilled water) and incubated in a thermostatic shaker (200 rpm, 25 °C) for 3 days. The mycelia were then collected for genomic DNA extraction.

Apple fruits (*Malus domestica* Borkh. cvs. Golden Delicious and Fuji) were harvested from a commercial orchard in Jingtai county, Gansu Province, China.

2.2. Sequence Alignments and Phylogenetic Analysis

Amino acid sequences of Erg4 protein in different fungal species were obtained through BLASTP searches on NCBI (http://www.ncbi.nlm.nih.gov/, accessed on 15 January 2020), and multiple sequence alignments were performed using DNAMAN 6.0 software. Phylogenetic analysis was performed using MEGA 7.0 software, and the neighbor-joining (NJ) tree was constructed with a bootstrap value of 1000. The conserved motif was predicted by the online MEME program (http://meme-suite.org/, accessed on 14 March 2021).

2.3. Gene Knockout and Complementation

The construction of *erg4A*, *erg4B* or *erg4C* knockout strains and their complementation strains was achieved by the homologous recombination strategy (Figures S1 and S2) [11]. Briefly, the genomic DNA of the WT strain was used as a template to obtain the upstream and downstream homologous recombination sequences (approximately 1 kb) of each gene by PCR amplification using the specified primer pairs. The upstream and downstream sequences of each gene were then inserted into the pCAMBIA1300-HPH vector to obtain the corresponding knockout vector. The knockout vector of each gene was transformed into the WT strain by *Agrobacterium tumefaciens*-mediated transformation (ATMT). Transformants were selected with 250 μg/mL hygromycin B and identified by PCR amplification. Complementation strains were obtained according to the method described by [12]. The DNA fragment of *erg4A*, *erg4B* or *erg4C* was inserted into the *XbalI* and *SacI* site of vector pCNEO, respectively. The vector was then transformed into the corresponding mutants using the ATMT method. Transformants were selected at 250 μg/mL G-418 (Solarbio Biotechnology Co., Ltd. Beijing, China) and confirmed by PCR amplification. All primers used to generate mutants and complementation strains are listed in Table S1.

2.4. Gene expression Analysis

The mycelium of *P. expansum* cultured for 3 days was collected from the CY liquid medium for the determination of gene expression. Total RNA was extracted using TRNzol Universal Reagent (Tiangen Biotech, Beijing, China) according to the manufacturer's instructions, and then reverse transcription was performed to generate cDNA using PrimeScriptTM RT Reagent Kit with gDNA Eraser. Real-time quantitative PCR (RT-qPCR) analyses were performed using SYBR Premix Ex Taq (Takara Biotechnology Co., Ltd., Dalian, China). The *β-tubulin* gene was used as an endogenous control for normalization. Relative expression levels were calculated using the $2^{-\Delta\Delta Ct}$ method [13]. Primer sequences are provided in Table S2.

2.5. Determination of Ergosterol Content

A 100-μL spore suspension containing 1×10^6 spores/mL of either the WT, knockout mutants or complementary strains was incubated in 100 mL of CY liquid medium at 200 rpm for 3 days at 25 °C. Fresh mycelia were then collected, filtered through sterile gauze, and then washed with sterilized water three times. Approximately 200 mg of dried mycelia from each strain were treated with 3 mL of 25% alcoholic potassium hydroxide and incubated at 85 °C for 1 h. Then, 1 mL of distilled water and 3 mL of pentane were added to the mixture and vortexed for 3 min, and then kept for 10 min. The top layer was transferred to a clear tube and evaporated at room temperature in a fume hood until dry. Before analysis, all samples were dissolved in 1 mL of methanol and then filtered through a 0.22 μm filter membrane. Ergosterol concentrations were quantified using a high-performance liquid chromatography (HPLC) system (Waters, Milford, MA, USA) and detected at 282 nm [14].

2.6. Colony Diameter and Colony Morphology

A 2 μL spore suspension containing 1×10^6 spores/mL of either the WT, knockout mutants or complementary strains was cultured on PDA medium for 7 days at 25 °C. Colony diameters were measured by the crossover method, and the colony morphology was recorded by photography [15].

2.7. Conidiophore Development and Spore Morphology

A 50 μL spore suspension containing 1×10^6 spores/mL of either the WT, knockout mutants or complementary strains was spread evenly on PDA plates and then a sterilized coverslip was inserted into the plates at an angle of approximately 45 degrees. Hyphae were allowed to grow along the junction of the coverslip and the medium to adhere to the coverslip. After incubation at 25 °C for 1.5 days, the conidiophore structure was observed under a microscope (Olympus Corporation, Tokyo, Japan) [16].

A 2 μL spore suspension containing 1×10^6 spores/mL of either the WT, knockout mutants or complementary strains was cultured on a PDA medium at 25 °C for 5 days, and then the spores were collected with sterile water and filtered through four layers of gauze to obtain a spore suspension. The spore suspension was centrifuged at 8000 rpm for 10 min, and then the supernatant was discarded. The spores were fixed in glutaraldehyde solution for 24 h, and then washed three times with PBS buffer. The spores were then dehydrated successively with different concentration gradients of aqueous ethanol (50%, 70%, 80% and

100%) for 15–20 min with each one. The samples were lightly adhered to a conductive adhesive, and then dried in a vacuum. After spraying with gold, the spore morphology was observed using a scanning electron microscope (JSM-5600LV, Tokyo, Japan) [17].

2.8. Determination of Sporulation and Spore Germination Rate

A 2 μL spore suspension containing 1×10^6 spores/mL of either the WT, knockout mutants or complementary strains was inoculated onto PDA plates and incubated at 25 °C for 7 days. Spores were obtained by adding 10 mL of sterile water to each plate. Sporulation was counted using a hemocytometer r [18].

A 10 μL spore suspension containing 1×10^6 spores/mL of either the WT, knockout mutants or complementary strains was inoculated onto PDA plates and incubated at 25 °C for 8 h. The spore germination rate of each strain was observed under a microscope (Olympus Corporation, Tokyo, Japan) [19].

2.9. Exogenous Stress Susceptibility Test

A 20 μL spore suspension containing 1×10^6 spores/mL of either the WT, knockout mutants or complementary strains was grown on the PDA plates supplemented with 1 M NaCl, 25 mg/mL Congo red (CR) or 2 mM H_2O_2. After incubation at 25 °C for 7 days, colony diameters were measured by the crossover method [20].

2.10. Pathogenicity Test

Apple fruits were soaked in a 1% sodium hypochlorite solution for 3 min and then dried at room temperature. After surface sterilization with alcohol, three wounds (1 mm in width, 2 mm in depth) were made on the equator of each apple fruit with a sterile nail. Subsequently, a 10 μL spore suspension containing 1×10^6 spores/mL of either the WT, knockout mutants or complementary strains was then inoculated into each wound. The inoculated fruits were placed in polyethylene bags and then stored at room temperature (22 ± 2 °C, RH 80–90%). After 7 days, the lesion diameter of the fruit was measured, and the lesion area was calculated according to the lesion diameter. Three replicates of each cultivar were performed, with six fruits inoculated per replicate [21].

2.11. Statistical Analysis

All the experiments were repeated at least three times. Excel 2020 was used to calculate means and standard errors for all data. OriginPro 2023 software (Northampton, MA, USA) was used for graphing. SPSS 26.0 software (SPSS, Inc., Chicago, IL, USA) was used to analyze the difference significance ($p < 0.05$).

3. Results

3.1. Sequence Alignment and Phylogenetic Analysis of Erg4A, Erg4B and Erg4C

The sequence alignment results showed that the amino acid sequence identity between Erg4A and Erg4B, Erg4A and Erg4C, and Erg4B and Erg4C was 21.9%, 61.7% and 19.9%, respectively (Figure 1A). The higher sequence identity of Erg4A and Erg4C proteins indicated that these two proteins may have similar biological functions in *P. expansum*.

The phylogenetic tree results showed that Erg4A, Erg4B and Erg4C proteins in *P. expansum* were separated into two branches. Among them, the Erg4A and Erg4C proteins were closely clustered, while the Erg4B protein and the Erg4 protein in *P. italicum* were located on the same branch, with the sequence identity of 97.4% (Figure 1B). These results showed that the Erg4A protein was closely related to the Erg4C protein, while the Erg4B protein was closely related to the Erg4 protein in *P. italicum*. In addition, the members of the *erg4* gene family were identified as 7 and 10 motifs, respectively. Among them, the Erg4A and Erg4C proteins contained 10 identical motifs, and the positional distribution of these motifs was uniform. The Erg4B protein contained 7 motifs (Figure 1C). These results indicated that Erg4A and Erg4C proteins may have similar functions in *P. expansum* due to their similar structure.

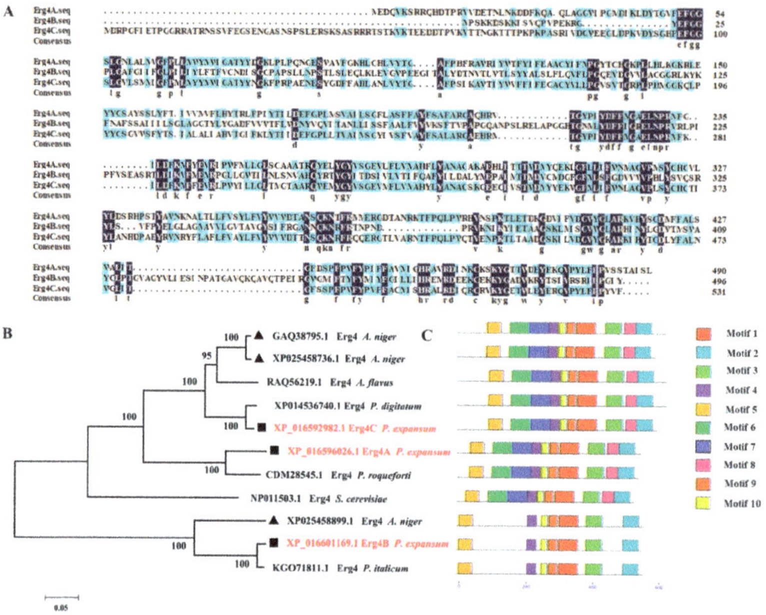

Figure 1. Bioinformatic analysis of Erg4A, Erg4B and Erg4C. (**A**) Alignment of amino acid sequences of Erg4A, Erg4B and Erg4C from selected fungi, including *A. niger*, *A. flavus*, *P. digitatum*, *P. expansum*, *P. roqueforti*, *P. italicum*. (**B**) The phylogenetic tree (neighbor-joining tree) was created using MEGA 7 software.(**C**) The conserved motif prediction of Erg4A, Erg4B and Erg4C by MEME program.

3.2. Transcript Levels of *erg4A*, *erg4B* and *erg4C* in WT and Mutants

In the WT strain, the expression levels of *erg4A*, *erg4B* and *erg4C* were significantly different, with the expression level of *erg4B* being the highest, followed by *erg4C*, and the lowest expression level was found in *erg4A* deletion strain. Compared to the expression of *erg4A* in the WT strain, the expression levels of *erg4B* and *erg4C* were 57.9-fold and 13.1-fold higher, respectively (Figure 2A). In the Δ*erg4A* mutant strain, the expression level of *erg4C* was 15.1-fold higher than that of *erg4B* (Figure 2B). In the Δ*erg4B* mutant strain, the expression levels of *erg4A* and *erg4C* were upregulated by 35.7% and 31.7%, respectively, compared to the WT. There was no significant difference between the expression levels of *erg4A* and *erg4C* in the Δ*erg4B* mutant strain (Figure 2C). In the Δ*erg4C* mutant strain, the expression level of *erg4A* was higher than that of *erg4B*, which increased by 49.5-fold (Figure 2D). These results indicate that *erg4A*, *erg4B* and *erg4C* had functional redundancy in *P. expansum*.

3.3. Effect of *erg4A*, *erg4B* and *erg4C* Deletion on Ergosterol Production

Compared with the WT strain, the ergosterol production in Δ*erg4A*, Δ*erg4B* and Δ*erg4C* was reduced by 40.6%, 48% and 27.9%, respectively, on day 7 of incubation, (Figure 3). The results indicated that the deletion of *erg4A*, *erg4B* or *erg4C* inhibited the ergosterol synthesis in *P. expansum*, with the *erg4B* knock-out having the most pronounced effect among them.

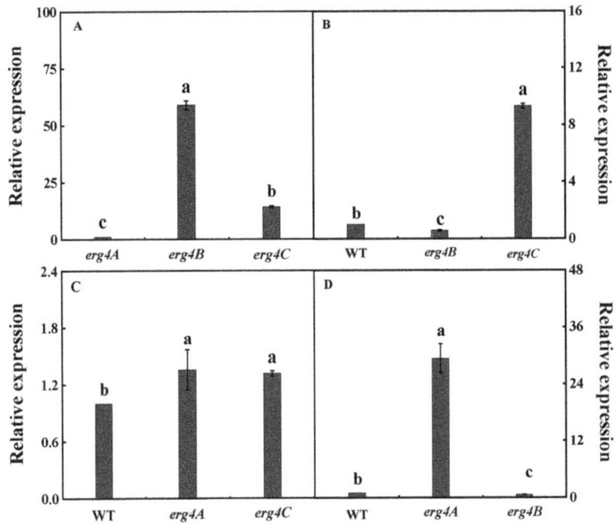

Figure 2. Transcript levels of *erg4A*, *erg4B* and *erg4C* in WT (**A**), and Δ*erg4A* (**B**), Δ*erg4B* (**C**) and Δ*erg4C* (**D**) mutants, respectively. Bars are the standard errors of the means. Different letters in the columns indicate significant differences ($p < 0.05$).

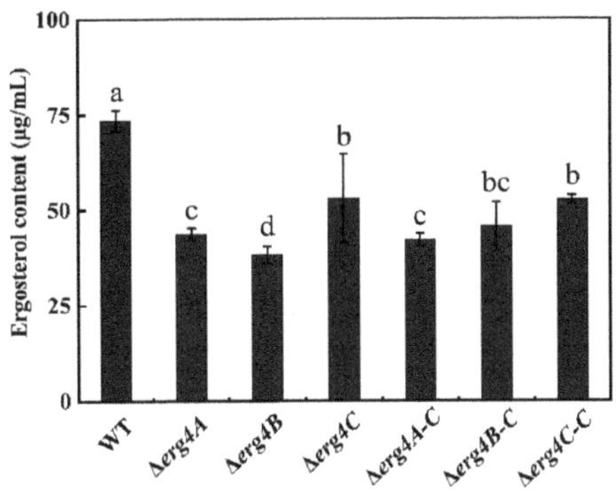

Figure 3. Ergosterol content of WT, *erg4* mutants and complementary strains cultured in CY liquid medium for 3 days. Bars are the standard errors of the means. Different letters in the columns indicate significant differences ($p < 0.05$).

3.4. Effect of erg4A, erg4B *and* erg4C *Deletion on Colony Morphology and Diameter*

The colony morphology and colony diameter of Δ*erg4A*, Δ*erg4B* and Δ*erg4C* strains were not significantly different from the WT strain. The spores of the WT strain were green, whereas the color of spores of the Δ*erg4A*, Δ*erg4B* and Δ*erg4C* strains were lighter in color. The Δ*erg4A-C*, Δ*erg4B-C* and Δ*erg4C-C* strains were similar in color to the WT strain (Figure 4A,B). These results indicated that the deletion of *erg4A*, *erg4B* or *erg4C* resulted in a lighter spore color of *P. expansum* but had no apparent effect on colony morphology and colony diameter.

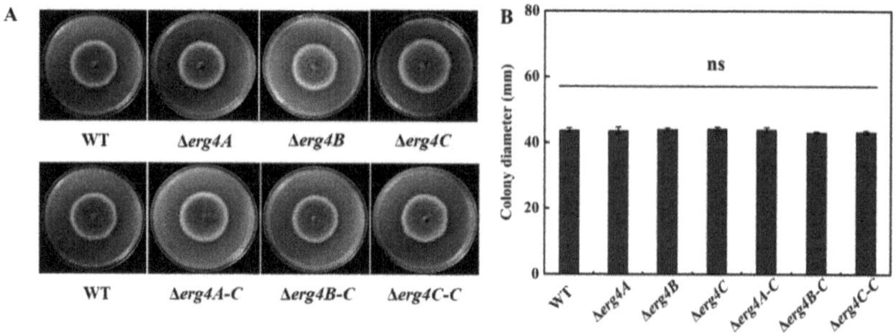

Figure 4. Colony morphology (**A**) and diameter (**B**) of WT, *erg4* mutants and complementary strains grown on PDA medium for 7 days. Bars are the standard errors of the means. ns indicates no significant difference ($p < 0.05$).

3.5. *Effect of* erg4A, erg4B *and* erg4C *Deletion on Sporulation and Spore Germination Rate*

Compared with the WT strain, the sporulation of Δ*erg4A*, Δ*erg4B* and Δ*erg4C* mutant strains was reduced by 40%, 44.7% and 47.1%, respectively, on day 7 of incubation. Sporulation of the Δ*erg4A-C* and Δ*erg4B-C* strains were recovered to some extent, but it was still lower than that of the WT strain. Sporulation was almost recovered in the Δ*erg4C-C* strain (Figure 5A). The spore germination rates of the Δ*erg4A*, Δ*erg4B* and Δ*erg4C* mutants and the corresponding complementary strains were not significantly different from the WT strain (Figure 5B). These results indicated that the deletion of *erg4A*, *erg4B* and *erg4C* significantly inhibited the sporulation of *P. expansum* but had no significant effect on spore germination.

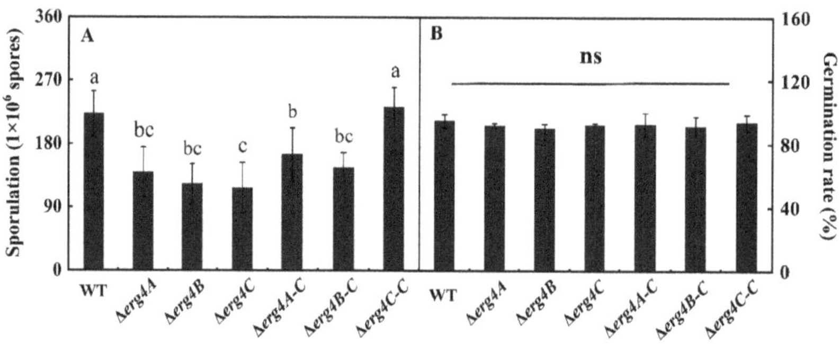

Figure 5. Sporulation (**A**) and spore germination rate (**B**) of WT, *erg4* mutants and complementary strains on PDA medium for 7 d and 8 h, respectively. Bars are the standard errors of the means. Different letters in the columns indicate significant differences ($p < 0.05$). ns indicates no significant difference.

3.6. *Effect of* erg4A, erg4B *and* erg4C *Deletion on Conidiophore Development and Spore Morphology*

The apical part of the sporangiophore of the WT strain expanded continuously and branched several times, producing several rounds of symmetrical or asymmetrical pedicels that produced clusters of greenish conidia at the tip. Compared with the WT strain, the sporulation structure of Δ*erg4A*, Δ*erg4B* and Δ*erg4C* strains was not significantly different, and the hyphae were able to form normal conidial heads and produce a large number of conidia (Figure 6A). The conidia of the WT strain were clustered, spherical or flattened. There was no significant difference between the conidia of Δ*erg4A* and the WT strains. However, the conidia of the Δ*erg4B* mutant showed obvious shrinkage, desiccation and water loss on the surface compared to the WT strain. A few of the Δ*erg4C* mutant conidia showed wrinkling compared to the WT strain (Figure 6B). These results indicate that among the three genes, the *erg4B* gene contributes most to the maintenance of the surface structure of *P. expansum*, followed by *erg4C*.

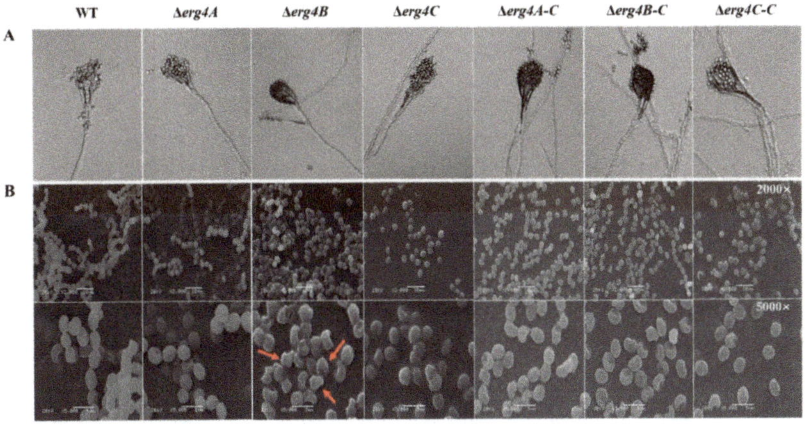

Figure 6. Conidiophore (**A**) and spore morphology (**B**) of WT, *erg4* mutants and complementary strains on the PDA medium for 1.5 days and 5 days, respectively. Red arrow indicates spore morphological shrinkage.

3.7. Effect of erg4A, erg4B and erg4C Deletion on Osmotic Stress, Cell Wall Integrity and Oxidative Stress

NaCl is an exogenous osmotic pressure reagent, CR is an inhibitor of cell wall synthesis, and H_2O_2 is an oxidative stress pressure reagent. On the NaCl medium, there was no significant difference showed between the WT and the three knockout strains (Figure 7A). On the medium containing CR, the colony diameter of the Δ*erg4A* strain was not different from that of the WT strain, whereas the colony diameters of the Δ*erg4B* and Δ*erg4C* strains were smaller than that of the WT strain (Figure 7B). On the medium containing H_2O_2, no significant difference in colony diameter was found between the WT and Δ*erg4A* mutant, while the deletion of *erg4B* and *erg4C* reduced the colony diameter of *P. expansum* (Figure 7C). These results suggest that both *erg4B* and *erg4C* are involved in the responses of *P. expansum* to cell wall integrity and oxidative stress.

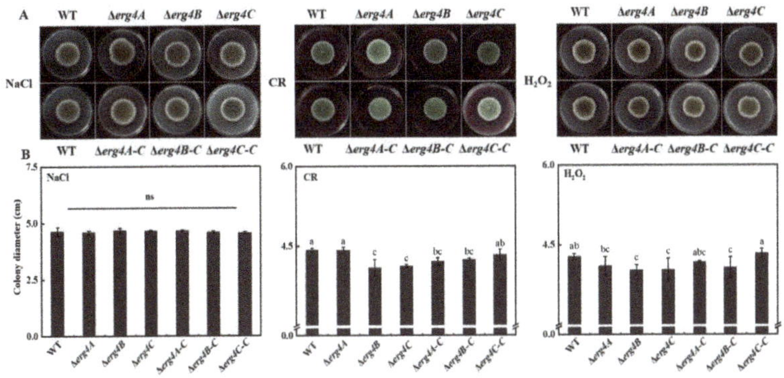

Figure 7. Colony morphology (**A**) and colony diameter (**B**) of WT, *erg4* mutants and complementary strains after 7 days of stress treatments. Bars are the standard errors of the means. ns indicates no significant difference. Different letters in the columns indicate significant differences ($p < 0.05$).

3.8. Effect of erg4A, erg4B and erg4C Deletion on Pathogenicity on Apple Fruit

Compared to the WT strain, the deletion of *erg4B* and *erg4C* slightly reduced the lesion area of Golden Delicious and Fuji fruit, but there was no significant difference between them (Figure 8). These results indicated that the deletion of *erg4A*, *erg4B* or *erg4C* had no significant effect on the pathogenicity of *P. expansum* on the apple fruits.

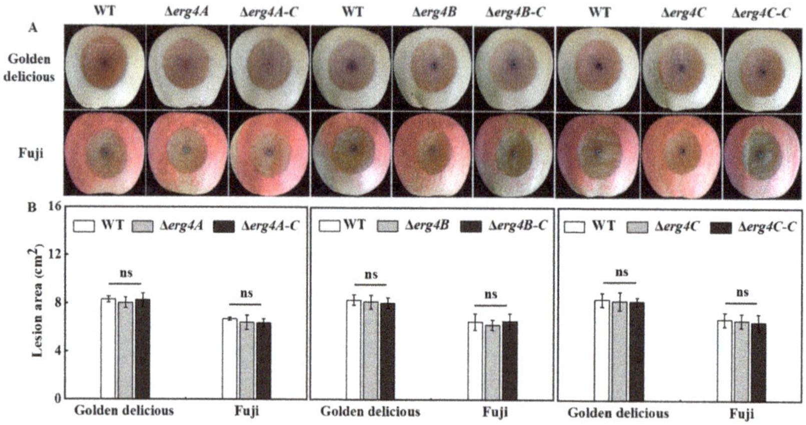

Figure 8. Disease symptoms (**A**) and lesion area (**B**) of WT, *erg4* mutants and complementary strains on two apple cultivars inoculated for 7 days. Bars are the standard errors of the means. ns indicates no significant difference.

4. Discussion

In this study, a higher amino acid sequence identity was found between Erg4A and Erg4C proteins (Figure 1A). Erg4A and Erg4C proteins contained the same number of motifs (Figure 1C), indicating that Erg4A protein is closely related to Erg4C protein. Therefore, we suggest that they may play similar roles in regulating ergosterol biosynthesis in *P. expansum*. Deletion of *erg4A* induced *erg4C* expression, deletion of *erg4B* induced *erg4A* and *erg4C* expression, and deletion of *erg4C* induced *erg4A* expression (Figure 2). These results suggest that a single *erg4* deletion is compensated for by other *erg4* genes, which is similar to the results in *A. fumigatus* [9]. Therefore, we propose that *erg4A*, *erg4B* and *erg4C* have redundant roles in ergosterol biosynthesis in *P. expansum*.

The biosynthesis of ergosterol is catalyzed by a series of enzymes encoded by the *erg* gene family, using acetyl-CoA as substrate, which is a highly conserved process in fungi. [22]. In the present study, the deletion of *erg4A*, *erg4B* or *erg4C* reduced ergosterol production in *P. expansum*. Among the three *erg4* genes, *erg4B* was more important for ergosterol biosynthesis in *P. expansum*. The result was similar to that of *F. graminearum* [8]. Since *erg4A*, *erg4B* and *erg4C* are key genes involved in the final step of ergosterol synthesis [23], deletion of all three genes could inhibit ergosterol synthesis. The expression of *erg4B* in the WT strain was higher than that in the Δ*erg4A*, Δ*erg4B* and Δ*erg4C* strains (Figure 2A), suggesting that *erg4B* plays a major function in the *erg4* gene family in *P. expansum*. Therefore, we propose that *erg4A*, *erg4B* and *erg4C* are all involved in ergosterol synthesis in *P. expansum*, and that these three genes independently regulate ergosterol synthesis, of which *erg4B* has the most obvious effect.

Conidia are the main mode of asexual reproduction in filamentous fungi, and their production is tightly regulated in cells [24]. In *A. nidulans*, the *BrlA*, *AbaA*, and *WetA* genes regulate the central pathway of conidiation [25]. The *BrlA* gene modulates the transcriptional regulation of genes involved in early sporulation development; AbaA activated by BrlA is required for the differentiation process in the middle stage of sporulation, and WetA activated by AbaA is essential for spore formation and maturation [26]. It has been shown that the G protein signaling pathway plays an important role in the regulation of sporulation and hyphal growth of *A. fumigatus*. Deletion of the Gβ-like protein gene *CpcB* resulted in delayed hyphal growth and reduced sporulation of *A. fumigatus* [27]. In this study, deletions of *erg4A*, *erg4B* and *erg4C* reduced sporulation of *P. expansum* (Figure 5A), which is consistent with the results of [8], who found that the *erg4* deletion mutant formed smaller and shorter conidia with less septation and showed a reduced conidiation. In addition, we also found that the sporulation of Δ*erg4A-C* and Δ*erg4B-C* strains was lower than that of the WT (Figure 5A). We constructed the complementary strains by transferring the target gene, together with its own promoter and terminator, into the corresponding mutant strains by *Agrobacterium*-mediated transformation. In this case, the target gene was integrated into the genome of the mutant strain as a random insertion. We speculate that the gene inserted by the target gene fragment may be one that has some effect on the sporulation of *P. expansum*, resulting in the sporulation of the complementary strains not fully reverting to the WT level. Previous studies have reported that the double deletion of *erg4A* and *erg4B* completely blocked the ergosterol synthesis in *A. fumigatus* and caused severe sporulation defects, while the complementary strain completely rescued the defects, indicating that ergosterol is required for the conidiation for *A. fumigatus* [9]. Furthermore, transcriptome analysis between the WT and *erg5* knockout strains of *A. fumigatus* showed that defects in ergosterol synthesis significantly downregulated central regulatory networks (BrlA, AbaA and WetA), the genes encoding heterotrimeric G-related proteins (GpaB, RicA, GpgA, RgsA) and the MAPK (MpkC, SakA) signaling pathway [28]. However, how *erg4* regulates the expression of genes involved in these signaling pathways has not been reported yet. Based on this information, we hypothesize that *erg4A*, *erg4B* and *erg4C* deletion reduces the sporulation of *P. expansum* by decreasing ergosterol levels and down-regulating the central regulatory network as well as the heterotrimer G-related protein and MAPK signaling pathways, whereas the details of the regulatory mechanism remain to be elucidated.

The cell wall is the first barrier of filamentous fungi to resist external stresses, and it plays an important role in maintaining cell morphology [29]. CR is an inhibitor of cell wall synthesis that binds mainly to chitin and β-1, 4-glucan in the cell wall [30]. In this study, Δ*erg4B* and Δ*erg4C* colony growth was inhibited under CR stress (Figure 7B), which is consistent with the results in *A. fumigatus* [9]. In *S. cerevisiae*, *erg* deletion inhibited ergosterol synthesis and reduced ergosterol content, leading to increased chitin synthesis and abnormal cell wall distribution [31]. Reduced septations in conidia and mycelia were found in *F. graminearum erg4* knockout mutants, suggesting that *erg4* affects cell wall formation [8]. In *S. cerevisiae*, *erg4* has been shown to be involved in cell wall assembly [32]. Additionally, *erg4A* may also be involved in the cell wall synthesis process in *A. fumigatus*, as the Δ*erg4A* mutant shows significant sensitivity to CR, indicating that *erg4* plays an important role in fungal cell wall integrity [9]. Disruption of ergosterol synthesis could also disrupt cell wall structure [33]. Therefore, we speculate that *erg4B* and *erg4C* could disrupt the cell wall integrity in *P. expansum* by inhibiting ergosterol synthesis. Furthermore, Δ*erg4B* and Δ*erg4C* strains were sensitive to H_2O_2 stress (Figure 7C), which is similar to the results of Liu et al. [8] and Long et al. [9]. They found that the deletion of *erg4* in *F. graminearum* and the deletion of *erg4A* in *A. fumigatus* increased the sensitivity of the fungi to H_2O_2 stress [8,9]. Deletion of *erg4B* and *erg4C* reduced the levels of ergosterol, which is an important component of the fungal cell membrane. Therefore, we hypothesized that the deletion of *erg4B and erg4C* would reduce ergosterol levels in *P. expansum*, leading to an increased sensitivity of cell membranes to H_2O_2 stress.

Deletion of *erg4A*, *erg4B* or *erg4C* did not attenuate the pathogenicity of *P. expansum* on apple fruit (Figure 8), which is similar to the finding that the deletion of *erg4A* or/and *erg4B* did not affect the virulence of *A. fumigatus* to mice [9]. However, in contrast, the deletion of *erg4* in *F. graminearum* reduced the pathogenicity of the fungus in wheat and tomato fruits, as only one *erg4* gene is present in *F. graminearum* [8]. In contrast, *A. fumigatus* had two *erg4* genes (*erg4A* and *erg4B*) with functional redundancy [9]. Therefore, we considered that the deletion of *erg4A*, *erg4B* or *erg4C* did not affect the pathogenicity of *P. expansum*, which was related to the functional redundancy of the family genes. Whether the double or triple knockout of *erg4A*, *erg4B* and/or *erg4C* affects the pathogenicity of *P. expansum* requires further verification.

A global transcriptome analysis is of great importance to elucidate the function of *erg4A*, *erg4B* and *erg4C* in *P. expansum*. In addition, we will construct double knockout mutants to further elucidate the mechanism of action of these three *P. expansum* genes in regulating ergosterol biosynthesis in our future study.

5. Conclusions

Deletion of *erg4A*, *erg4B* or *erg4C* inhibited the ergosterol levels in *P. expansum*, whereas the *erg4B gene* had a greater effect on ergosterol biosynthesis. Deletion of *erg4A*, *erg4B* or *erg4C* also reduced the sporulation of *P. expansum*. Furthermore, *erg4B* and *erg4C* were involved in the spore morphology, cell wall integrity and oxidative stress response of *P. expansum*. However, deletion of *erg4A*, *erg4B* or *erg4C* had no significant effect on colony growth, osmotic stress of *P. expansum* or pathogenicity on apple fruit.

Supplementary Materials: The following supporting information can be downloaded at: https://www.mdpi.com/article/10.3390/jof9050568/s1, Figure S1: Homologous recombination strategy; Figure S2: PCR results of gene knockout mutants of erg4s in P. expansum; Table S1: The primers used for the construction and identification of gene deletion mutants and complementation strains; Table S2: Primers sequences used for RT-qPCR.

Author Contributions: Formal analysis, writing original draft: Z.H. Data curation: X.Z. Software: B.W. Writing—review and editing: Y.Z. and D.G. Project administration, supervision: D.P. and E.S. Supervision, validation, project management: H.X. Conceptualization, project management, fundraising, supervision: Y.B. All authors have read and agreed to the published version of the manuscript.

Funding: This work was supported by the Israel-China Project of National Natural Science Foundation of China (grant no. 31861143046).

Institutional Review Board Statement: Not applicable.

Informed Consent Statement: Not applicable.

Data Availability Statement: Not applicable.

Conflicts of Interest: The authors declare no conflict of interest.

References

1. Wang, K.L.; Ngea, G.L.N.; Godana, E.A.; Shi, Y.; Lanhuang, B.; Zhang, X.Y.; Zhao, L.N.; Yang, Q.Y.; Wang, S.Y.; Zhang, H.Y. Recent advances in *Penicillium expansum* infection mechanisms and current methods in controlling *P. expansum* in postharvest apples. *Crit. Rev. Food Sci. Nutr.* **2021**, *20*, 2598–2611. [CrossRef] [PubMed]
2. Jordá, T.; Puig, S. Regulation of ergosterol biosynthesis in *Saccharomyces cerevisiae*. *Genes* **2020**, *11*, 795. [CrossRef]
3. Liu, J.F.; Xia, J.J.; Nie, K.L.; Wang, F.; Deng, L. Outline of the biosynthesis and regulation of ergosterol in yeast. *World J. Microbiol. Biotechnol.* **2019**, *35*, 98. [CrossRef] [PubMed]
4. Hu, Z.H.; He, B.; Ma, L.; Sun, Y.L.; Niu, Y.L.; Zeng, B. Recent advances in ergosterol biosynthesis and regulation mechanisms in *Saccharomyces cerevisiae*. *Indian J. Microbiol.* **2017**, *57*, 270–277. [CrossRef] [PubMed]
5. Zweytick, D.; Hrastnik, C.; Kohlwein, S.D.; Daum, G. Biochemical characterization and subcellular localization of the sterol C-24(28) reductase, erg4p, from the yeast *saccharomyces cerevisiae*. *FEBS Lett.* **2000**, *470*, 83–87. [CrossRef] [PubMed]
6. He, X.P.; Zhang, B.R.; Tan, H.R. Overexpression of a sterol C-24(28) reductase increases ergosterol production in *Saccharomyces cerevisiae*. *Biotechnol. Lett.* **2003**, *25*, 773–778. [CrossRef]
7. Venegas, M.; Barahona, S.; González, A.M.; Sepúlveda, D.; Zúñiga, G.E.; Baeza, M.; Cifuentes, V.; Alcaíno, J. Phenotypic analysis of mutants of ergosterol biosynthesis genes (ERG3 and ERG4) in the red yeast *Xanthophyllomyces dendrorhous*. *Front. Microbiol.* **2020**, *11*, 1312. [CrossRef]
8. Liu, X.; Jiang, J.H.; Yin, Y.N.; Ma, Z.H. Involvement of FgERG4 in ergosterol biosynthesis, vegetative differentiation and virulence in *Fusarium graminearum*. *Mol. Plant Pathol.* **2013**, *14*, 71–83. [CrossRef] [PubMed]
9. Long, N.B.; Xu, X.L.; Zeng, Q.Q.; Sang, H.; Lu, L. Erg4A and Erg4B are required for conidiation and azole resistance via regulation of ergosterol biosynthesis in *Aspergillus fumigatus*. *Appl. Environ. Microbiol.* **2017**, *83*, e02924-16. [CrossRef]
10. Han, Z.H.; Zong, Y.Y.; Zhang, X.M.; Wang, B.; Prusky, D.; Bi, Y. Bioinformatic, subcellular localization and expression analysis of *erg4* in *Penicillium expansum*. *Biotechnol. Bull.* **2021**, *37*, 60–70. (In Chinese) [CrossRef]
11. Li, B.Q.; Zong, Y.Y.; Du, Z.L.; Chen, Y.; Zhang, Z.Q.; Qin, G.Z.; Zhao, W.M.; Tian, S.P. Genomic characterization reveals insights into patulin biosynthesis and pathogenicity in *Penicillium* species. *Mol. Plant Microbe Interact.* **2015**, *28*, 635–647. [CrossRef] [PubMed]

12. Chen, Y.; Li, B.Q.; Xu, X.D.; Zhang, Z.Q.; Tian, S.P. The pH-responsive PacC transcription factor plays pivotal roles in virulence and patulin biosynthesis in *Penicillium expansum*. *Environ. Microbiol.* **2018**, *20*, 4063–4078. [CrossRef] [PubMed]
13. Livak, K.J.; Schmittgen, T.D. Analysis of relative gene expression data using real-time quantitative PCR and the $2^{-\Delta\Delta CT}$ method. *Methods* **2001**, *25*, 402–408. [CrossRef] [PubMed]
14. Ouyang, Q.; Liu, Y.; Oketch, O.R.; Zhang, M.; Shao, X.; Tao, N. Citronellal exerts its antifungal activity by targeting ergosterol biosynthesis in *Penicillium digitatum*. *J. Fungi* **2021**, *7*, 432. [CrossRef] [PubMed]
15. Zong, Y.Y.; Li, B.Q.; Tian, S.P. Effects of carbon, nitrogen and ambient pH on patulin production and related gene expression in *Penicillium expansum*. *Int. J. Food Microbiol.* **2015**, *206*, 102–108. [CrossRef]
16. Prakash, P.Y.; Bhargava, K.A. A modified micro chamber agar spot slide culture technique for microscopic examination of filamentous fungi. *J. Microbiol. Methods* **2016**, *123*, 126–129. [CrossRef]
17. Yahyazadeh, M.; Omidbaigi, R.; Zare, R.; Taheri, H. Effect of some essentialoils on mycelial growth of *Penicillium digitatum* Sacc. *World J. Microbiol Biotechnol.* **2008**, *24*, 1445–1450. [CrossRef]
18. Xu, X.D.; Chen, Y.; Li, B.Q.; Tian, S.P. Arginine methyltransferase PeRmtC regulates development and pathogenicity of *Penicillium expansum* via mediating key genes in conidiation and secondary metabolism. *J. Fungi* **2021**, *7*, 807. [CrossRef]
19. Ma, D.Y.; Ji, D.C.; Liu, J.L.; Xu, Y.; Chen, T.; Tian, S.P. Efficacy of methyl thujate in inhibiting *Penicillium expansum* growth and possible mechanism involved. *Postharvest Biol. Technol.* **2020**, *3*, 111070. [CrossRef]
20. Zhang, X.M.; Zong, Y.Y.; Gong, D.; Yu, L.R.; Sionov, E.; Bi, Y.; Prusky, D. NADPH oxidase regulates the growth and pathogenicity of *Penicillium expansum*. *Front. Plant Sci.* **2021**, *12*, 696210. [CrossRef]
21. Chen, Y.; Li, B.Q.; Zhang, Z.Q.; Tian, S.P. Pathogenicity assay of *Penicillium expansum* on apple fruits. *Bio-Protocol.* **2017**, *7*, e2264. [CrossRef] [PubMed]
22. Alcazar-Fuoli, L.; Mellado, E. Ergosterol biosynthesis in *Aspergillus fumigatus*: Its relevance as an antifungal target and role in antifungal drug resistance. *Front. Microbiol.* **2012**, *3*, 439. [CrossRef] [PubMed]
23. Dhingra, S.; Cramer, R.A. Regulation of sterol biosynthesis in the human fungal pathogen *Aspergillus fumigatus*: Opportunities for therapeutic development. *Front. Microbiol.* **2017**, *8*, 92. [CrossRef]
24. Luciano-Rosario, D.; Keller, N.P.; Jurick, W.M. *Penicillium expansum*: Biology, omics, and management tools for a global postharvest pathogen causing blue mould of pome fruit. *Mol. Plant Pathol.* **2020**, *21*, 1391–1404. [CrossRef] [PubMed]
25. Wu, M.Y.; Mead, M.E.; Lee, M.K.; Ostrem Loss, E.M.; Kim, S.C.; Rokas, A.; Yu, J.H. Systematic dissection of the evolutionarily conserved wetA developmental regulator across a genus of filamentous fungi. *mBio.* **2018**, *70*, 317–343. [CrossRef]
26. Chen, J.F.; Liu, Y.; Tang, G.R.; Jin, D.; Chen, X.; Pei, Y.; Fan, Y.H. The secondary metabolite regulator, BbSmr1, is a central regulator of conidiation via the BrlA-AbaA-WetA pathway in *Beauveria bassiana*. *Environ. Microbiol.* **2021**, *23*, 810–825. [CrossRef] [PubMed]
27. Cai, Z.D.; Chai, Y.F.; Zhang, C.Y.; Qiao, W.R.; Sang, H.; Lu, L. The Gβ-like protein CpcB is required for hyphal growth, conidiophore morphology and pathogenicity in *Aspergillus fumigatus*. *Fungal Genet. Biol.* **2015**, *81*, 120–131. [CrossRef] [PubMed]
28. Long, N.B.; Zhong, G. The C-22 sterol desaturase Erg5 is responsible for ergosterol biosynthesis and conidiation in *Aspergillus fumigatus*. *J. Microbiol.* **2022**, *60*, 620–626. [CrossRef]
29. Yin, Z.; Tang, W.; Wang, J.; Liu, X.; Yang, L.; Gao, C.; Zhang, J.; Zhang, H.; Zheng, X.; Wang, P.; et al. Phosphodiesterase MoPdeH targets MoMck1 of the conserved mitogen-activated protein (MAP) kinase signalling pathway to regulate cell wall integrity in rice blast fungus *Magnaporthe oryzae*. *Mol. Plant Pathol.* **2016**, *17*, 654–668. [CrossRef]
30. Ram, A.F.; Klis, F.M. Identification of fungal cell wall mutants using susceptibility assays based on calcofluor white and congo red. *Nat. Protoc.* **2006**, *1*, 2253–2256. [CrossRef]
31. Marisco, G.; Saito, S.T.; Ganda, I.S.; Brendel, M.; Pungartnik, C. Low ergosterol content in yeast adh1 mutant enhances chitin maldistribution and sensitivity to paraquat-induced oxidative stress. *Yeast.* **2011**, *28*, 363–373. [CrossRef] [PubMed]
32. Tiedje, C.; Holland, D.G.; Just, U.; Höfken, T. Proteins involved in sterol synthesis interact with Ste20 and regulate cell polarity. *J. Cell Sci.* **2007**, *120*, 3613–3624. [CrossRef] [PubMed]
33. Lesage, G.; Bussey, H. Cell wall assembly in *Saccharomyces cerevisiae*. *Microbiol. Mol. Biol. Rev.* **2016**, *70*, 317–343. [CrossRef] [PubMed]

Journal of
Fungi

MDPI

Article

Ena Proteins Respond to PacC-Mediated pH Signaling Pathway and Play a Crucial Role in Patulin Biosynthesis

Ruiling Zhuo [1,2,3], Yong Chen [1,2], Mengyang Xing [1,2,3], Zhanquan Zhang [1,2], Shiping Tian [1,2,3] and Boqiang Li [1,2,4,*]

1 Key Laboratory of Plant Resources, Institute of Botany, Chinese Academy of Sciences, Beijing 100093, China
2 China National Botanical Garden, Beijing 100093, China
3 University of Chinese Academy of Sciences, Beijing 100049, China
4 Key Laboratory of Post-Harvest Handling of Fruits, Ministry of Agriculture, Beijing 100093, China
* Correspondence: bqli@ibcas.ac.cn

Abstract: *Penicillium expansum* is a main producer of patulin that causes severe postharvest decay and food safety issues in the fruit industry. Development, pathogenicity, and patulin production of *P. expansum* are strongly influenced by the PacC-pH signaling pathway. Global transcription factor PacC regulates various fungal biological processes through a complicated molecular network. In the present study, three Ena family genes (*PeEnas*), *PeEnaA*, *PeEnaB*, and *PeEnaC*, as important downstream targets of PePacC, were identified in *P. expansum*. Deletion of *PeEnaA*, *PeEnaB*, and *PeEnaC* showed little effect on mycelial growth under alkaline or high salinity conditions, but double and triple deletion of these genes impaired the virulence of *P. expansum* on apple fruit. Notably, patulin biosynthesis of *P. expansum* was distinctly inhibited in the deletion mutants of *PeEnas*. *PeEnas* regulated expressions of the patulin gene cluster, *AP1*, *CreA*, *Sge1*, and *Hog1* at the transcriptional level and played roles in maintaining membrane potential. Overexpression of *PeEnaC* in Δ*PePacC* restored the patulin production defect of Δ*PePacC*. Our results indicated that, as downstream targets of PePacC, the PeEna family proteins play a crucial role in patulin biosynthesis in *P. expansum*.

Keywords: Ena family; fruit; mycotoxin; *Penicillium expansum*; blue mold

1. Introduction

Penicillium expansum, a saprophytic phytopathogen, infects numerous fruit and vegetable hosts and causes severe blue mold rot. It also contaminates hosts with mycotoxin patulin, which causes food safety issues [1]. Understanding the regulatory mechanisms of pathogenicity and patulin biosynthesis will lay the foundation for the management of blue mold [2].

As one of the most important environmental factors, ambient pH significantly affects pathogen development and pathogenicity [3]. *P. expansum* can survive over a broad range of pH, with pH 4.0–5.0 being a conducive condition for spore germination and mycelial growth [4]. To sense and respond to ambient pH, fungi evolve a fungal-specific Rim/Pal signaling pathway to modulate gene expression through the activation of a key transcription factor PacC [5]. In *Aspergillus nidulans*, PacC was initially identified in three forms: $PacC^{72}$, $PacC^{53}$, and $PacC^{27}$ [6]. Among them, $PacC^{27}$ is considered the active form and is produced by the two proteolytic cleavages of the entire length form, $PacC^{72}$. It contains three Cys_2His_2 zinc finger structures, and the core binding motif is 5′-GCCARG-3′ [7]. PacC regulates a variety of biological processes as a global transcription factor in fungi [8]. In *P. expansum*, PePacC was required for virulence, patulin production, conidiation, and vegetative growth [9]. PacC activates multiple alkaline-expressed genes and represses acid-expressed genes involved in transport, secondary metabolism, and cell wall degradation under neutral to alkaline conditions in *P. expansum* [9–11].

Ambient pH directly affects the charge of inorganic or organic acid ions. To maintain optimal cation homeostasis, cells employ diverse ion transporters. Upregulation of transporters may restore cation homeostasis in fungi at varying pH conditions [12–14]. The fungal cation pump is a large superfamily of plasma membrane P-type ATPases divided into five families (Types I-V) [15,16]. The Ena family proteins (Enas), corresponding to typical P-type ATPases of Group IID, couple ATP hydrolysis to transport cations against electrochemical gradients. The Ena ATPases have been recognized to be present in bryophytes, protozoa, and fungi but not in flowering plants [17,18]. Ena1/2 was originally named in *Saccharomyces cerevisiae* for Latin exitus natru (sodium exit) and was shown to mediate cellular tolerability to Na^+, Li^+, and alkaline pH [19]. In *S. cerevisiae*, Ena1 plays the dominant role in Na^+ export [18]. In *A. nidulans*, three Ena orthologues (EnaA, EnaB, and EnaC) were identified, of which EnaA and EnaB were necessary in response to ions and alkaline pH, and EnaC was a putative pseudogene [20]. Enas were regulated by PacC/Rim101 pathway, Crz1 pathway, nutrient availability, and HOG pathway at transcriptional or post-transcriptional levels to adapt to high pH and salt stress [13,17,20]. PacC/Rim101 pathway was significantly involved in regulating the gene expression of *Enas* in response to alkaline pH stress in *S. cerevisiae* and *A. nidulans* [21].

Moreover, Enas significantly influenced the virulence of some fungal pathogens. The absence of Ena1 decreased virulence in pathogenic fungi, including mammalian pathogens *Cryptococcus neoformans* [22], insect pathogens *Beauveria bassiana* [23], and *Metarhizium acridum* [14]. However, functional studies of Enas in phytopathogens have rarely been reported. The aim of the present study is to investigate the functions of Enas in development, pathogenicity, and mycotoxin production in *P. expansum*. Regulation of transcriptional factor PePacC on *Enas* genes was also explored. Three orthologues (*EnaA*, *EnaB*, and *EnaC*) of the PeEna family (PeEnas) were identified in *P. expansum*. *PeEnas* were found in response to ambient pH and were directly regulated by the PePacC. *PeEnas* were involved in mycelial growth under alkaline or high salinity conditions and virulence on apple fruit. Particularly, the crucial role of PeEnas in patulin biosynthesis was revealed for the first time.

2. Materials and Methods

2.1. Fungal Strains and Culture Conditions

P. expansum T01 strain isolated from infected apple fruit was taken as wild-type (*WT*) throughout this study [24]. Δ*PePacC* was constructed in our previous study [9]. The strains were cultured on potato dextrose agar (PDA) medium under dark conditions for 7–10 d at 25 °C. The conidia were collected and counted using the automated cell counter (Countstar IY1200, Shanghai, China).

2.2. Phylogenetic Relationships and Conserved Domain Analysis

The amino acid sequences of EnaA (CBF71175), EnaB (CBF85251), and EnaC (CBF79858) in *A. nidulans* were used as bait for PeEnaA (PEG01338), PeEnaB (PEG09496), PeEnaC (PEG10401) in *P. expansum*. The initial protein sequences of Ena homologous from other fungi were obtained from the NCBI database (http://www.ncbi.nlm.nih.gov/ (accessed on 12 August 2021)) by BLASTP. Representation of domain organization and extension of PeEnaA, PeEnaB, and PeEnaC were based on Pfam databases (http://pfam-legacy.xfam.org/ (accessed on 12 August 2021)). Multiple protein sequences mentioned above were aligned using Clustal W. With MEGA 7, a Neighbor-Joining (NJ) tree was constructed, and 1000 bootstrap replicates were performed.

2.3. Mutant Generation and Complementation

P. expansum transformation was performed by *Agrobacterium tumefaciens*-mediated transformation method (ATMT) [24]. The hygromycin phosphotransferase gene *hph* was applied as a resistance marker for single gene deletion, the neomycin resistance gene *neo* was used to construct double gene deletion and complementary strains, and the nourseothricin resistance gene *nat* was used to construct triple gene deletion strains. Homologous sequences on both sides of the target gene (5′ flanking and 3′ flanking) were amplified from DNA in the genome and inserted into multiple cloning sites of the modified pCAMBIA1300 vector with *HPH*, *NEO*, or *NAT*, respectively. The primers used for gene replacement and mutant identification are listed in Table S1. Positive transformants of *PeEnaA*, *PeEnaB*, and *PeEnaC* were screened by PCR assay and further confirmed by Southern blotting assay for single gene deletion strains (Figure S1). Full-length *PeEnaA*, *PeEnaB*, and *PeEnaC* fragments were transformed into deletion mutant strains with pCAMBIA1300-NEO in constructing complement strains.

2.4. Chromatin Immunoprecipitation Assay

To test whether PacC directly controls *Enas* expression, chromatin immunoprecipitation (ChIP) was performed using an antibody to PacC-GFP [25,26]. Mycelia from *WT-GFP* and *ΔPePacC::PePacC-GFP* mutant strains were collected and immersed in 1% formaldehyde for 10 min under a vacuum. Genomic DNA and protein cross-linking were performed in nuclear extraction buffer. A final concentration of 0.125 M glycine was added to the reaction system for 5 min to stop the cross-linking reaction. The fixed material was collected for nuclei extraction, as described by Wang et al. (2021) [27]. Enriched nuclei were sonicated by fragmenting the nuclear membrane and cutting gDNA to an average size of 500–1000 bp. A portion of the supernatant was set aside and reversed, cross-linked as input DNA. The remaining fraction was used as immunoprecipitation (IP) by incubating the anti-GFP antibody (ab290, dilution 1:500) with pre-blocked Dynabeads™ Protein G (Invitrogen; 10003D) overnight at 4 °C, followed by incubating chromatin samples with the antibody for 4 h, and subsequently with low salt buffer, high salt buffer, lithium chloride buffer, and TE buffer to wash the magnetic beads. IP complexes were then eluted from the magnetic beads with freshly prepared elution buffer and reversed cross-linking. The samples were then separated from the magnetic beads by elution and reverse cross-linking. IP DNA was extracted by the TIAN Quick Midi Purification Kit (Tiangen; DP204). The PacC binding sites (5′-GCCARG-3′ containing elements) in the promoters of the Ena genes were analyzed with SnapGene Viewer version 6.0.2 (http://www.snapgene.com (accessed on 29 November 2021)). Regions A and B of each gene were selected as representative regions for chromatin immunoprecipitation with qPCR (ChIP-qPCR). Specific primers were designed to amplify promoter regions surrounding 5′-GCCARG-3′ containing elements of immunoprecipitated DNA. The relative enrichment of each gene was calculated with quantitative PCR determination and normalization of IP samples to input [27].

2.5. Phenotypic Analysis

Phenotypic analysis was based on Chen et al. (2018) and Xu et al. (2023) methods with minor modifications [9,25]. All strains were cultured on a minimal medium (MM) and adjusted to different pH values (pH 5, pH 8) using citrate–phosphate buffer (Table S2). The colony diameters of the strains were measured by the crossover method using Vernier calipers, and the colony morphology was recorded by photographs. The virulence assay was performed on apple fruit (*Malus domestica* cv. Fuji). Four wounds (2 mm in diameter and 5 mm in depth) were placed uniformly on the equator of each apple fruit. 5 μL of conidial suspension (2×10^5 conidia mL^{-1}) was transferred to each wound. Inoculated fruit were kept at a constant temperature of 25 °C in high humidity. Lesion diameters were measured every 2 d. To determine patulin production, 1 μL of conidial suspension (10^6 conidia mL^{-1}) of the described strain was inoculated onto a PDA medium pre-covered with cellophane sheets (square with 1 cm sides). After 24 h of incubation, the cellophane sheets with mycelia were transferred and floated on 1 mL Czapek yeast extract (CY) medium buffered at pH 3, pH 5, and pH 8 with citrate–phosphate buffer (Table S2) for 36 h on a 24-well cell culture plate at 25 °C. Mycelia were collected for RNA extraction, and filtrates were collected for patulin determination using HPLC. HPLC detection was performed according to Li et al. [2]. The mobile phase consisted of acetonitrile and water, with a flow rate of 1 mL min^{-1} (10:90, v/v), in isocratic elution mode, and the detection wavelength was 276 nm.

2.6. Reverse Transcription and Quantitative PCR Analysis

Total RNA was extracted from aspirated mycelia using TRNzol universal reagent (Tiangen; DP424). The quality of RNA was assessed by utilizing 1% agarose gel electrophoresis and staining with StarStain Red Plus Nucleic Acid Dye (GenStar, China, E110-01). Moreover, the OD260/OD280 ratio of extracted RNA was quantified between 1.8 and 2.0 using a NanoDrop N-1000 spectrophotometer (NanoDrop Technologies, Wilmington, DE, USA). Reverse transcription and quantitative PCR (RT-qPCR) were carried out as previously described [24]. The data obtained were evaluated by the ΔΔCt method, using the β-tubulin gene as an internal reference. A summary of qPCR primers is provided in Table S3. A heatmap showing expression changes was generated by TBtools-II (Toolbox for Biologists) v1.120 (https://github.com/CJ-Chen/TBtools (accessed on 2 July 2022)).

2.7. Membrane Potential Assay

To analyze the fungal membrane potential, protoplasts were first prepared. The conidial suspension of the indicated strains in CY (15 mL, 5×10^7 conidia mL^{-1}) was shaken for 20 h at 25 °C and the germinating mycelia were incubated in enzymatic digestion buffer (0.8 M $MgSO_4$, 1 % w/v Lysing Enzymes from Trichoderma (L1412, Sigma), 0.1 % w/v Snailase (S8280, Solarbio, Beijing, China)) for 2 h with gentle shaking (100 rpm) in dark. Protoplasts were collected by centrifugation for 5 min (1500× g) and then shifted to pH 5 or pH 8 conditions and continually shaken (100 rpm) for another 1 h. 2 mM bis (1,3-dibutylbarbituric acid) trimethine oxonol (DiBAC4(3), Invitrogen, Carlsbad, CA, USA) was added to the sample and incubated for 10 min at 4 °C in the dark. Fluorescence was examined using a confocal laser scanning microscope with 488 nm excitation and 509 nm emission using a confocal Zeiss 980 laser scanning microscope with Elyra7 (Zeiss, Oberkochen, Germany) [28,29]. Images and fluorescence measurements of the confocal data were captured with ZEISS ZEN 3.2 (blue edition) software (Zeiss, Oberkochen, Germany) under the same parameters.

2.8. Subcellular Localization of EnaC

Subcellular localization of PeEnaC proteins was observed as previously described [2], and FM4-64 (Thermo Fisher Scientific, Waltham, MA, USA) was used to stain the plasma membrane. Briefly, 5×10^7 conidia mL^{-1} were incubated in CY medium for 15 h. Prechilled FM4-64 (1 mg mL^{-1} master mix, 1:40 dilution) was added to the 50 μL system, and samples were incubated on ice for 15–20 min, followed by confocal imaging. 488 nm/540 nm and 516 nm/640 nm excitation/emission wavelengths were used to detect the fluorescence of GFP and FM4-64, respectively.

2.9. Statistical Analysis

The software SPSS version 20.0 (SPSS Inc., Chicago, IL, USA) was used. Differences among multiple groups of means were analyzed by one-way ANOVA followed by Duncan's multiple range test. The significance was considered when $p < 0.05$. For comparisons in gene expression of *PeEnaA*, *PeEnaB*, and *PeEnaC* between pH 3 and 8, and DNA fragments enrichment, a Student's t-test was used.

3. Results

3.1. Sequence Features of Ena ATPases in P. expansum

A phylogenetic evolutionary tree was constructed to analyze the Ena homologs of six species, including *P. expansum*, *A. nidulans*, *B. bassiana*, *M. acridum*, *Colletotrichum gloeosporioides*, and *S. cerevisiae*. Conserved domains of Ena homologs were further analyzed to denote their roles as P-type ATPases.

The PeEna family has four conserved structural domains, namely the Cation_ATPase_N domain (I) (pfam00690), the E1-E2_ATPase domain (II) (pfam00122), the haloacid dehalogenase-like hydrolase domain (III) (pfam00702), and the Cation_ATPase_C domain (IV) (pfam00689) (Figure 1A). As shown in Figure 1B, the evolutionary tree constructed based on PeEna protein and its homologs was divided into four groups, with the PeEnas in *P. expansum* being most closely related to orthologous proteins in *A. nidulans* (Figure 1B). PeEna homologous proteins were predicted to have ten transmembrane regions and be localized in the plasma membrane using the TOPCONS web server (Figure S2).

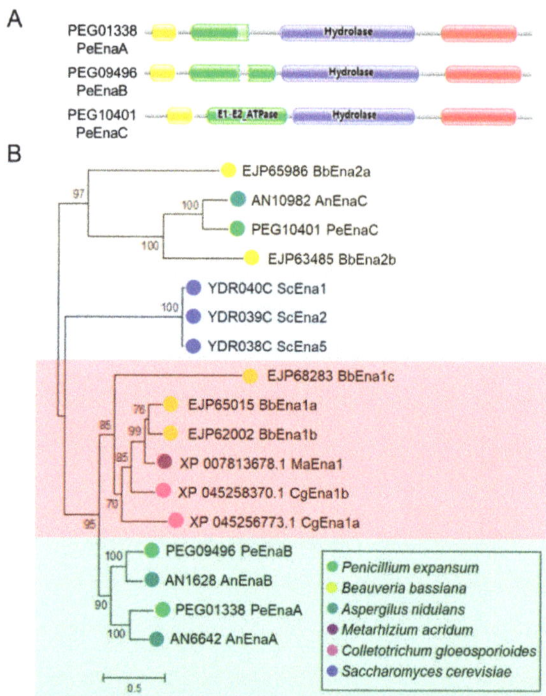

Figure 1. Functional conserved domain and phylogenetic tree analysis of Ena family proteins. (**A**). Functional conserved domain analysis of Ena family via Pfam database. The cation-ATPase-N (pfam00690) domain is marked in red, the E1-E2_ATPase (pfam00122) domain is marked in green, the halo acid dehalogenase-like hydrolase (HAD) (COG4087. pfam00702) domain is marked in blue, and Cation_ATPase_C (pfam00689) domain is marked in yellow. (**B**). Neighbor-Joining phylogenetic analysis of Ena family proteins in *P. expansum* and five other fungal species. MEGA 7 was used to construct the phylogenetic tree. Bootstrap values (1000 replicates) are shown for each branch. Enas of *P. expansum* are indicated in bright green.

3.2. PePacC Directly Binds to the Promoter Regions of PeEnas and Activates Transcription

As shown in Figure 2A, the expression of three *PeEnas* was up-regulated at pH 8. The relative expression levels of *PeEnaA*, *PeEnaB*, and *PeEnaC* were extensively increased by 7285-, 51-, and 41-fold in *WT* at pH 8 compared to pH 3. The deletion of PePacC markedly suppressed the expression of these genes at pH 8. Sequence analysis of 1000 bp up-stream promoter regions of *PeEna* sequences revealed the presence of 6, 5, and 2 PacC binding motif (5′-GCCARG-3′ Box) in *PeEnaA*, *PeEnaB*, and *PeEnaC* promoter regions, respectively (Figure 2B), suggesting that the expression of *PeEnaA*, *PeEnaB*, and *PeEnaC* may be regulated by PePacC. To further confirm this hypothesis, we performed ChIP-qPCR analysis to detect whether the 5′-GCCARG-3′ motif was enriched in the promoter region. The degree of PacC binding to promoters was expressed as the percentage of DNA fragments that coimmunoprecipitated with anti-GFP antibodies relative to the input DNAs. The results showed that promoter regions of *PeEnaA*, *PeEnaB*, and *PeEnaC* were significantly enriched by the anti-GFP antibody in the Δ*PePacC::PacC-GFP* strain (Figure 2C). Together, it was suggested that PePacC could bind to the promoter region and transcriptionally activate *PeEnaA*, *PeEnaB*, and *PeEnaC*.

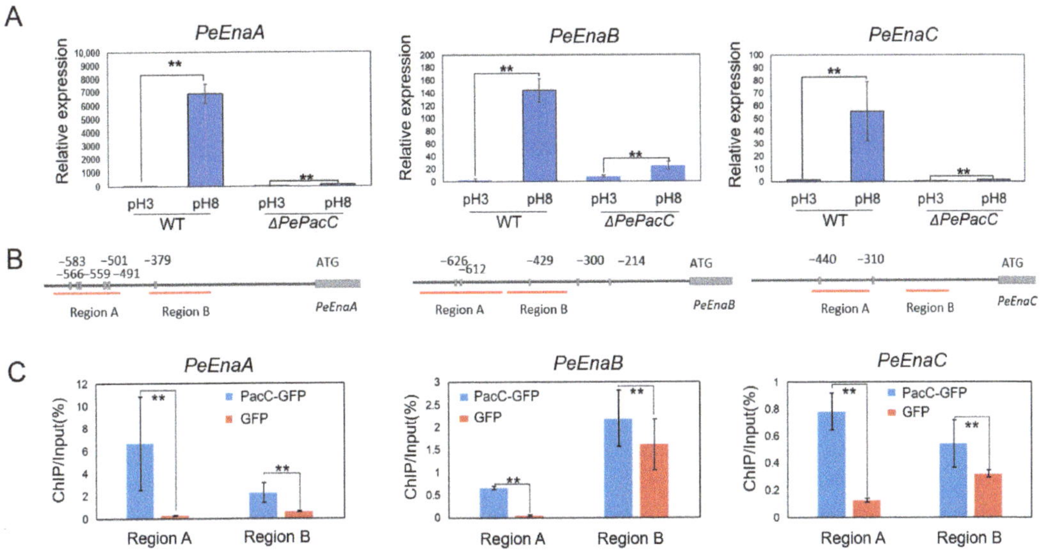

Figure 2. PePacC binds to the *PeEna* genes' promoter and activates their transcription. (**A**). RT-qPCR analysis of the expression patterns of the *PeEnas* in *WT* and Δ*PePacC* at pH 3 and pH 8. Error bars represent the standard deviation of three replicates. Asterisks (**) indicate significant differences according to Student's *t*-test ($p < 0.01$). (**B**). Analysis of the promoter regions of PacC target genes. Boxes represent elements of PacC binding motif 5′-GCCARG-3′, and numbers indicate the position of these motifs relative to the translation start site. Red lines with capital letters represent regions used for ChIP-qPCR. (**C**). ChIP-qPCR detection of the percentage of DNA fragments enriched by anti-GFP antibody in specific regions of *PeEnaA*, *PeEnaB*, and *PeEnaC* relative to input DNAs. Error bars represent standard deviation of three replicates. Asterisks (**) indicate significant differences according to Student's *t*-test ($p < 0.01$).

3.3. PeEnas Are Involved in the Growth and Virulence of P. expansum

Mycelial growth among single-deletion, double-deletion, and triple-deletion strains of *PeEnaA*, *PeEnaB*, *PeEnaC*, *WT*, and Δ*PePacC* strains on MM at pH 5 and pH 8 were compared (Figure 3). The colony diameter of Δ*PeEnaABC* was reduced by about 14.3% compared to that of the WT strain when incubated at pH 8 with 1.5 M NaCl for 4 d. However, there was no significant difference between WT and *PeEnas* single or double-deletion strains under high sodium stress or alkaline pH (Figure 3C,D).

In vivo, assays on apple fruit were conducted to assess the virulence of *P. expansum*. At 5 d post inoculation, lesion diameters of Δ*PeEnaBC*, Δ*PeEnaAC*, and Δ*PeEnaABC* were significantly reduced by 10.2–14.8% compared to *WT* (Figure 4A,B).

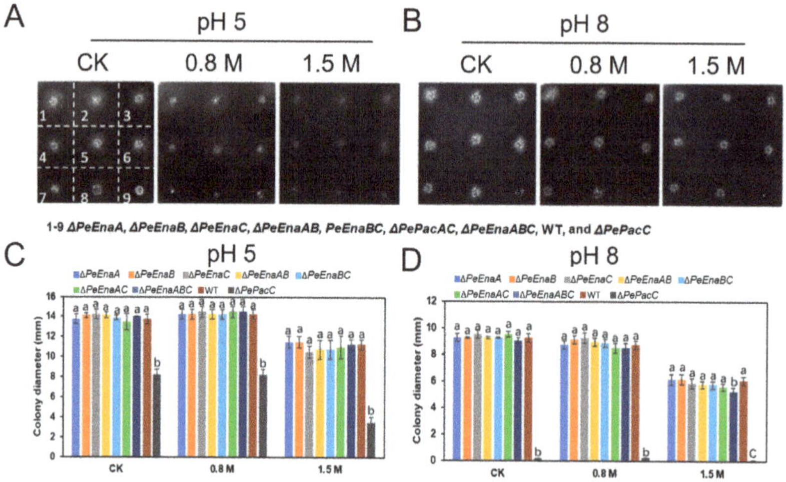

Figure 3. Mycelial growth of *WT*, Δ*PePacC*, and *PeEnas* knockout strains under different pH and high sodium stress of *P. expansum*. (**A**,**B**). Colony morphology of Δ*PeEnaA*, Δ*PeEnaB*, Δ*PePacC*, Δ*PeEnaAB*, Δ*PeEnaBC*, Δ*PePacAC*, Δ*PeEnaABC*, *WT*, and Δ*PePacC* (number 1–9) strains on 4 d after inoculation at MM adjusted to pH 5 and pH 8 conditions, supplemented with 0.8 M NaCl, 1.5 M NaCl. (**C**,**D**). Colony diameters of the indicated strains on MM media for 4 d at pH 5 and pH 8 conditions. Error bars represent the standard deviation of three replicates. Different letters on bars indicate significance according to One-way ANOVA followed by Duncan's multiple range test ($p < 0.05$).

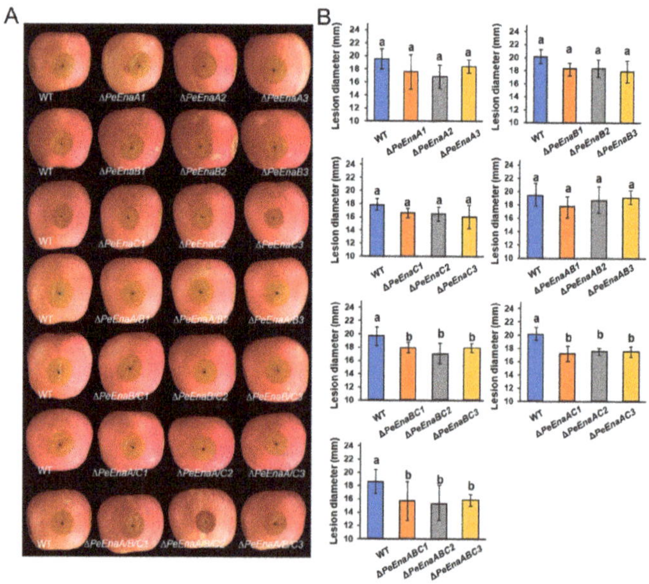

Figure 4. Virulence of *WT* and *PeEnas* knockout strains on apple fruit. (**A**). Disease symptoms on apple inoculated with conidia of *WT*, Δ*PeEnaA*, Δ*PeEnaB*, Δ*PeEnaC*, Δ*PeEnaAB*, Δ*PeEnaBC*, Δ*PePacAC*, and Δ*PeEnaABC* strains after inoculation. (**B**). Lesion diameters of the indicated strains after 5 d of inoculation. Error bars represent the standard deviation of three independent biological replicates. Different letters on bars indicate significance according to One-way ANOVA followed by Duncan's multiple range test ($p < 0.05$).

3.4. PeEnas Affect Patulin Biosynthesis in P. expansum

The patulin production of *WT*, single-deletion, double-deletion, and triple-deletion strains of *PeEnaA*, *PeEnaB*, and *PeEnaC* was assessed (Figure 5A). Compared to *WT*, patulin production in all deletion mutants was significantly reduced (Figure 5B). Among Δ*PeEnaA*, Δ*PeEnaB*, and Δ*PeEnaC*, the patulin production in Δ*PeEnaC* was reduced the most by around 56% when compared to the *WT*. Patulin biosynthesis was further impaired in double-deletion and triple-deletion strains. Patulin production in Δ*PeEnaAC* and Δ*PeEnaABC* was reduced to 30.1% and 23.6% of *WT*, respectively (Figure 5B). In addition, expression levels of all 15 patulin cluster genes were detected by RT-qPCR assay after incubation for 2 d in the strains (Figure 5C). The results suggested that the expression levels of all genes in the gene cluster were downregulated in *PeEnas* deficient mutants. Furthermore, the expression levels of several known secondary metabolism regulators, including 3 Velvet complex components, four global transcription factors, and 4 HOG pathway members, were also evaluated. The gene expression levels of transcription factor AP1, CreA, AreB, and Sge1, as well as vital members Hog1 and Pbs2 in the HOG pathway, were significantly decreased in *PeEnas* deletion mutants in contrast to *WT* (Figure 5C).

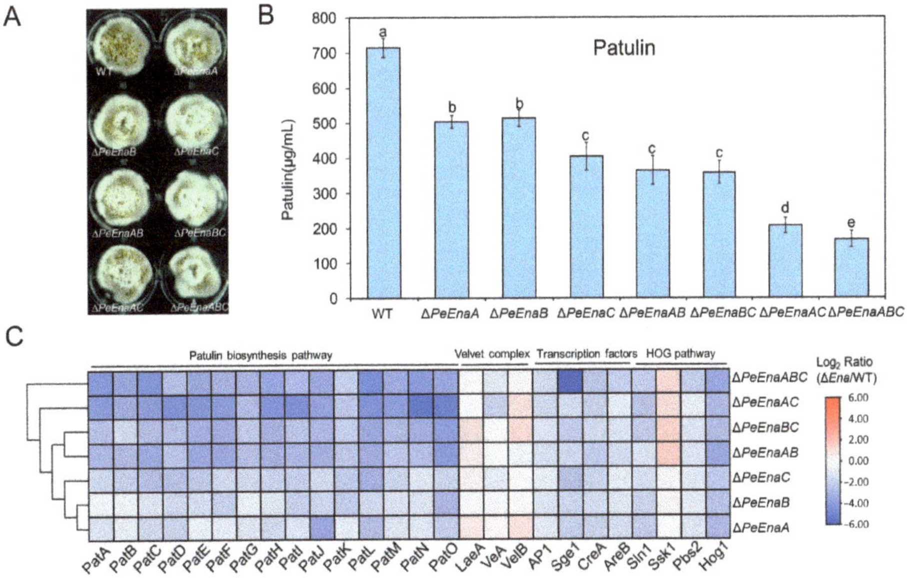

Figure 5. PeEnas play a vital role in patulin production. (**A**). Morphologies of the *WT*, Δ*PeEnaA*, Δ*PeEnaB*, Δ*PeEnaC*, and the relevant complementation strains after 2 d of culture in CY. (**B**). Patulin production of the indicated strains. Error bars represent the standard deviation of three independent biological replicates. Different letters on bars indicate significance according to One-way ANOVA followed by Duncan's multiple range test ($p < 0.05$). (**C**). Heatmap showing expression changes of 15 genes in the patulin cluster (*PatA-PatO*), 3 coding genes in Velvet Complex, 4 coding genes of the global transcription factor, and 4 coding genes in the HOG pathway in the indicated strains. The change fold was based on the $\log_2$ scale of relative expression ratio and was expressed as a color gradient. Each column in the heatmap represented Δ*PeEnas* to *WT*.

3.5. PeEnaC Rescues the Defective of Patulin Biosynthesis in ΔPePacC

PeEnaC as a representative of the *Enas* was selected to construct a Δ*PePacC::PeEnaC-GFP* strain. The overexpression of *PeEnaC* was validated by RT-qPCR assay (Figure 6D). Subcellular localization assay showed that PeEnaC protein was localized in the plasma membrane as TOPCONS predicted (Figure S2) in both Δ*PeEnaC::PeEnaC-GFP* and

Δ*PePacC::PeEnaC-GFP* strains (Figure 6B). Patulin production of the indicated strains was further quantified. Patulin biosynthesis was significantly increased in Δ*PePacC::PeEnaC-GFP* compared to Δ*PePacC* under both pH 5 and pH 8 conditions (Figure 6C). Moreover, expression levels of all 15 patulin cluster genes were up-regulated in *PeEnaC*-complementary mutant compared to Δ*PePacC* (Figure 6E).

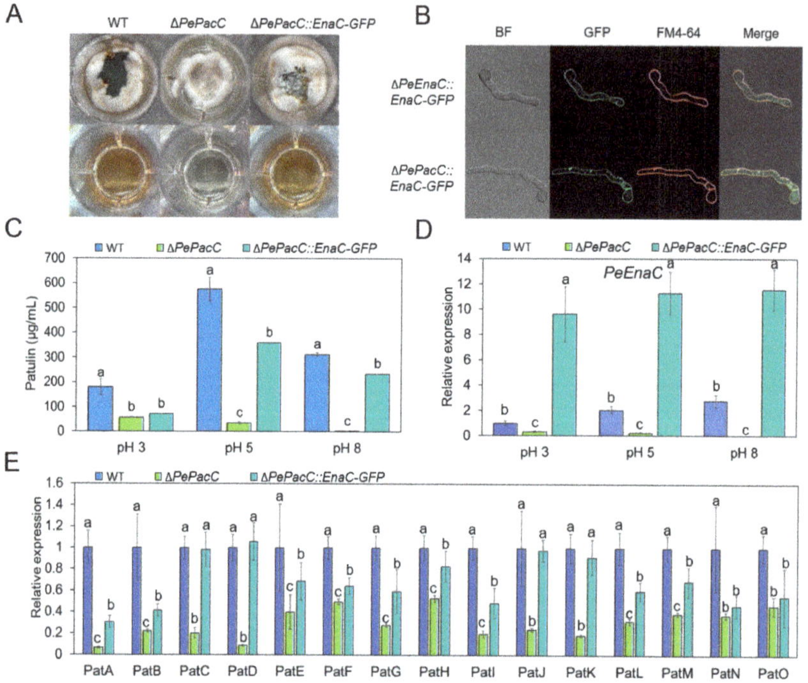

Figure 6. PeEnaC rescues the detective of patulin biosynthesis in Δ*PePacC*. (**A**). Morphologies of the *WT*, Δ*PePacC*, and Δ*PePacC::PeEnaC-GFP* in CY media at pH 5 for 2 d. (**B**). Subcellular localization of PeEnaC in Δ*PeEnaC::PeEnaC-GFP* and Δ*PePacC::PeEnaC-GFP* strains. (**C**). Patulin production of the indicated strains at pH 3, 5, and 8. (**D**). The gene expression analysis of *PeEnaC* in *WT*, Δ*PePacC*, and Δ*PePacC::PeEnaC-GFP* strains in CY media at pH 3, 5, and 8. (**E**). The gene expression analysis of patulin cluster genes of *WT*, Δ*PePacC*, and Δ*PePacC::PeEnaC-GFP* strains in CY media buffered at pH 5. Error bars represent standard deviation of three independent biological replicates. Different letters on bars indicate significance according to One-way ANOVA followed by Duncan's multiple range test ($p < 0.05$).

3.6. PeEnas Involve in Maintaining Membrane Potential in P. expansum

The permeability of ions in the cell can be monitored by observing the cellular accumulation of anionic voltage-sensitive green fluorescent oxonol DiBAC4(3) [29]. DiBAC4(3) is highly voltage-sensitive and enters depolarized cells, where it binds to lipid-rich intracellular components [30]. Compared to WT, Δ*PeEnaA*, Δ*PeEnaB*, Δ*PeEnaC*, Δ*PeEnaABC*, and Δ*PePacC* strains demonstrated a higher degree of DiBAC4(3)-binding at both pH 5 and 8 conditions (Figure 7). These results indicated that the degree of depolarization was stronger in *Enas* deletion mutants and Δ*PePacC*, which has a low expression of *Enas*.

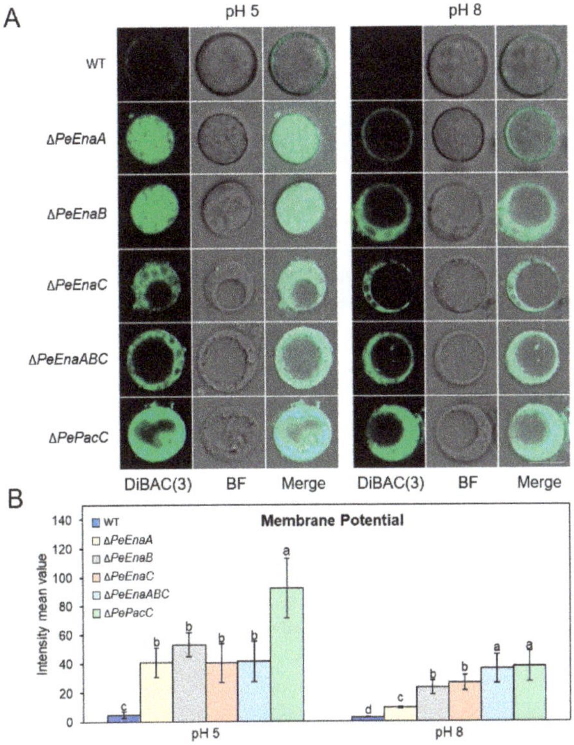

Figure 7. *PeEnas* involve in plasma membrane potential maintenance of *P. expansum*. (**A**). Membrane potential assay using a fluorescent indicator, DiBAC4(3), in protoplast at 20 h (control) and shifted to pH 5 or pH 8 for 1 h. Bar = 5 μm. (**B**). Fluorescence measurements were calculated with the ZEISS ZEN 3.2 (blue edition) software (Zeiss, Oberkochen, Germany). Error bars represent the standard deviation of three independent biological replicates. Different letters on bars indicate significance according to One-way ANOVA followed by Duncan's multiple range test ($p < 0.05$).

4. Discussion

In the present study, putative sodium ATPases of the Ena family were identified in *P. expansum* and characterized. The *P. expansum* genome encodes three PeEna ATPases, PeEnaA, PeEnaB, and PeEnaC, which is similar to that of *A. nidulans* [20]. It has been reported that AnEnaA and AnEnaB play important roles in cellular $Na^+/K^+/H^+$ homeostasis, environmental adaptation, and virulence, while AnEnaC is a pseudogene in *A. nidulans* [20]. We found that AnEnaC did not have a complete ATPase (pfam00122) domain and HAD (pfam00702) domain (Figure S3). However, PeEnaC has a complete predicted ATPase structure (PF00122), which indicates the properties of PeEnaC were more similar to the function of typical P-type ATPases. In contrast to previous studies of Ena function, the growth of single, double, and triple-deletion mutants of *PeEnaA*, *PeEnaB*, and *PeEnaC* were unaffected in control and high salinity conditions (Figure 3). Only the triple-deletion mutant, Δ*PeEnaABC*, showed a slight decrease in colony diameter under 1.5 M Na^+ and pH 8 conditions. *P. expansum* may have evolved other mechanisms to cope with high salt stress. In addition, we found that double and triple-deletion of *PeEnaA*, *PeEnaB*, and *PeEnaC* impaired the virulence of *P. expansum* on apple fruit, indicating that PeEnas may be involved in various biological processes.

According to the previous study, gene expression of Enas was extensively regulated to adapt to ambient pH [14,20,23,31]. PacC is the key transcription factor in the fungal-specific pH signaling pathway [8,32,33]. By analyzing the transcription pattern of the PeEnas, it was found that the gene expression levels of *PeEnaA*, *PeEnaB*, and *PeEnaC* were significantly increased at alkaline pH, and no gene transcription of any PeEnas was detected in the absence of PePacC. Further, analysis of the PePacC binding motif on the gene promoter region and ChIP-qPCR experiments demonstrated that PacC is an important positive regulator of *PeEnas* gene expression (Figure 2). This result is consistent with the previous report that Ena protein is regulated by PacC in *A. nidulans* (*AnEnaA* and *AnEnaB*), *F. graminearum* (*FgENA5*), and *A. fumigatus* (*AfEna1*) [20,34,35]. Nevertheless, Rim101(PacC) does not interact directly with the ENA1 promoter in *S. cerevisiae* but instead acts as a repressor of Nrg1 expression, which directly represses the ENA1 transcription [17]. The contrary conclusion between yeast and filamentous fungi may be attributed to the variable cis-elements of Ena orthologues generated through evolution. Fungi have evolved to use various sensors and signaling pathways to regulate cation pump expression as each fungus is exposed to different cation concentrations [35].

Few studies have been reported on secondary metabolism regulated by Ena ATPase in fungi. Interestingly, our data demonstrated that biosynthesis of vital mycotoxin in *P. expansum*, patulin, was distinctly decreased by the single deletion of three *PeEnas* (Figure 5B). Double and triple-deletion of *PeEnaA*, *PeEnaB*, and *PeEnaC* further impaired patulin production, indicating that the three PeEna family genes have a synergistic effect in regulating patulin biosynthesis. Analysis of gene expression on the patulin gene cluster suggested that PeEnas could regulate patulin biosynthesis at the transcriptional level (Figure 5C). In addition, the expressions of several well-known regulators, *AP1*, *CreA*, *Sge1*, and *Hog1* genes, were significantly downregulated in *PeEnas* deletion mutants compared to *WT*, while global Velvet complex transcription factors were not affected. In fungi, AP1-like bZIP factor AP1 regulates oxidative stress-responsive genes and is involved in secondary metabolism [36,37]. Hog1 functions in oxidative stress tolerance and is involved in trichothecene biosynthesis [38,39]. CREA ensures preferential glucose utilization by blocking the expression of other genes required for carbon source metabolism and is a transcriptional repressor of the carbon catabolite (CCR) [40]. Loss of function creA strains of *P. expansum* do not produce patulin on apple fruit [41]. Sge1 is a homologous protein of Wor1 and is involved in lifestyle switching, effector expression, and regulation of secondary metabolite biosynthesis [42]. The results suggested that PeEnas may modulate patulin biosynthesis through a broad network in *P. expansum*.

PacC, as a globally regulated transcription factor, not only mediates the pH signaling pathway but is also involved in the regulation of mycotoxin biosynthesis [8]. In *A. nidulans*, a PacC binding motif is present in the promoter region of the aflatoxin biosynthetic pathway-specific transcription factor AFLR and key enzyme IpnA encoding genes [43]. In *P. expansum*, patulin production and expression of the biosynthetic cluster were significantly down-regulated in the Δ*PePacC* strain under both acidic and alkaline conditions, indicating that PePacC positively regulates the expression of the biosynthetic gene cluster, thereby affecting patulin production [9]. PacC binding motifs were found in the promoter regions of nine patulin cluster genes [9], indicating that PePacC may directly regulate the expression of patulin biosynthetic genes. In the present study, we found that PePacC could directly regulate the expression of *PeEnas*, while the latter affected patulin production. In the three *PeEnas*, *PeEnaC* demonstrated a stronger effect on patulin production compared to *PeEnaA* and *PeEnaB* (Figure 5B). Interestingly, *PeEnaC* complementary experiment in Δ*PePacC* strain showed that overexpression of *PeEnaC* could partly restore gene expression of the patulin cluster and patulin production of Δ*PePacC* (Figure 6), indicating that PeEnas may play important roles in the regulation of PePacC on patulin biosynthesis.

5. Conclusions

As P-type plasma membrane ATPases, Ena family proteins are important in environmental adaptation in fungi. In the present study, three Ena family genes, *PeEnaA*, *PeEnaB*, and *PeEnaC*, were identified in *P. expansum*. All the genes responded to ambient pH and were directly regulated by PePacC, the key transcription factor of the fungal pH signaling pathway. PeEnas have little effect on mycelial growth under alkaline or high salinity conditions but are involved in virulence in fruit and maintenance of cell membrane potential. Notably, a crucial role of PeEnas in regulating the biosynthesis of secondary metabolites, patulin, was reported for the first time. PeEnas affect gene expressions of patulin cluster, AP1, CreA, Sge1, and Hog1, and play a function in the regulation of PePacC on patulin biosynthesis. The PeEna proteins may be used as potential targets for patulin contamination control. In addition, our results provide new insights for elucidating the complicated regulatory network of the global transcription factor PacC.

Supplementary Materials: The following supporting information can be downloaded at: https://www.mdpi.com/article/10.3390/jof9080806/s1, Figure S1: Knockout of *PeEnaA*, *PeEnaB*, and *PeEnaC* in *P. expansum*; Figure S2: The transmembrane topology of the PeEna proteins; Figure S3: Functional conserved domain analysis of Ena family proteins via Pfam database; Figure S4: The melting curves of representative genes in this study; Table S1: The primers used for construction and identification of gene deletion, complementation and eGFP tag strains in this study; Table S2: Formulation of the citrate–phosphate buffer; Table S3: The primers used for RT-qPCR in this study.

Author Contributions: Conceptualization, B.L.; Methodology, R.Z., Y.C. and M.X.; Investigation, R.Z., Y.C. and M.X.; Writing—Original draft preparation, R.Z.; Writing—Review and editing, B.L., Y.C., Z.Z. and S.T.; visualization, R.Z. and B.L.; supervision, B.L.; funding acquisition, B.L. and S.T. All authors have read and agreed to the published version of the manuscript.

Funding: This research was funded by the National Natural Science Foundation of China (grant number: 32261133624; 32072273), National Key R&D Program of China (grant number: 2021YFD2100501/05), and Youth Innovation Promotion Association, CAS (grant number: Y201919).

Institutional Review Board Statement: Not applicable.

Informed Consent Statement: Not applicable.

Data Availability Statement: Not applicable.

Conflicts of Interest: The authors declare no conflict of interest.

References

1. Li, B.; Chen, Y.; Zhang, Z.; Qin, G.; Chen, T.; Tian, S. Molecular basis and regulation of pathogenicity and patulin biosynthesis in *Penicillium expansum*. *Compr. Rev. Food Sci. Food Saf.* **2020**, *19*, 3416–3438. [CrossRef] [PubMed]
2. Li, B.; Chen, Y.; Zong, Y.; Shang, Y.; Zhang, Z.; Xu, X.; Wang, X.; Long, M.; Tian, S. Dissection of patulin biosynthesis, spatial control and regulation mechanism in *Penicillium expansum*. *Environ. Microbiol.* **2019**, *21*, 1124–1139. [CrossRef]
3. Barad, S.; Sela, N.; Kumar, D.; Kumar-Dubey, A.; Glam-Matana, N.; Sherman, A.; Prusky, D. Fungal and host transcriptome analysis of pH-regulated genes during colonization of apple fruits by *Penicillium expansum*. *BMC Genom.* **2016**, *17*, 330. [CrossRef]
4. Tannous, J.; Atoui, A.; El Khoury, A.; Francis, Z.; Oswald, I.P.; Puel, O.; Lteif, R. A study on the physicochemical parameters for *Penicillium expansum* growth and patulin production: Effect of temperature, pH, and water activity. *Food Sci. Nutr.* **2016**, *4*, 611–622. [CrossRef]
5. Penalva, M.A.; Tilburn, J.; Bignell, E.; Arst, H.N., Jr. Ambient pH gene regulation in fungi: Making connections. *Trends Microbiol.* **2008**, *16*, 291–300. [CrossRef]
6. Diez, E.; Alvaro, J.; Espeso, E.A.; Rainbow, L.; Suarez, T.; Tilburn, J.; Arst, H.N., Jr.; Penalva, M.A. Activation of the *Aspergillus* PacC zinc finger transcription factor requires two proteolytic steps. *EMBO J.* **2002**, *21*, 1350–1359. [CrossRef]
7. Rollins, J.A.; Dickman, M.B. pH signaling in *Sclerotinia sclerotiorum*: Identification of a *pacC/RIM1* homolog. *Appl. Environ. Microbiol.* **2001**, *67*, 75–81. [CrossRef] [PubMed]
8. Li, B.; Chen, Y.; Tian, S. Function of pH-dependent transcription factor PacC in regulating development, pathogenicity, and mycotoxin biosynthesis of phytopathogenic fungi. *FEBS J.* **2021**, *289*, 1723–1730. [CrossRef]
9. Chen, Y.; Li, B.; Xu, X.; Zhang, Z.; Tian, S. The pH-responsive PacC transcription factor plays pivotal roles in virulence and patulin biosynthesis in *Penicillium expansum*. *Environ. Microbiol.* **2018**, *20*, 4063–4078. [CrossRef] [PubMed]
10. Chen, Y.; Zhang, Z.; Tian, S.; Li, B. Application of -omic technologies in postharvest pathology: Recent advances and perspectives. *Curr. Opin. Food Sci.* **2022**, *45*, 100820. [CrossRef]
11. Zhuo, R.; Li, G.; Peng, H.; Zong, Y.; Wang, X.; Lu, S.; Chen, Y.; Zhang, Z.; Tian, S.; Li, B. Extensive regulation of pH-responsive transcription factor PacC on secondary metabolism contributes to development and virulence of *Botrytis cinerea*. *Postharvest Biol. Technol.* **2023**, *197*, 112219. [CrossRef]

12. Lamb, T.M.; Mitchell, A.P. The transcription factor Rim101p governs ion tolerance and cell differentiation by direct repression of the regulatory genes *NRG1* and *SMP1* in *Saccharomyces cerevisiae*. *Mol. Cell. Biol.* **2003**, *23*, 677–686. [CrossRef] [PubMed]
13. Serra-Cardona, A.; Petrezselyova, S.; Canadell, D.; Ramos, J.; Arino, J. Coregulated expression of the Na^+/phosphate Pho89 transporter and Ena1 Na^+-ATPase allows their functional coupling under high-pH stress. *Mol. Cell. Biol.* **2014**, *34*, 4420–4435. [CrossRef]
14. Ma, Q.; Jin, K.; Peng, G.; Xia, Y. An ENA ATPase, MaENA1, of *Metarhizium acridum* influences the Na^+-, thermo- and UV-tolerances of conidia and is involved in multiple mechanisms of stress tolerance. *Fungal Genet. Biol.* **2015**, *83*, 68–77. [CrossRef]
15. Dyla, M.; Kjaergaard, M.; Poulsen, H.; Nissen, P. Structure and mechanism of P-Type ATPase ion pumps. *Annu. Rev. Biochem.* **2020**, *89*, 583–603. [CrossRef]
16. Palmgren, M.G.; Nissen, P. P-type ATPases. *Annu. Rev. Biophys.* **2011**, *40*, 243–266. [CrossRef]
17. Arino, J.; Ramos, J.; Sychrova, H. Monovalent cation transporters at the plasma membrane in yeasts. *Yeast* **2019**, *36*, 177–193. [CrossRef] [PubMed]
18. Arino, J.; Ramos, J.; Sychrova, H. Alkali metal cation transport and homeostasis in yeasts. *Microbiol. Mol. Biol. Rev.* **2010**, *74*, 95–120. [CrossRef] [PubMed]
19. Garciadeblas, B.; Rubio, F.; Quintero, F.J.; Banuelos, M.A.; Haro, R.; Rodrigueznavarro, A. Differential expression of two genes encoding isoforms of the ATPase involved in sodium effiux in *Saccharomyces cerevisiae*. *Mol. Genet. Genom.* **1993**, *236*, 363–368. [CrossRef]
20. Markina-Inarrairaegui, A.; Spielvogel, A.; Etxebeste, O.; Ugalde, U.; Espeso, E.A. Tolerance to alkaline ambient pH in *Aspergillus nidulans* depends on the activity of ENA proteins. *Sci. Rep.* **2020**, *10*, 14325. [CrossRef] [PubMed]
21. Platara, M.; Ruiz, A.; Serrano, R.; Palomino, A.; Moreno, F.; Arino, J. The transcriptional response of the yeast Na^+-ATPase ENA1 gene to alkaline stress involves three main signaling pathways. *J. Biol. Chem.* **2006**, *281*, 36632–36642. [CrossRef] [PubMed]
22. Idnurm, A.; Walton, F.J.; Floyd, A.; Reedy, J.L.; Heitman, J. Identification of *ENA1* as a virulence gene of the human pathogenic fungus *Cryptococcus neoformans* through signature-tagged insertional mutagenesis. *Eukaryot. Cell* **2009**, *8*, 315–326. [CrossRef] [PubMed]
23. Mou, Y.N.; Gao, B.J.; Ren, K.; Tong, S.M.; Ying, S.H.; Feng, M.G. P-type Na^+/K^+ ATPases essential and nonessential for cellular homeostasis and insect pathogenicity of *Beauveria bassiana*. *Virulence* **2020**, *11*, 1415–1431. [CrossRef] [PubMed]
24. Li, B.; Zong, Y.; Du, Z.; Chen, Y.; Zhang, Z.; Qin, G.; Zhao, W.; Tian, S. Genomic characterization reveals insights into patulin biosynthesis and pathogenicity in *Penicillium* species. *Mol. Plant Microbe Interact.* **2015**, *28*, 635–647. [CrossRef]
25. Xu, X.; Chen, Y.; Li, B.; Tian, S. Histone H3K4 methyltransferase PeSet1 regulates colonization, patulin biosynthesis, and stress responses of *Penicillium expansum*. *Microbiol. Spectr.* **2023**, *11*, e0354522. [CrossRef] [PubMed]
26. Gu, Q.; Wang, Y.; Zhao, X.; Yuan, B.; Zhang, M.; Tan, Z.; Zhang, X.; Chen, Y.; Wu, H.; Luo, Y.; et al. Inhibition of histone acetyltransferase GCN5 by a transcription factor FgPacC controls fungal adaption to host-derived iron stress. *Nucleic Acids Res.* **2022**, *50*, 6190–6210. [CrossRef]
27. Wang, W.; Wang, P.; Li, X.; Wang, Y.; Tian, S.; Qin, G. The transcription factor SlHY5 regulates the ripening of tomato fruit at both the transcriptional and translational levels. *Hort. Res.* **2021**, *8*, 83. [CrossRef]
28. Veerana, M.; Yu, N.N.; Bae, S.J.; Kim, I.; Kim, E.S.; Ketya, W.; Lee, H.Y.; Kim, N.Y.; Park, G. Enhancement of fungal enzyme production by radio-frequency electromagnetic fields. *J. Fungi* **2022**, *8*, 1187. [CrossRef]
29. Molina-Hernandez, J.B.; Capelli, F.; Laurita, R.; Tappi, S.; Laika, J.; Gioia, L.; Valbonetti, L.; Chaves-López, C. A comparative study on the antifungal efficacy of cold atmospheric plasma at low and high surface density on *Aspergillus chevalieri* and mechanisms of action. *Innov. Food Sci. Emerg. Technol.* **2022**, *82*, 103194. [CrossRef]
30. Schuster, M.; Steinberg, G. The fungicide dodine primarily inhibits mitochondrial respiration in *Ustilago maydis*, but also affects plasma membrane integrity and endocytosis, which is not found in *Zymoseptoria tritici*. *Fungal Genet. Biol.* **2020**, *142*, 103414. [CrossRef]
31. Kane, P.M. Proton transport and pH control in fungi. *Adv. Exp. Med. Biol.* **2016**, *892*, 33–68. [CrossRef] [PubMed]
32. Barda, O.; Maor, U.; Sadhasivam, S.; Bi, Y.; Zakin, V.; Prusky, D.; Sionov, E. The pH-responsive transcription factor PacC governs pathogenicity and ochratoxin A biosynthesis in *Aspergillus carbonarius*. *Front. Microbiol.* **2020**, *11*, 210. [CrossRef]
33. Zhang, M.; Wei, Q.; Xia, Y.; Jin, K. MaPacC, a pH-responsive transcription factor, negatively regulates thermotolerance and contributes to conidiation and virulence in *Metarhizium acridum*. *Curr. Genet.* **2020**, *66*, 397–408. [CrossRef]
34. Loss, O.; Bertuzzi, M.; Yan, Y.; Fedorova, N.; McCann, B.L.; Armstrong-James, D.; Espeso, E.A.; Read, N.D.; Nierman, W.C.; Bignell, E.M. Mutual independence of alkaline- and calcium-mediated signalling in *Aspergillus fumigatus* refutes the existence of a conserved druggable signalling nexus. *Mol. Microbiol.* **2017**, *106*, 861–875. [CrossRef] [PubMed]
35. Son, H.; Park, A.R.; Lim, J.Y.; Lee, Y.W. Fss1 is involved in the regulation of an *ENA5* homologue for sodium and lithium tolerance in *Fusarium graminearum*. *Environ. Microbiol.* **2015**, *17*, 2048–2063. [CrossRef] [PubMed]
36. Perez-Perez, W.D.; Carrasco-Navarro, U.; Garcia-Estrada, C.; Kosalkova, K.; Gutierrez-Ruiz, M.C.; Barrios-Gonzalez, J.; Fierro, F. bZIP transcription factors PcYap1 and PcRsmA link oxidative stress response to secondary metabolism and development in *Penicillium chrysogenum*. *Microb. Cell Fact.* **2022**, *21*, 50. [CrossRef] [PubMed]
37. Keller, N.P. Translating biosynthetic gene clusters into fungal armor and weaponry. *Nat. Chem. Biol.* **2015**, *11*, 671–677. [CrossRef]
38. Li, Y.; He, P.; Tian, C.; Wang, Y. CgHog1 controls the adaptation to both sorbitol and fludioxonil in *Colletotrichum gloeosporioides*. *Fungal Genet. Biol.* **2020**, *135*, 103289. [CrossRef]

39. Chen, Y.; Kistler, H.C.; Ma, Z. *Fusarium graminearum* trichothecene mycotoxins: Biosynthesis, regulation, and management. *Annu. Rev. Phytopathol.* **2019**, *57*, 15–39. [CrossRef]
40. Reijngoud, J.; Arentshorst, M.; Ruijmbeek, C.; Reid, I.; Alazi, E.D.; Punt, P.J.; Tsang, A.; Ram, A.F.J. Loss of function of the carbon catabolite repressor CreA leads to low but inducer-independent expression from the feruloyl esterase B promoter in *Aspergillus niger*. *Biotechnol. Lett.* **2021**, *43*, 1323–1336. [CrossRef]
41. Tannous, J.; Kumar, D.; Sela, N.; Sionov, E.; Prusky, D.; Keller, N.P. Fungal attack and host defence pathways unveiled in near-avirulent interactions of *Penicillium expansum creA* mutants on apples. *Mol. Plant Pathol.* **2018**, *19*, 2635–2650. [CrossRef] [PubMed]
42. Gurdaswani, V.; Ghag, S.B.; Ganapathi, T.R. *FocSge1* in *Fusarium oxysporum* f. sp. *cubense* race 1 is essential for full virulence. *BMC Microbiol.* **2020**, *20*, 255. [CrossRef] [PubMed]
43. Ehrlich, K.C.; Cary, J.W.; Montalbano, B.G. Characterization of the promoter for the gene encoding the aflatoxin biosynthetic pathway regulatory protein AFLR. *Acta Biochim. Biophys. Sin.* **1999**, *1444*, 412–417. [CrossRef] [PubMed]

Journal of
Fungi

MDPI

Article

Inhibitory Mechanisms of *trans*-2-Hexenal on the Growth of *Geotrichum citri-aurantii*

Qiuli Ouyang *, Shiwei Shi, Yangmei Liu, Yanqin Yang, Yonghua Zhang, Xingxing Yuan, Nengguo Tao * and Lu Li

School of Chemical Engineering, Xiangtan University, Xiangtan 411105, China; luli9003@163.com (L.L.)
* Correspondence: ouyang199006@126.com (Q.O.); nengguotao@126.com (N.T.); Tel.: +86-731-58292246 (N.T.)

Abstract: *Geotrichum citri-aurantii* (*G. citri-aurantii*) is one of the most important postharvest pathogens leading to a postharvest loss of citrus by causing sour rot. In this study, the antifungal activity of *trans*-2-hexenal, a natural component of essential oil, against *G. citri-aurantii* was evaluated. *Trans*-2-hexenal treatment inhibited the mycelia growth of *G. citri-aurantii* with a minimum inhibitory concentration and minimum fungicidal concentration of *trans*-2-hexenal at 0.50 and 1.00 μL/mL, respectively. Moreover, *trans*-2-hexenal efficiently reduced the incidence of sour rot of Satsuma fruit inoculated with *G. citri-aurantii*. Ultrastructural observations and Fourier transform infrared (FT−IR) results showed that *trans*-2-hexenal treatment affected the cell wall and cell membrane instructions of *G. citri-aurantii*. The content of β-1,3-glucan was significantly decreased after *trans*-2-hexenal treatment, but the cell wall permeability was not changed. The decrease in lipid and ergosterol contents might be responsible for this antifungal activity. Several important genes, *FKS1*, *ERG1*, *ERG7*, and *ERG11*, showed decreasing expression levels after *trans*-2-hexenal treatment. Molecule-docking results also indicated that *trans*-2-hexenal could join with the protein of FKS1, ERG1, ERG7, and ERG11 to impact enzyme activities. These results demonstrated that *trans*-2-hexenal is a promising fungicide for controlling sour rot of harvested citrus fruit by damaging the membrane integrity of *G. citri-aurantii*.

Keywords: *G. citri-aurantii*; *trans*-2-hexeanl; antifungal mechanism; cell wall; cell membrane

1. Introduction

G. citri-aurantii is a necrotrophic fungal pathogen that infects citrus fruit, and it has the characteristics of strong infectivity, fast infectivity, and being difficult to control [1,2]. Synthetic fungicides are exclusively used to control this disease but cause serious hazardous effects on the fruit rind, the environment, and human health [3,4]. Thus, it is necessary to focus on developing alternatives to synthetic fungicides for handling and maintaining the quality of citrus fruit [5–7].

Previous studies have shown that plant essential oils and their antimicrobial components have significant inhibitory effects on many postharvest pathogenic fungi of citrus, such as *G. citri-aurantii*, *Penicillium digitatum*, and *P. italicum*, and that they have the advantages of safety, high efficiency, and low residue, which means they have the potential to control postharvest diseases in citrus [8,9]. *Trans*-2-hexenal is a volatile component of plant essential oil, which naturally exists in citronella oil, camphor oil, apples, and grapes. It is an important signal molecule for plants to respond to and defend against their external environment. Some studies have shown that *trans*-2-hexenal has good inhibitory effects against *Colletotrichum acutatum*, *Alternaria alternata*, *P. cyclopium*, *P. expansum*, and *Botrytis cinerea* [10–13]. In addition, *trans*-2-hexenal was also successfully applied in order to control postharvest diseases such as gray mold in tomato fruits, green mold in citrus fruits, and black rot in 'Zaosu' pears [9,14,15].

As mentioned above, several studies have shown that *trans*-2-hexenal is a potential biological alternative to other preservative methods of controlling postharvest diseases, but there are few studies related to the investigation of its exact mechanism of antifungal action. Studies have shown that *trans*-2-hexenal interferes with the cell wall structure of fungi. For example, Arroyo et al. [10] pointed out that treatment with *trans*-2-hexenal can cause obvious cracks in the cell wall, disorder cell components to a high degree, and cause the organelle morphology to disappear and the cells to crack. Research by Zhang et al. [13] found that the membrane permeability of *P. cyclopium* increased with an increase of *trans*-2-hexenal concentration, resulting in the release of cell components and the leakage of potassium ions. At the same time, the integrity of the cell membrane of *P. cyclopium* was destroyed due to the decrease in the lipid content. Ma et al. [16] showed that 1.0 μL/mL *trans*-2-hexenal inhibited spore germination by disrupting the mitochondrial energy metabolism of *Aspergillus flavus*; the *trans*-2-hexenal treatment also decreased the acetyl-CoA and ATP contents, and the mitochondrial dehydrogenases activity increased by 65.7 ± 3.7%, 53.9 ± 4.0%, and 23.8 ± 2.2%, respectively. However, the antifungal activity of *trans*-2-hexenal against *G. citri-aurantii* has not been determined, and the antifungal mechanism of *trans*-2-hexenal against *G. citri-aurantii* has not been studied.

Thus, this study aimed to (1) study the antifungal properties of *trans*-2-hexenal against *G. citri-aurantii* in vitro and in vivo, (2) investigate the effect of the cell wall and cell membrane of *G. citri-aurantii* in the presence of *trans*-2-hexenal, and (3) further explore the possible mechanism through RT-qPCR and molecular docking.

2. Materials and Methods

2.1. Pathogen

G. citri-aurantii was provided by the Department of Biotechnology and Food Engineering, Xiangtan University, Xiangtan, China. The fungus was purified and preserved at 28 ± 2 °C on potato dextrose agar (PDA). A spore suspension (5×10^6 spores/mL) in potato dextrose broth (PDB) was prepared using a hemocytometer.

2.2. Fruit

Satsuma mandarin fruits (*Citrus unshiu* Marc. cv. Miyagawa Wase) were harvested on 18 October 2018 from an orchard in Xiangtan, Hunan, China. Healthy fruits of uniform size and without scars were selected for the experiments.

2.3. Chemicals

Trans-2-hexenal (98%) was obtained from Aladdin (Shanghai, China). All the chemicals were analytical grade.

2.4. Antifungal Activity of *trans*-2-Hexenal against *G. citri-aurantii*

The inhibition of *trans*-2-hexenal on the growth of *G. citri-aurantii* mycelia was tested in vitro through the agar dilution method [17]. Briefly, *trans*-2-hexenal solutions were prepared by dissolving the requisite amount in Tween-80 (0.5%, Aladdin, Shanghai, China) and adding it to PDA (20 mL) to achieve the desired concentrations (0, 0.25, 0.50, 1.0, 2.0, and 4.0 μL/mL). A 6 mm diameter mycelial disk of inoculate was cut from the actively growing culture of the PDA plates. Then, they were placed at the center of each new Petri plate (90 mm in diameter). The culture plates were then incubated at 28 ± 2 °C for 2 d. Each treatment was performed in triplicate. The percentage of inhibition of mycelial growth (MGI) was calculated according to the following formula:

$$\text{MGI}\ (\%) = [(\text{dc} - \text{dt})/(\text{dc} - 6)] \times 100\%$$

where dc (cm) is the average diameter of the control and dt (cm) is the average diameter of the treatment. The lowest concentration that completely inhibited the growth of *G. citri-aurantii* after 2 d of incubation was considered to be the minimum inhibitory concentration (MIC). The minimum fungicidal concentration (MFC) was regarded as the lowest concentration that prevented the growth of the pathogen after 4 d of incubation at 28 ± 2 °C, indicating that more than 99.5% of the original inocula were killed.

2.5. In Vivo Experiments of trans-2-Hexenal against G. citri-aurantii

The effect of the *trans*-2-hexenal on the incidence of sour rot was determined as described previously by Dou et al. [18]. All fresh citrus fruit were surface-sterilized by immersing in 2% sodium hypochlorite solution (v/v) for 2 min, then washed with distilled water, wounded (depth of 3 mm and width of 3 mm) with a sterile needle, inoculated with 20 μL of *G. citri-aurantii* spore suspension (10^5 spores mL^{-1}), and left to air-dry. After being inoculated with *G. citri-aurantii*, the fruit were soaked in wax amended with *trans*-2-hexenal at 1× MFC and 10× MFC. The fruit with wax and inoculated with the pathogen inoculation was used as a control. The inoculated fruit was kept in sealed incubators at 25 ± 2 °C to ensure a high relative humidity (80–85% relative humidity). Each treatment was performed in triplicate, and each replicate contained 20 Satsuma fruits. The incidence rate of disease (measured by counting the number of green-mold-infected wounds) was calculated as follows:

$$Diseaseincidence = \frac{Number\ of\ rotten\ wounds}{Total\ number\ of\ wounds} \times 100\%$$

2.6. Scanning Electron Microscopy (SEM) of trans-2-Hexenal against G. citri-aurantii

The mycelia treated with *trans*-2-hexenal for 30 min, as described above, were directly examined using a JEOL JSM-6360LV SEM instrument (JEOL, Tokyo, Japan). The hyphae grown on PDA without *trans*-2-hexenal were used as a control. The procedures for the SEM observation were described in our previous study [19].

2.7. Transmission Electron Microscopy (TEM) of trans-2-Hexenal against G. citri-aurantii

The mycelia treated with *trans*-2-hexenal for 30 min, as described above, were directly examined using a transmission electron microscope (JEM-1230; JEOL Ltd., Tokyo, Japan) operated at an accelerating voltage of 80 kV. The hyphae grown on PDA without *trans*-2-hexenal were used as a control. The procedures for the TEM observation were described in our previous study [19].

2.8. Fourier Transform Infrared (FT−IR) Spectroscopy of trans-2-Hexenal against G. citri-aurantii

The effect of *trans*-2-hexenal on the mycelia composition of *G. citri-aurantii* was analyzed using Fourier transform infrared spectroscopy (FT−IR) (Thermo Fisher Scientific, Waltham, MA, USA) [20]. The mycelia treated with 1/2 MIC *trans*-2-hexenal for 30 min were collected, frozen with liquid nitrogen, and then vacuum freeze-dried. Subsequently, the mycelia were ground (100 mesh) to obtain uniform dried powder. The samples were prepared using the potassium bromide-disk technique for the FT−IR detection. The scanning range was 4000–400 cm^{-1} with the resolution of 4 cm^{-1} and 128 separate scans. The infrared spectrum was analyzed using Unscrambler X (Version 10.4).

2.9. Effect of trans-2-Hexenal on the Cell Wall of G. citri-aurantii

The effects of *trans*-2-hexenal on the cell wall integrity of *G. citri-aurantii* were analyzed using calcofluor white (Sigma, St. Louis, MO, USA) staining coupled with fluorescence microscopy. The mycelia treated with 1/2 MIC *trans*-2-hexenal for 0, 30, 60, and 120 min were centrifuged at 4000× g for 10 min. The collected mycelia were stained with 10 μL of calcofluor white stain after the addition of 10 μL KOH (10%) following the manufacturer's instructions. The samples were observed with a fluorescence microscope (Nikon ECLIPSE TS100, Tokyo, Metropolis, Japan). The fungal culture in PDB without *trans*-2-hexenal was used as a control.

2.10. Effect of trans-2-Hexenal on the Cell Wall of G. citri-aurantii

2.10.1. Effect of trans-2-Hexenal on the Cell Wall Integrity of G. citri-aurantii

The effects of *trans*-2-hexenal on the cell wall integrity of *G. citri-aurantii* were analyzed using calcofluor white (Sigma, St. Louis, MO, USA) staining coupled with fluorescence microscopy. The mycelia treated with 1/2 MIC *trans*-2-hexenal for 0, 30, 60, and 120 min were centrifuged at 4000× *g* for 10 min. The collected mycelia were stained with 10 μL of calcofluor white stain after the addition of 10 μL KOH (10%, Aladdin, Shanghai, China) following the manufacturer's instructions. The samples were observed with a fluorescence microscope (Nikon ECLIPSE TS100, Tokyo Metropolis, Japan). The fungal culture in PDB without *trans*-2-hexenal was used as a control.

2.10.2. Effect of *trans*-2-Hexenal on the Chitin Content of *G. citri-aurantii*

The chitin contents of the *G. citri-aurantii* treated with the 1/2 MIC *trans*-2-hexenal treatments in PDB were determined using the method of Francois [21]. A total of 0.5 g of dried mycelia was soaked in 4 mL of concentrated HCl (Soleibao, Beijing, China) at 25 °C for 24 h, then diluted with distilled water until the HCl reached a concentration of 8.5 mol/L, and the solution was further digested in a boiling water bath. After cooling, it was adjusted to neutral with 1 mol/L of NaOH (Aladdin, Shanghai, China) solution, the volume was made constant to 100 mL, then filtered with filter paper, and the supernatant was shaken to obtain the sample to be tested. Then, 200 μL supernatant was mixed with 400 μL acetylacetone reagent and placed in a 90 °C water bath for 1 h. After cooling to room temperature, 4 mL absolute ethyl alcohol and 400 μL 4-dimethylaminobenzaldehyde were added to the supernatant, the volume was fixed to 5 mL with absolute ethyl alcohol, and it was left to stand at room temperature for 1 h. The absorbance of the solution was measured at a wavelength of 530 nm, and glucosamine hydrochloride was used as the standard curve.

2.10.3. Effect of *trans*-2-Hexenal on the β-1,3-Glucan Content of *G. citri-aurantii*

The method of Fortwendel et al. [22] was used for the determination of β-1,3-glucan. The mycelia were washed with 0.1 mol/L NaOH solution and then freeze-dried and ground into powder. A certain amount of powder was added to 1 mol/L NaOH solution, ultrasonicated for 30 s, placed in a water bath at 52 °C for 30 min, and centrifuged for 5 min after cooling. A total of 50 μL of the supernatant was taken, 185 μL of aniline blue solution was added, and it was placed in a water bath at 52 °C for 30 min and left to stand for 30 min. The fluorescence value (excitation wavelength: 405 nm; emission wavelength: 460 nm) was measured using a fluorescence spectrophotometer (Lengguang Technology Co., Ltd., Shanghai, China).

2.10.4. Effect of *trans*-2-Hexenal on the Extracellular Alkaline Phosphatase (AKP) Activities of *G. citri-aurantii*

The extracellular AKP activities of the *G. citri-aurantii* mycelia that received different *trans*-2-hexenal treatments in PDB, as described above, were assayed with a UV-2450 UV/V spectrophotometer (Shimadzu (China) Co., Ltd., Shanghai, China) using a commercially available kit, following the instructions. The fungal culture in PDB without *trans*-2-hexenal was used as a control. Each experiment was repeated three times. The enzyme activity is expressed as U/g prot.

2.11. *Effect of trans-2-Hexenal on the Cell Membrane of G. citri-aurantii*

2.11.1. Effect of *trans*-2-Hexenal on the Cell Membrane Integrity of *G. citri-aurantii*

The cell membrane integrity of the *G. citri-aurantii* that received different *trans*-2-hexenal treatments in PDB, as described above, was analyzed using propidium iodide (PI) staining coupled with an ECLIPSE TS100 microscope (Nikon, Tokyo Metropolis, Japan) and F97 PRO fluorescence spectrophotometer (Lengguang Technology, Shanghai, China) [23].

2.11.2. Effect of *trans*-2-Hexenal on the Total Lipid Content of *G. citri-aurantii*

The total lipid content of *G. citri-aurantii* cells with *trans*-2-hexenal at various concentrations (0 and 1/2 MIC) for 0, 30, 60, and 120 min was determined using the phosphovanillin method [19]. The fungal culture in PDB without *trans*-2-hexenal was used as a control.

2.11.3. Effect of *trans*-2-Hexenal on the Ergosterol Contents of *G. citri-aurantii*

The ergosterol contents of *G. citri-aurantii* cells that received different *trans*-2-hexenal treatments in PDB, as described above, were determined using the HPLC method [23]. The fungal culture in PDA without *trans*-2-hexenal was used as a control.

2.12. Real-Time Fluorescence Quantitative PCR (RT-qPCR) Analysis

The effects of *trans*-2-hexenal on the transcriptional profiles of *FKS1* (the key gene that synthesizes β-1,3-glucan synthase) and genes related to ergosterol synthesis (*ERG1*, *ERG7*, and *ERG11*) in *G. citri-aurantii* were evaluated, and the sequences were obtained from a previous RNA-Seq of *G. citri-aurantii*. RNA was extracted from *G. citri-aurantii* cells exposed to *trans*-2-hexenal at concentrations of 0 and 1/2 MIC for 0, 30, 60, and 120 min using Trizol reagent (Invitrogen, Carlsbad, CA, USA) following the manufacturer's instructions. Two micrograms of DNA-free RNA were used for the reverse transcription using M-MLV (Promega, Madison, WI, USA) with oligo dT18. The RT-qPCR was performed on a BIO-RAD CFX Connect Thermal Cycler using FastStart Universal SYBR Green Master (Roche, Basel, Switzerland). All primer pairs for the expression assays are listed in Table 1. The RT-qPCR was programmed as follows: initial denaturation at 95 °C for 10 min, followed by 40 cycles of denaturation at 95 °C for 15 s and a combined annealing and extension step at 60 °C for 1 min. The $2^{-\Delta\Delta CT}$ method was used to quantify the value of every sample using the actin gene as an internal reference [24].

Table 1. Primer pair sequences designed for validation of differentially expressed genes in control and 1/2 MIC *trans*-2-hexenal treatment of *G. citri-aurantii* using RT-qPCR.

Gene ID	Genes	Primer Sequence (5′-3′)
CL1729.Contig1_All	*FKS1*-F *FKS1*-R	AGGTTGAAGGCAAGCGTACTCT CAGGAAGTGGCTCAGGAATAGGT
Unigene4828_All	*ERG1*-F *ERG1*-R	AAGTCCTACACCTCCAAGGCTAC GAATATCGGCGTCAGTGAGAACC
Unigene3819_All	*ERG7*-F *ERG7*-R	TAACGCATATCCAGGACGACCAA CGCACAATCTCAATTCGCTCTTC
Unigene3920_All	*ERG11*-F *ERG11*-R	CGCCGTAAGGAAGGAAACATTGA AAGACGAAGTAGCAGCCGAAGT
CL313.Contig2	*Actin1*-F *Actin1*-R	TTACGCCGGTTTCTCCCTCC GACGATTTCACGCTCGGCAG

2.13. Molecular Docking

Selecting FKS1, ERG1, ERG7, and ERG11 as the receptors and using *Alphafold2* (https://colab.research.google.com/github/sokrypton/ColabFold/blob/main/AlphaFold2ipynb; accessed on 8 July 2022) for the homology modeling, the structure of the models were optimized using *ModRefiner* (https://zhanggroup.org/ModRefiner/; accessed on 8 July 2022), and the obtained models were evaluated with *SAVES v6.0* (https://saves.mbi.ucla.edu/; accessed on 20 July 2022). In this way, the three-dimensional protein structure models of ERG1, ERG7, ERG11, and FKS1 were obtained. The structure of the ligand small molecule *trans*-2-hexenal (CID: 5281168; MF: $C_6H_{10}O$) was obtained from the chemical structure database of the *PubChem* (https://pubchem.ncbi.nlm.nih.gov/; accessed on 20 July 2022) website. Using *PyMol* (Version 2.5.2) to process the receptor protein, the water molecules and metal ions were deleted. The *Openbabel* module in *PyRx* (Version 0.8) was used to minimize the ability of the ligands. The *Autodock Vina* module in *PyRx* was used for the molecular docking. The optimal docking model was selected according to the binding energy. *PyMol* and *Ligplot+* (Version 2.2.8) were used to connect the three-dimensional and two-dimensional visual analysis of the model.

2.14. Statistical Analyses

All data are expressed as the mean ± SD (standard deviation), and they were measured using three independent replicates and analyzed using one-way analysis of variance (ANOVA) followed by Duncan's test. A value of $p < 0.05$ was considered statistically significant using SPSS statistical software package release 16.0 (SPSS Inc., Chicago, IL, USA).

3. Results

3.1. Antifungal Activity of trans-2-Hexenal against G. citri-aurantii

Table 2 shows the effect of *trans*-2-hexenal on the mycelial growth of *G. citri-aurantii* in vitro. The results show that the mycelial growth considerably decreased with an increasing *trans*-2-hexenal concentration and incubation time. Mycelia growth was inhibited to different degrees at 0.25 μL/mL to 0.50 μL/mL of *trans*-2-hexenal. At a concentration of 0.50 μL/mL, the growth of *G. citri-aurantii* was completely inhibited after 2 d of incubation. As the duration of the culture was prolonged to 4 d, 72.9 ± 3.9% and 100.0 ± 0.0% of the mycelial growth was inhibited by 0.50 and 1.00 μL/mL of *trans*-2-hexenal, respectively. Thus, the MIC and MFC of *trans*-2-hexenal were 0.50 and 1.00 μL/mL, respectively.

Table 2. Antifungal activity of *trans*-2-hexenal against *G. citri-aurantii.*

Concentration (μL/mL)	Inhibitory Rate (%)				
	1 d	2 d	3 d	4 d	5 d
0.25	100.0 ± 0.0 a	53.7 ± 5.2 b	24.8 ± 1.7 c	27.1 ± 3.0 c	23.2 ± 1.4 c
0.50	100.0 ± 0.0 a	100.0 ± 0.0 a	90.5 ± 1.7 b	72.9 ± 3.9 b	64.0 ± 8.7 b
1.00	100.0 ± 0.0 a	100.0 ± 0.0 a	100.0 ± 0.0 a	100.0 ± 0.0 a	100.0 ± 0.0 a
2.00	100.0 ± 0.0 a	100.0 ± 0.0 a	100.0 ± 0.0 a	100.0 ± 0.0 a	100.0 ± 0.0 a
4.00	100.0 ± 0.0 a	100.0 ± 0.0 a	100.0 ± 0.0 a	100.0 ± 0.0 a	100.0 ± 0.0 a

Note: Mean values ± SD (standard deviation) followed by different letters (a–c) represent significantly different scores in the same phase ($p < 0.05$).

3.2. In Vivo Experiments of trans-2-Hexenal against G. citri-aurantii

trans-2-Hexenal (1× and 10× MFC) effectively reduced the decay of citrus fruit inoculated with *G. citri-aurantii* (Table 3), and the disease progression in the inoculated citrus fruit treated with *trans*-2-hexenal is presented in Figure 1. The control group began to decay within 2 d with 11 ± 4% decay, while the citrus fruit remained healthy after treatment with *trans*-2-hexenal. The citrus fruit in the 1× and 10× MFC *trans*-2-hexenal groups began to rot after 3 d and 5 d of treatment, respectively. After 7 d of storage, the incidence of fruit in the fruit wax control group reached 100%, while that in the 1× and 10× MFC treatment groups were only 85 ± 4% and 33 ± 12%, respectively.

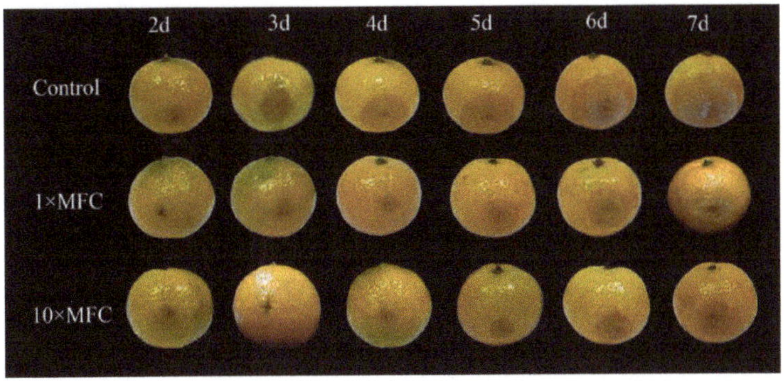

Figure 1. Disease progression in inoculated citrus fruit treated with *trans*-2-hexenal (0×, 1× and 10× MFC) during storage at 25 ± 2 °C and 80–85% RH. The data presented are the means of pooled data. Error bars indicate the SDs of the means ($n = 20$).

Table 3. Effect of *trans*-2-hexenal on the incidence of Satsuma fruit inoculated with *G. citri-aurantii.*

Treatments	Incidence Rate (%)						
	1 d	2 d	3 d	4 d	5 d	6 d	7 d
Control	0 ± 0 a	11 ± 4 a	22 ± 4 a	49 ± 4 a	73 ± 0 a	89 ± 4 a	100 ± 0 a
1× MFC *trans*-2-hexenal	0 ± 0 a	0 ± 0 b	16 ± 4 b	22 ± 8 b	60 ± 7 b	80 ± 0 b	85 ± 4 b
10× MFC *trans*-2-hexenal	0 ± 0 a	0 ± 0 b	0 ± 0 c	9 ± 0 c	22 ± 8 c	29 ± 8 b	33 ± 12 c

Note: "a–c" indicates the difference among different treatment groups with the same storage time ($p < 0.05$).

3.3. Scanning Electron Microscopy (SEM) of trans-2-Hexenal against G. citri-aurantii

The effects of *trans*-2-hexenal on the surface morphology of *G. citri-aurantii* are shown in Figure 2. The mycelia in the control group were regular in shape, uniform in thickness, smooth on the surface, healthy, and full (Figure 2A,B). However, after the 1/2 MIC *trans*-2-hexenal treatment, the surface of the mycelia became wrinkled and severely twisted, shrank, and collapsed (Figure 2C,D). The results show that *trans*-2-hexenal can change the morphology of mycelia of *G. citri-aurantii.*

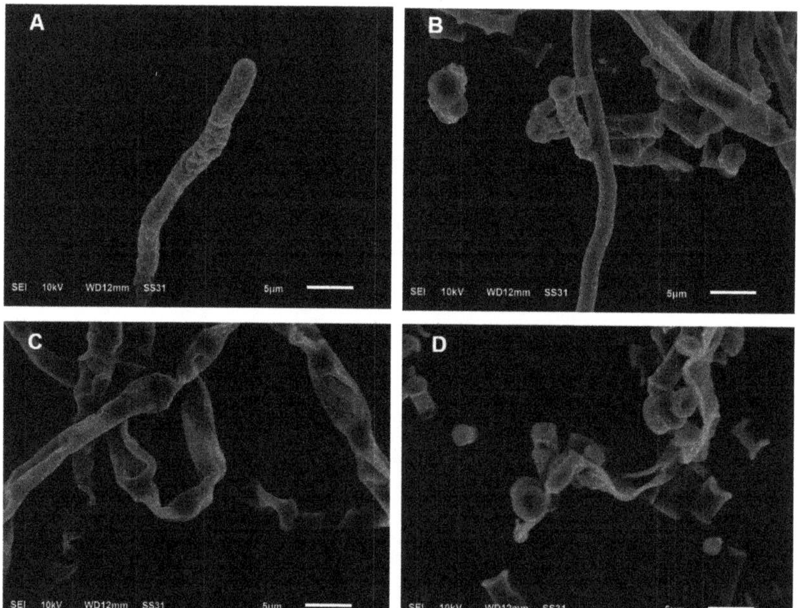

Figure 2. SEM images of (**A**,**B**) untreated control culture of *G. citri-aurantii;* (**C**,**D**) culture after incubation with the *trans*-2-hexenal.

3.4. Transmission Electron Microscopy (TEM) of trans-2-Hexenal against G. citri-aurantii

The effects of *trans*-2-hexenal on the internal morphology of *G. citri-aurantii* are shown in Figure 3. In the control group, the cells were composed of uniform cell walls, cell membranes, and cytoplasm, with uniform organelles and a complete structure (Figure 3A). After the 1/2 MIC *trans*-2-hexenal treatment, the internal morphology and ultrastructure of the *G. citri-aurantii* cells were destroyed, the cell walls became thicker, the cell membranes were irregularly twisted and, in some areas, the organelle structures were disordered, and the mitochondria were enlarged with irregular distribution (Figure 3B).

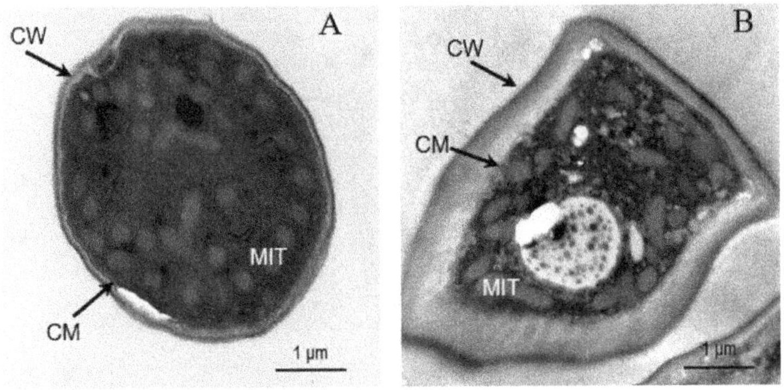

Figure 3. TEM images of (**A**) untreated control culture of *G. citri-aurantii*; (**B**) culture after incubation with the *trans*-2-hexenal. CW: cell wall; CM: cell membrane; MIT: mitochondrion.

3.5. FT−IR of *trans*-2-Hexenal against *G. citri-aurantii*

In the region of >3000 cm^{-1}, after the *trans*-2-hexenal treatment, the O–H expansion (3775 cm^{-1}) of alcohols in the carbohydrates shifted, suggesting that the *trans*-2-hexenal treatment may change the components of the cell walls and cell membranes of *G. citri-aurantii* (Figure 4). The N-H stretching (3415 cm^{-1}) of the amino groups and amide groups also shifted, which may be related to the Michael addition reaction between the α,β-unsaturated carbonyl groups in *trans*-2-hexenal and the amino groups in protein, resulting in the degradation of the cell wall proteins. In the 2800–3100 cm^{-1} region, *trans*-2-hexenal caused a significant shift of the C-H asymmetric stretching (2928 cm^{-1}) in the methyl and acyl chains, which might be related to the changes in the fatty acids caused by the condensation of the active methylene-containing compounds with aldehydes. The amides in the polypeptides and the protein (1584 cm^{-1} and 1637 cm^{-1}) in the 1500–1700 cm^{-1} region all shifted, which might be related to the peroxidation and degradation of protein. In the range of 900–1500 cm^{-1}, the peak positions at 1400 cm^{-1} and 1077 cm^{-1}, corresponding to fatty acids, proteins, polysaccharides, and nucleic acid, shifted after the treatment with *trans*-2-hexenal, suggesting that the cell wall's polysaccharides and nucleic acid were affected by *trans*-2-hexenal.

3.6. Effect of *trans*-2-Hexenal on the Cell Wall of *G. citri-aurantii*

Figure 5A shows the effect of *trans*-2-hexenal on the chitin content of *G. citri-aurantii*. There was no significant difference in the chitin content between the *trans*-2-hexenal-treated group and the control group, suggesting that *trans*-2-hexenal had no effect on the chitin content of *G. citri-aurantii*.

The content of β-1,3-glucan in the *trans*-2-hexenal-treated group was significantly lower than that in the control group (Figure 5B). At 30 min, the *trans*-2-hexenal-treated group decreased by 17.31% compared with the control group ($p < 0.05$), which indicates that the content of β-1,3-glucan significantly reduced after *trans*-2-hexenal treatment.

The activity of extracellular AKP after *trans*-2-hexenal treatment had no significant difference with the control group ($p < 0.05$, Figure 5C), which further indicates that *trans*-2-hexenal did not damage the cell wall integrity of *G. citri-aurantii*.

3.7. Effect of *trans*-2-Hexenal on the Cell Membrane of *G. citri-aurantii*

According to the results of the PI staining (Figure 6A), the fluorescence value of the mycelia treatment group after the 1/2 MIC *trans*-2-hexenal treatment for 60 min was 1.24 times that of the control group, which was significantly higher than that of the control group ($p < 0.05$), indicating that the integrity of the cell membrane might begin to be destroyed after 60 min.

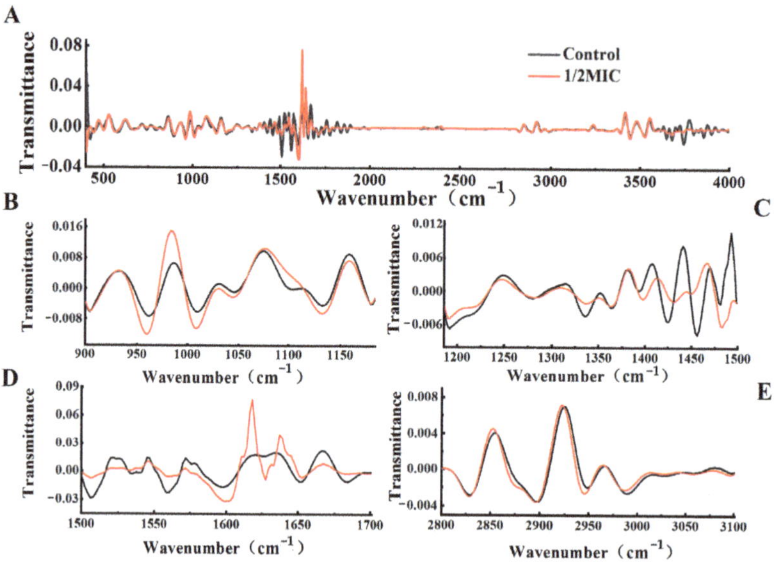

Figure 4. FT−IR spectra analysis of *G. citri-aurantii* under *trans*-2-hexenal treatment. (**A**) FT−IR spectra of the second derivative (400–4000 cm^{-1}); (**B**) the second derivative of FT−IR spectra of the nucleic acid and polysaccharide region (900−1200 cm^{-1}); (**C**) the second derivative of FT−IR spectra of the mixed region of fatty acid, protein, and polysaccharide (1200−1500 cm^{-1}); (**D**) the second derivative of FT−IR spectra of the mixed region (1500−1700 cm^{-1}); (**E**) the second derivative of FT−IR spectra of the lipid region (2800−3100 cm^{-1}).

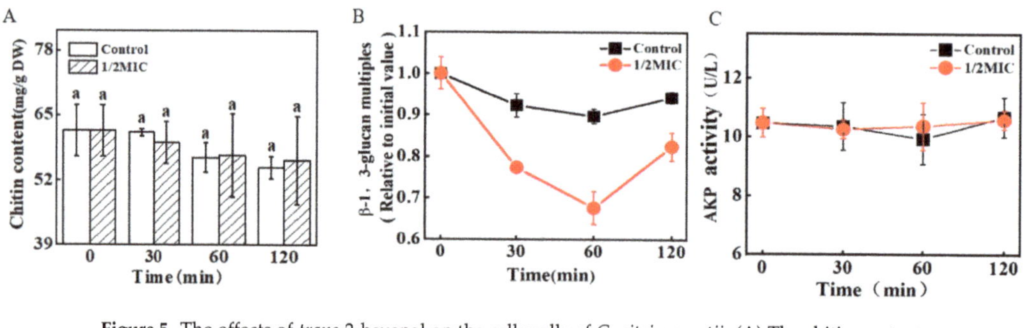

Figure 5. The effects of *trans*-2-hexenal on the cell walls of *G. citri-aurantii*. (**A**) The chitin content of *G. citri-aurantii*; (**B**) the *β*-1,3-glucan content of *G. citri-aurantii*; (**C**) the extracellular AKP activity of *G. citri-aurantii*. The data presented are the means of pooled data. Error bars indicate the SDs of the means (n = 3). "a" indicated that there was no difference between different treatment groups ($p < 0.05$).

The effect of *trans*-2-hexenal on the total lipid content of *G. citri-aurantii* is shown in Figure 6B. After the 1/2 MIC *trans*-2-hexenal treatment for 60 min, the total lipid content was 123.8 ± 1.6 mg/g DW, which was significantly lower than that of the control group (185.4 ± 13.3 mg/g DW) ($p < 0.05$), suggesting that the integrity of the cell membrane of *G. citri-aurantii* was destroyed, and the cells were damaged after the 1/2 MIC *trans*-2-hexenal treatment for 60 min.

The ergosterol content (3.54 ± 0.07 mg/g DW) of *G. citri-aurantii* was significantly lower than that of the control group (5.89 ± 0.24 mg/g DW) ($p < 0.05$) after 30 min of treatment with *trans*-2-hexenal, suggesting that *trans*-2-hexenal could affect the synthesis of ergosterol (Figure 6C).

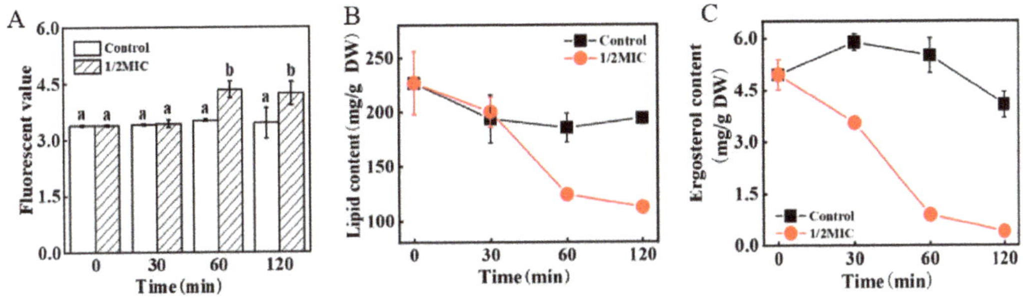

Figure 6. The effects of *trans*-2-hexenal on the cell membranes of *G. citri-aurantii*. (**A**) The plasma membrane integrity of *G. citri-aurantii* mycelia and mycelia fluorescence times; (**B**) the total lipids contents and (**C**) ergosterol contents of *G. citri-aurantii* mycelia. The data presented are the means of pooled data. Error bars indicate the SDs of the means ($n = 3$). "a,b" indicates the difference among different treatment groups ($p < 0.05$).

3.8. RT-qPCR

In order to determine the effect of *trans*-2-hexenal on β-1,3-glucan and ergosterol, the key biosynthesis genes related to β-1,3-glucan biosynthesis and ergosterol biosynthesis were selected for RT-qPCR analysis. As revealed in Figure 7, the expression levels of *FKS1*, *ERG1*, *ERG7*, and *ERG11* were significantly lower than the control samples in the whole *trans*-2-hexenal treatment period.

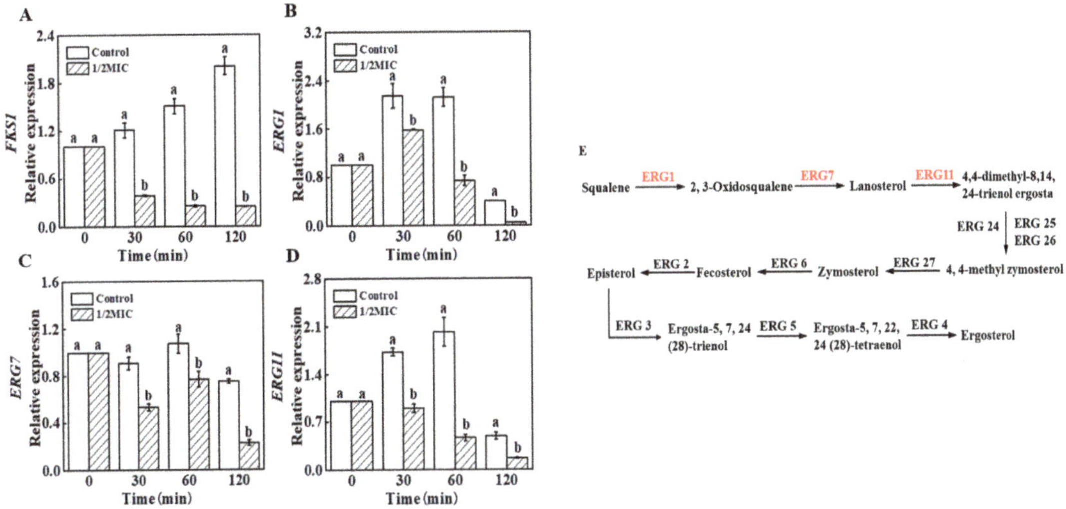

Figure 7. Changes in the expression of the β-1,3-glucan and ergosterol biosynthesis genes of *G. citri-aurantii* by control and 1/2 MIC of *trans*-2-hexenal for 0, 30, 60, and 120 min ((**A**) *FKS1*; (**B**) *ERG1*; (**C**) *ERG7*; (**D**) *ERG11*). (**E**) Ergosterol biosynthesis pathway. The data presented are the means of pooled data. Error bars indicate the SDs of the means ($n = 3$). "a,b" indicates the difference among different treatment groups ($p < 0.05$).

3.9. Molecular Docking

Using *Alphafold2*, FKS1, ERG1, ERG7, and ERG11 were homology modeled; then, *trans*-2-hexenal was docked against the FKS1, ERG1, ERG7, and ERG11 protein, respectively (Figure 8). The binding energies of *trans*-2-hexenal to the best conformations of FKS1, ERG1, ERG7, and ERG11 were −4.9 kcal/mol, −4.6 kcal/mol, −4.1 kcal/mol, and −3.9 kcal/mol, respectively. Because of the different binding conformations of the ligand molecules in the conjugates, the amino acid residues around the active sites of their interactions and the strength of their interactions were different. *trans*-2-Hexenal formed a hydrophobic interaction with Ile1260, Gly1261, Ile1262, Leu1263, Leu1348, Tyr1587, Pro1589, Phe1868, and Leu1867 in FKS1, while the aldehyde group (-COH) of *trans*-2-hexenal acted as an acceptor to form a hydrogen bond with a bond length of 2.88 Å with the side chain of Gly1264 in FKS1; the binding of *trans*-2-hexenal to FKS1 with that hydrophobic interaction and hydrogen bonding promote the binding of *trans*-2-hexenal to FKS1. trans-2-Hexenal formed hydrophobic interactions with amino acid residues Pro399, Leu363, Leu403, Tyr81, Leu367, Val211, Ile83, Leu92, Phe37, and Phe389 in ERG1; His724, Ile725, Glu726, Glu736, Gln127, Tyr722, and Pro405 in ERG7; and Leu371, Val494, Gln490, Tyr493, Asn378, and Gln372 in ERG11. Hydrophobic interactions provide the main driving force for the combination of *trans*-2-hexenal with ERG1, ERG7, and ERG11, which results in altered interactions among amino acid residues within the protein molecule and plays an important role in maintaining protein conformation.

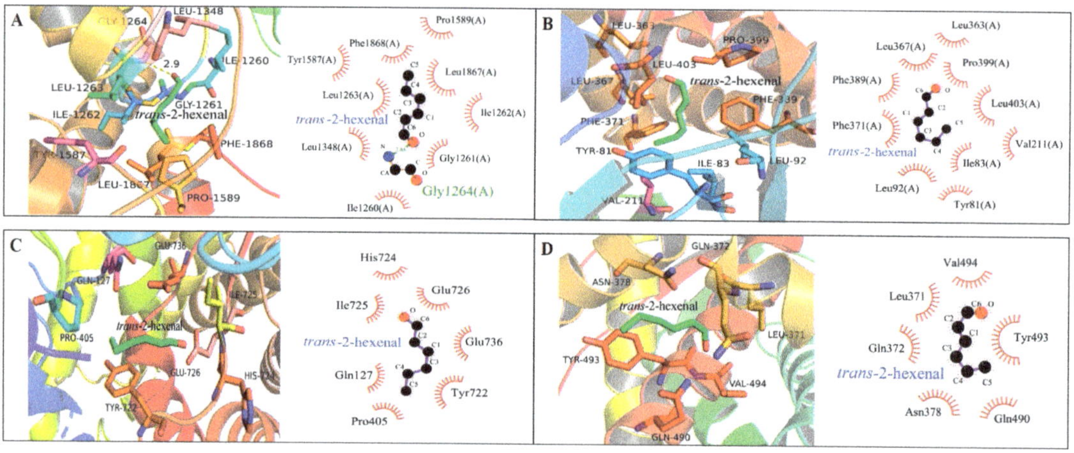

Figure 8. The 3D images of FKS1 (**A**), ERG1 (**B**), ERG7 (**C**), and ERG11 (**D**) proteins and the 2D interaction diagram of *trans*-2-hexenal with FKS1 (**A**), ERG1 (**B**), ERG7 (**C**), and ERG11 (**D**) proteins. The dark green rod-like structure in the diagram is *trans*-2-hexenal, and the other color rod-like structures are amino acid residues.

4. Discussion

G. citri-aurantii is considered one of the most important pathogens in terms of the economic losses that hinder citrus worldwide [2,18,19]. In the present study, *trans*-2-hexenal showed pronounced antifungal efficacy against *G. citri-aurantii* in vitro and in vivo. These results are consistent with those of previous studies with regard to the antifungal activity of *trans*-2-hexenal [9,10,13,15], indicating that the application of *trans*-2-hexenal is a potent method for controlling *G. citri-aurantii*.

The fungal cell wall is a dynamic structure that protects cells from osmotic pressure and other environmental stresses. It is of great significance to the growth of fungal cells. The destruction of the cell wall's structure easily leads to the cell's dissolution and death. Plant essential oils, such as tea tree essential oil, *Hyssopus officinalis* essential oil, cinnamaldehyde, and thymol, can destroy the cell wall of fungi [10,25,26]. Cinnamaldehyde mainly inhibits the growth of *G. citri-aurantii* by inhibiting chitin synthesis, accelerating chitin hydrolysis, and destroying the integrity of the cell wall [19]. In this study, *trans*-2-hexenal treatment did not change the chitin content and cell wall integrity, but the β-1,3-glucan content of *G. citri-aurantii* decreased significantly. These results are similar to the results of Li et al. [27] in which o-vanillin mainly reduced the β-1,3-glucan content in the cell wall of *A. flavus*, but the content of chitin did not change. FKS family genes (*FKS1*, *FKS2*, and *FKS3*) are the key genes that synthesize β-1,3-glucan synthase, and *FKS1* is an essential gene, in most, that can regulate β-1,3-glucan synthase activity [28]. García et al. [29] found that the deletion of *FKS1* in yeast would lead to a decrease in the glucan content and slow the growth of the strain. The RT-qPCR results show that *trans*-2-hexenal decreased the expression of *FKS1*, which is consistent with the decrease in the β-1,3-glucan content, indicating that the β-1,3-glucan synthesis pathway was inhibited. Inhibitors of β-1,3-glucan synthase affect gene expression by targeting key groups of synthases, which, in turn, leads to a decrease in the content of β-1,3-glucan in the cell wall, resulting in cell rupture and death [30]. Douglas et al. [31] found that echinocandin binds to the catalytic subunit of β-1,3-glucan synthase to reduce the integrity of the fungal cell wall. Molecular docking showed that the hydrogen bond and hydrophobic interaction between *trans*-2-hexenal and FKS1 lead to a change in the FKS1 protein's structure, rotatable bond dihedral angle, and amino acid residue side chain, which reduces its expression and interferes with the normal function of FKS1. *Trans*-2-hexenal inhibited the synthesis of the cell wall by combining different amino acid residues with the active site of FKS1 in the cell wall of *G. citri-aurantii*, which is consistent with the antibacterial mechanism of eugenol and citral against *P. roqueforti* and *A. niger* [32]. The decrease in β-1,3-glucan content in the cell wall may lead to an increase in fungal cell wall permeability, as evidenced by the increase in extracellular AKP activity of fungi [26]. However, despite the decrease in β-1,3-glucan content, the cell wall became thicker and the activity of extracellular AKP activity remained unchanged, suggesting that *trans*-2-hexenal did not destroy the cell wall permeability of *G. citri-aurantii* but induced its cell wall components to restructure, resulting in a decrease in the β-1,3-glucan content.

The integrity and fluidity of the cell membrane are crucial for the survival and growth of fungi, and they are key points in many kinds of drug treatment [33]. The fungal cell membrane is the protective barrier of fungal cells, and it is selective for substances to enter and leave the cells. Therefore, damage to the integrity of the cell membrane will cause the leakage of components, leading to cell apoptosis [34,35]. Lipids are an important component of the cell membrane, and they play important roles in cell membrane integrity and material transportation [36,37]. Ergosterol, as a unique component of fungal cell membranes, plays important roles in cell membrane fluidity and membrane protein function [38–40]. It was found that the synthesis of ergosterol in *A. flavus* cell membranes was significantly inhibited after being treated with *Artemisia annua* essential oil; thus, the fungal cell membranes were the main site [41]. After being treated with *trans*-2-hexenal, the lipid content of *G. citri-aurantii* decreased at 60 min. The ergosterol content in the treatment group decreased at 30 min, indicating that ergosterol was the primary target of *trans*-2-hexenal in inhibiting *G. citri-aurantii*. *ERG1*, *ERG7*, and *ERG11* are key genes in the sterol biosynthesis pathway, encoding squalene epoxidase, lanosterol synthase, and lanosterol-14-α-demethylase in the sterol biosynthesis pathway, respectively [19,42]. *ERG1*, *ERG7*, and *ERG11* in *G. citri-aurantii* were significantly downregulated after *trans*-2-hexenal treatment, which is consistent with the results of the ergosterol content, further indicating that the ergosterol synthesis pathway was attacked. Meanwhile, it is worth noting that the molecular docking analysis of the formed complexes shows that the formation of hydrophobic interactions may lead to a longer residence time of *trans*-2-hexenal in ERG1, ERG7, and ERG11, and the hydrophobic interactions may enhance the effect of *trans*-2-hexenal on the ergosterol synthesis pathway. This result is consistent with that of Song et al. [43], who found that *trans*-2-hexenal fumigation reduced the ergosterol content of *Botrytis cinerea* and influenced ergosterol biosynthetic gene expression levels.

5. Conclusions

In summary, *trans*-2-hexenal reduced the β-1,3-glucan content and ergosterol content of *G. citri-aurantii* by inhibiting the expressions of related genes. With the prolongation of the treatment time, the total lipid content decreased, and the cell membrane integrity was further destroyed.

Author Contributions: Conceptualization, Q.O.; data curation, Q.O.; formal analysis, Q.O.; funding acquisition, Q.O.; investigation, Q.O., S.S., Y.L. and Y.Y.; methodology, S.S., Y.L., Y.Z., X.Y. and N.T.; supervision, Q.O.; writing—original draft, Q.O.; writing—review & editing, Q.O., N.T. and L.L. All authors have read and agreed to the published version of the manuscript.

Funding: This study was supported by the National Natural Science Foundation of China, grant number 32001755, Scientific Research Fund of Hunan Provincial Education Department, grant number 22B0165, and Hunan Provincial Natural Science Foundation of China, grant number 2021JJ40533.

Institutional Review Board Statement: Not applicable.

Informed Consent Statement: Not applicable.

Data Availability Statement: The datasets generated during and/or analyzed during the current study are available from the corresponding author upon reasonable request.

Conflicts of Interest: The authors declare no conflict of interest.

References

1. Ferraz, L.P.; Cunha, T.D.; da Silva, A.C.; Kupper, K.C. Biocontrol ability and putative mode of action of yeasts against *Geotrichum citri-aurantii* in citrus fruit. *Microbiol. Res.* **2016**, *188–189*, 72–79. [CrossRef] [PubMed]
2. Xu, L.X.; Feng, L.Y.; Sun, J.; Mao, L.T.; Li, X.J.; Jiang, Y.M.; Duan, X.W.; Li, T.T. Antifungal activities of a natural trisaccharide ester against sour rot in mandarin fruit. *Postharvest Biol. Technol.* **2022**, *191*, 111981. [CrossRef]
3. McKay, A.H.; Förster, H.; Adaskaveg, J.E. Efficacy and application strategies for propiconazole as a new postharvest fungicide for managing sour rot and green mold of citrus fruit. *Plant Dis.* **2012**, *96*, 235–242. [CrossRef] [PubMed]
4. Zhao, J.; Zhang, D.Y.; Wang, Z.; Tian, Z.H.; Fan, Y.; Lu, X.J.; Long, C.A. Genome sequencing and transcriptome analysis of *Geotrichum citri-aurantii* on citrus reveal the potential pathogenic- and guazatine-resistance related genes. *Genomics* **2020**, *112*, 4063–4071. [CrossRef] [PubMed]
5. Cheng, X.; Yang, Y.; Zhu, X.; Yuan, P.; Shan, Y. Inhibitory mechanisms of cinnamic acid on the growth of *Geotrichum citri-aurantii*. *Food Control* **2022**, *131*, 108459. [CrossRef]
6. Li, Y.H.; Shao, X.F.; Xu, J.Y.; Wei, Y.Y.; Xu, F.; Wang, H.F. Tea tree oil exhibits antifungal activity against *Botrytis cinerea* by affecting mitochondria. *Food Chem.* **2017**, *234*, 62–67. [CrossRef] [PubMed]
7. Wuryatmo, E.; Able, A.J.; Ford, C.M.; Scott, E.S. Effect of volatile citral on the development of blue mould, green mould and sour rot on navel orange. *Australas. Plant Path.* **2014**, *43*, 403–411. [CrossRef]
8. Bazioli, J.M.; Belinato, J.R.; Costa, J.H.; Akiyama, D.Y.; Fill, T.P. Biological control of citrus postharvest phytopathogens. *Toxins* **2019**, *11*, 460. [CrossRef]
9. Zhang, Y.H.; Ouyang, Q.L.; Duan, B.; Reymick, O.O.; Chen, Y.; Tan, Y.Z.; Zhu, X.R.; Su, D.L.; Li, G.Y.; Tao, N.G. Trans-2-hexenal/beta-cyclodextrin effectively reduces green mold in citrus fruit. *Postharvest Biol. Technol.* **2022**, *187*, 111871. [CrossRef]
10. Arroyo, F.T.; Moreno, J.; Daza, P.; Boianova, L.; Romero, F. Antifungal activity of strawberry fruit volatile compounds against *Colletotrichum acutatum*. *J. Agric. Food Chem.* **2007**, *55*, 5701–5707. [CrossRef]
11. Myung, K.; Hamilton-Kemp, T.R.; Archbold, D.D. Interaction with and Effects on the profile of proteins of *Botrytis cinerea* by C6 aldehydes. *J. Agric. Food Chem.* **2007**, *55*, 2182–2188. [CrossRef] [PubMed]
12. Wang, X.H.; Fu, M.R.; Qu, X.Q.; Liu, J.J.; Bu, J.W.; Feng, S.R.; Zhao, H.D.; Jiao, W.X.; Sun, F. (E)-2-Hexenal-based coating induced acquired resistance in apple and its antifungal effects against *Penicillium expansum*. *LWT-Food Sci. Technol.* **2022**, *163*, 113536. [CrossRef]
13. Zhang, J.; Tian, H.; Sun, H.; Wang, X. Antifungal activity of trans-2-hexenal against *Penicillium cyclopium* by a membrane damage mechanism. *J. Food Biochem.* **2017**, *41*, e12289. [CrossRef]
14. Dong, Y.P.; Li, C.Y.; Long, H.T.; Liu, Z.T.; Huang, Y.; Zhang, M.; Wang, T.L.; Liu, Y.X.; Bi, Y.; Prusky, D.B. Preparation and use of trans-2-hexenal microcapsules to preserve 'Zaosu' pears. *Sci. Hortic.* **2021**, *283*, 110091. [CrossRef]
15. Guo, M.; Feng, J.; Zhang, P.; Jia, L.; Chen, K. Postharvest treatment with trans-2-hexenal induced resistance against *Botrytis cinerea* in tomato fruit. *Australas. Plant Path.* **2015**, *44*, 121–128. [CrossRef]
16. Ma, W.; Zhao, L.; Zhao, W.; Xie, Y.L. (E)-2-Hexenal, as a potential natural antifungal compound, inhibits *Aspergillus flavus* spore germination by disrupting mitochondrial energy metabolism. *J. Agric. Food Chem.* **2019**, *67*, 1138–1145. [CrossRef]
17. Wu, Y.L.; Ouyang, Q.L.; Tao, N.G. Plasma membrane damage contributes to antifungal activity of citronellal against *Penicillium digitatum*. *J. Food Sci. Technol.* **2016**, *53*, 3853–3858. [CrossRef]

18. Dou, S.W.; Ouyang, Q.L.; You, K.Y.; Qian, J.J.; Tao, N.G. An inclusion complex of thymol into β-cyclodextrin and its antifungal activity against *Geotrichum citri-aurantii*. *Postharvest Biol. Technol.* **2018**, *138*, 31–36. [CrossRef]
19. Ouyang, Q.L.; Duan, X.F.; Li, L.; Tao, N.G. Cinnamaldehyde exerts its antifungal activity by disrupting the cell wall integrity of *Geotrichum citri-aurantii*. *Front. Microbiol.* **2019**, *10*, 55. [CrossRef]
20. Wang, Y.; Feng, K.; Yang, H.; Yuan, Y.H.; Yue, T.L. Antifungal mechanism of cinnamaldehyde and citral combination against *Penicillium expansum* based on FT-IR fingerprint, plasma membrane, oxidative stress and volatile profile. *RSC Adv.* **2018**, *8*, 5806–5815. [CrossRef]
21. Francois, J.M. A simple method for quantitative determination of polysaccharides in fungal cell walls. *Nat. Protoc.* **2016**, *1*, 2995–3000. [CrossRef]
22. Fortwendel, J.R.; Juvvadi, P.R.; Pinchai, N.; Perfect, B.Z.; Alspaugh, J.A.; Perfect, J.R.; Steinbach, W.J. Differential effects of inhibiting chitin and 1,3-β-D-glucan synthesis in ras and calcineurin mutants of *Aspergillus fumigatus*. *Antimicrob. Agents Chemother.* **2009**, *53*, 476–482. [CrossRef] [PubMed]
23. Ouyang, Q.L.; Okwong, R.O.; Chen, Y.P.; Tao, N.G. Citronellal exerts its antifungal activity by targeting ergosterol biosynthesis in *Penicillium digitatum*. *J. Fungi* **2021**, *7*, 432. [CrossRef] [PubMed]
24. Livak, K.J.; Schmittgen, T.D. Analysis of relative gene expression data using real-time quantitative PCR and the $2^{-\Delta\Delta CT}$ method. *Methods* **2021**, *25*, 402–408. [CrossRef]
25. Ghfir, B.; Fonvieille, J.L.; Dargent, R. Influence of essential oil of *Hyssopus officinalis* on the chemical composition of the walls of *Aspergillus fumigatus* (Fresenius). *Mycopathologia* **1997**, *138*, 7–12. [CrossRef] [PubMed]
26. Shao, X.; Cheng, S.; Wang, H.; Yu, D.; Mungai, C. The possible mechanism of antifungal action of tea tree oil on *Botrytis cinerea*. *J. Appl. Microbiol.* **2013**, *11114*, 1642–1649. [CrossRef]
27. Li, Q.; Zhu, X.M.; Xie, Y.L.; Zhong, Y. o-Vanillin, a promising antifungal agent, inhibits *Aspergillus flavus* by disrupting the integrity of cell walls and cell membranes. *Appl. Microbiol. Biotechnol.* **2021**, *105*, 5147–5158. [CrossRef]
28. Healey, K.R.; Paderu, P.; Hou, X.; Ortigosa, C.J.; Bagley, N.; Patel, B.; Zhao, Y.N.; Perlin, D.S. Differential regulation of echinocandin targets Fks1 and Fks2 in *Candida glabrataby* the Post-Transcriptional Regulator Ssd1. *J. Fungi* **2020**, *4*, 143. [CrossRef]
29. García-Rodriguez, L.J.; Trilla, J.A.; Castro, C.; Valdivieso, M.H.; Durán, A.; Roncero, C. Characterization of the chitin biosynthesis process as a compensatory mechanism in the fks1 mutant of *Saccharomyces cerevisiae*. *FEBS Lett.* **2000**, *478*, 84–88. [CrossRef]
30. Ma, L.; Salas, O.; Bowler, K.; Bar-Peled, M.; Sharon, A. UDP-4-keto-6-deoxyglucose, a transient antifungal metabolite, weakens the fungal cell wall partly by inhibition of UDP-galactopyranose mutase. *MBio* **2017**, *8*, e01559-17. [CrossRef]
31. Douglas, C.M.; Foor, F.; Marrinan, J.A.; Morin, N.; Nielsen, J.B.; Dahl, A.M.; Mazur, P.; Baginsky, W.; Li, W.; el-Sherbeini, M. The *Saccharomyces cerevisiae* FKS1 (ETG1) gene encodes an integral membrane protein which is a subunit of 1,3-β-D-glucan synthase. *Proc. Natl. Acad. Sci. USA* **1994**, *91*, 12907–12911. [CrossRef] [PubMed]
32. Ju, J. *Study on the Synergistic Inhibitory Mechanism of Eugenol and Citral against Penicillium Roqueforti and Aspergillus niger*; Jiangnan University: Wuxi, China, 2021; pp. 27–35.
33. Li, L.; Tang, X.; Ouyang, Q.L.; Tao, N.G. Combination of sodium dehydroacetate and sodium silicate reduces sour rot of citrus fruit. *Postharvest Biol. Technol.* **2019**, *151*, 19–25. [CrossRef]
34. Liu, K.; Zhou, X.; Fu, M. Inhibiting effect of epsilon-poly-lysine (ε-PL) on *Pencillium digitatum* and its involved mechanism. *Postharvest Biol. Technol.* **2017**, *123*, 94–101. [CrossRef]
35. Sant, D.G.; Tupe, S.G.; Ramana, C.V.; Deshpande, M.V. Fungal cell membrane-promising drug target for antifungal therapy. *J. Appl. Microbiol.* **2016**, *121*, 1498–1510. [CrossRef]
36. Segawa, K.; Nagata, S. An apoptotic 'eat me'signal: Phosphatidylserine exposure. *Trends Cell Biol.* **2015**, *25*, 639–650. [CrossRef]
37. Vartabedian, V.F.; Savage, P.B.; Teyton, L. The processing and presentation of lipids and glycolipids to the immune system. *Immunol. Rev.* **2016**, *272*, 109–119. [CrossRef]
38. Dupont, S.; Lemetais, G.; Ferreira, T.; Cayot, P.; Gervais, P.; Beney, L. Ergosterol biosynthesis: A fungal pathway for life on land? *Evolution* **2012**, *66*, 2961–2968. [CrossRef]
39. Krumpe, K.; Frumkin, I.; Herzig, Y.; Rimon, N.; Oezbalci, C.; Bruegger, B.; Rapaport, D.; Schuldiner, M. Ergosterol content specifies targeting of tail-anchored proteins to mitochondrial outer membranes. *Mol. Biol. Cell* **2012**, *23*, 3927–3935. [CrossRef]
40. Zhang, Y.Q.; Gamarra, S.; Garcia-Effron, G.; Park, S.; Perlin, D.S.; Rao, R. Requirement for ergosterol in V-ATPase function underlies antifungal activity of azole drugs. *PLoS Pathog.* **2010**, *6*, e1000939. [CrossRef]
41. Kumar, M.; Dwivedy, A.K.; Sarma, P.; Dkhar, M.S.; Kayang, H.; Raghuwanshi, R.; Dubey, N.K. Chemically characterised *Artemisia nilagirica* (Clarke) Pamp. essential oil as a safe plant-based preservative and shelf-life enhancer of millets against fungal and aflatoxin contamination and lipid peroxidation. *Plant Biosyst.* **2020**, *154*, 269–276. [CrossRef]
42. Bhattacharya, S.; Esquivel, B.D.; White, T.C. Overexpression or deletion of ergosterol biosynthesis genes alters doubling time, response to stress agents, and drug susceptibility in *Saccharomyces cerevisiae*. *mBio* **2018**, *9*, e01291-18. [CrossRef] [PubMed]
43. Song, G.; Du, S.L.; Sun, H.L.; Liang, Q.W.; Wang, H.H.; Yan, L.M.; Zhang, J.H. Antifungal mechanism of (E)-2-hexenal against *Botrytis cinerea* growth revealed by transcriptome analysis. *Front. Microbiol.* **2022**, *13*, 951751. [CrossRef] [PubMed]

Journal of *Fungi*

MDPI

Article

Insights into the Isolation, Identification, and Biological Characterization Analysis of and Novel Control Strategies for *Diaporthe passiflorae* in Postharvest Passion Fruit

Huiling Wang [1,2], Hongbin Chen [1,*], Yu Lin [3], Meiling Li [2], Qingqing Liu [2], Yuzhao Lin [1], Xuanjing Jiang [1] and Yihui Chen [2,*]

1 College of Oceanology and Food Science, Quanzhou Normal University, Quanzhou 362000, China
2 Institute of Postharvest Technology of Agricultural Products, College of Food Science, Fujian Agriculture and Forestry University, Fuzhou 350002, China
3 Department of Intelligent Manufacturing, MinXi Vocational and Technical College, Longyan 364021, China
* Correspondence: yummyway@163.com (H.C.); harris2197395@163.com (Y.C.)

Abstract: Postharvest diseases seriously restrict developments in the passion fruit industry. In this study, we aimed to identify the postharvest pathogen affecting passion fruit, investigate its pathogenicity, and explore relevant control methods. The pathogen was isolated from rotting passion fruit and identified using morphological characteristics, ITS sequences, and phylogenetic tree analyses. Additionally, preliminary studies were conducted to assess the biological characteristics of the pathogen and evaluate the efficacy of various treatments for disease control. The fungus on the passion fruit called B4 was identified as *Diaporthe passiflorae*. Optimal conditions for mycelial growth were observed at 25–30 °C and pH 5–6, with starch as the carbon source and peptone as the nitrogen source. Infection by *D. passiflorae* accelerated fruit decay, reduced the $h°$ value of the peel, and increased the peel cell membrane permeability when compared to the control. Notably, treatments with appropriate concentrations of ε-poly-l-lysine, salicylic acid, and melatonin showed inhibitory effects on the pathogen's growth in vitro and may thus be potential postharvest treatments for controlling brown rot caused by *D. passiflorae* in passion fruit. The results provide a scientific basis for the development of strategies to control postharvest decay and extend the storage period of passion fruit.

Keywords: *Passiflora caerulea*; postharvest diseases; *Diaporthe passiflorae*; biological characteristics; antifungal

1. Introduction

Passion fruit (*Passiflora cerulean* L.) is a perennial, evergreen, climbing woody vine that originates from South America but is now widely cultivated in tropical and subtropical regions worldwide [1]. Brazil is the world's largest producer of passion fruit, with an annual production of 683,000 tons in 2021, covering an area of about 45,000 hectares and generating an income of about USD 286 million for farmers [2]. In China, passion fruit is mainly grown in the provinces of Fujian, Hainan, Guangdong, and Guangxi [3]. In 2019, the cultivation area of passion fruit in China was about 30,000 hectares, and the annual production was about 600,000 tons [4]. Passion fruit produces a unique aroma for a tropical fruit, as it contains the scents of apple, mango, pineapple, banana, lemon, and litchi fruit. Its juice is sweet and sour, with a unique flavor, and it is known colloquially as the "king of fruit juices" [5–7]. Passion fruit is also rich in various phenolic compounds, dietary fiber, pectin, carotene, vitamins, and other active substances required by the human body, and it has antioxidant, antihypertensive, antitumor, hypolipidemic, and other medicinal values [8–10]. Owing to its unique flavor and high nutritional value, passion fruit is sought after by consumers, and the market demand for passion fruit is increasing, thereby leading to its high commercial value; thus, the planting area of passion fruit is increasing. For example, the planted area of passion fruit in China increased from less than 700 hectares in 2011 to more than 30,000 hectares in 2019, with an annual production of about 600,000 tons [4,11].

However, as passion fruit ripens during the high-temperature season and has a high level of respiratory action [12,13], the peel can easily lose its luster and experience water loss, shrinkage, and rotting, which makes storage challenging [14,15]. Additionally, passion fruit is susceptible to pathogen infections that lead to fruit rot. These problems seriously affect the quality and commercial value of passion fruit. Globally, several fungal pathogens have been identified as causing passion fruit decays, including *Diaporthe infercuda* [16], *Phytophthora drechsleri* [17], *Phytophthora nicotianae* [18], *Colletotrichum brevisporum* [19], *Lasiodiplodia theobromae* [20], and *Trichothecium roseum* [21].

To improve the shelf life of passion fruit and control the occurrence of postharvest diseases, chemical preservation treatments are currently widely used in the market [4,12,22]. However, the prolonged use of chemical fungicides can lead to a series of problems, including the development of pathogen resistance, the contamination of the environment, and harm to human health [23,24]. Therefore, it is necessary to find new safety strategies to control the occurrence of postharvest diseases. ε-poly-l-lysine (ε-PL) is a natural preservative with easy solubility and nontoxic, harmless, and antifungal characteristics, and it has been widely used in the food industry [25,26]. Both salicylic acid (SA) and melatonin (MT) treatments can delay fruit senescence and help maintain the postharvest quality [27–30]. Therefore, this study aimed to isolate and identify the pathogen causing brown rot in passion fruit, as well as its biological characteristics, and to explore the effects of ε-PL, SA, and MT treatments on the fungal pathogen in vitro.

2. Materials and Methods

2.1. Materials and Reagents

The 'Fujian Passion Fruit No. 3' passion fruits were collected from an orchard in Nan'an City, Fujian Province, China, and shipped to Quanzhou on the same day. Fruits of a uniform size, similar color, consistent maturity, no disease, no damage, and that were apparently healthy were selected for the experiment. A total of 390 passion fruits were selected after cleaning and subjected to the following treatments. Among them, 30 passion fruits were used for measuring the initial fruit indicators (day 0). The remaining 360 passion fruits were randomly divided into two groups, with 180 fruits in each group. One group was punched, and the other group was punched and placed with 8 mm *D. passiflorae* plugs. After treatment, 10 fruits were packed in polyethylene film bags (0.015 mm thick) and stored at 25 ± 1 °C and 85% relative humidity for 6 days. In the experiment, 30 passion fruits (3 bags) from each treatment were randomly selected every day for physiological index analysis.

2.2. Isolation and Purification of Pathogens

Postharvest passion fruits were packed in 0.015 mm polyethylene cling film bags and stored at 28 ± 1 °C and 85% relative humidity (RH) until the onset of decay. The margins (4 × 4 mm) between the symptomatic and healthy tissue were cut from the rotten fruit. Then, the surfaces were disinfected and cleaned with 75% ethanol [31]. The isolated tissue blocks were transferred to potato glucose agar (PDA) medium. Finally, the plates were cultured at constant temperature (28 °C) and humidity (85% RH) in an artificial climate chamber. When the colonies reached a diameter of approximately 3 cm, the mycelia at the edges of the colonies were placed into new PDA medium for purification, and this was repeated 3–4 times to obtain purified strains.

2.3. Morphological Identification of the Pathogen

The fungus plugs (Ø = 5 mm) were removed from the edge of the colony, inoculated on the PDA medium, and cultured in the artificial climate chamber with constant temperature (28 °C), humidity (85%), and darkness condition. The morphology of the pathogen colonies was observed, photographed, and recorded daily until the colonies covered the Petri dish on the fifth day. The prepared Sabouraud agar plate was cut into several squares (1 cm^2) using a sterile inoculant needle via an aseptic procedure and placed on the agar plate [32,33]. The pathogens to be tested were inoculated on the upper part of the surrounding edge of the agar block, and then a sterile cover glass was placed on the agar with sterile tweezers and they were cultured at 28 °C under a plate lid. After the spores had grown on a sterile slide, a drop of normal saline was added, and the cultured cover slide was placed on the slide using sterile tweezers. Finally, the characteristics of the spores were observed and photographed with a light microscope (DM2000 LED, Leica, Wetzlar, Germany). Morphological characteristics were identified using relevant guidelines [34,35].

2.4. Pathogenicity Detection of the Pathogen

Pathogenicity testing was performed according to the method described by Chen et al. [31]. Healthy, disease-free, mature, and uniform-sized passion fruits were selected and washed with deionized water and dried naturally. The selected fruits were divided into three groups of 10 fruits each. A small hole (Ø = 8 mm) was punched on the equatorial surface of the passion fruit using a sterile puncher. After the wound was dried, the fungal plug (Ø = 8 mm) was inoculated, and the treatments without fungal plug inoculation were used as the controls. The fruits were then placed in an artificial climate box with constant conditions of 28 °C and 85% RH for preservation and photographed every day.

2.5. Molecular Biological Identification of Pathogens

DNA extraction, PCR amplification, and pathogen gene sequencing were performed by Qingdao Yixin Testing Technology Service (Qingdao, China). The ITS, TUB, and TEF-1α sequencing results of strain B4 were analyzed using the Basic Local Alignment Search Tool (BLAST) in the GenBank database. Subsequently, by comparing the similarity between the B4 strain sequence and the existing sequences in the database, the sequences with high similarity were obtained. Finally, DNAMAN software (Version 9.0, LynnonBiosoft, Chicago, IL, USA) was used for sequence comparison and homology analysis, and MEGA11.0 (Mega Limited, Auckland, New Zealand) was used for neighbor-joining analysis to construct the phylogenetic tree [36].

2.6. Biological Characteristics of the Pathogens

The biological properties were determined as described by Gui et al. [37] with slight modifications. The pathogen was inoculated on different carbon source (D-fructose, glucose, sucrose, and starch) and nitrogen source (peptone, KNO_3, and $NaNO_3$) media. The pathogens were also inoculated on PDA media with different pH values (4, 5, 6, 7, 8, 9, 10, and 11). The PDA medium inoculated with the pathogen was incubated in an artificial climate chamber at 85% RH and different temperatures (20 °C, 25 °C, 28 °C, 30 °C, and 35 °C). After 4 days, the diameters of the colonies on the medium were measured using the crossover method and then photographed and recorded.

2.7. In Vivo Assay

2.7.1. Assay of Wound Inoculation with *D. passiflorae*

Passion fruits of uniform size that were healthy, disease-free, ripe, and consistent were selected. The samples were rinsed with deionized water and dried naturally. A small wound (Ø = 8 mm) was made in the passion fruit with an equatorial plane using a sterile puncher. After the wound had dried, a *D. passiflorae* cake with a diameter of 8 mm was inoculated [20]. Ten fruits were randomly selected from the treatment and control groups each day for observation. The diameters of the lesions were then measured using the crossover method and photographed for record.

2.7.2. Effects of *D. passiflorae* Inoculation on the Fruit Hue Angle

The hue angle of the fruit was determined as described by Zhao et al. [38]. Ten fruits were randomly selected from the fruits inoculated with *D. passiflorae* and the control group each day. A colorimeter (CR400, Konica Minolta, Tokyo, Japan) was used to measure 4 points on the equatorial surface of the fruit. The h° value was recorded and averaged.

2.7.3. Effects of *D. passiflorae* Inoculation on Cell Membrane Permeability

The determination of the cell membrane permeability for the fruit peel followed the method described by Shi et al. [39] with slight modifications. Thirty peels (5 mm diameter) from ten fruits were sampled and rinsed with distilled water. Then, the peels were transferred to a scale test tube containing 20 mL of distilled water. The peels were then allowed to stand for 1 h, and the initial electrolyte leakage (C1) was measured using a conductivity meter (STARTER3100C, Ohaus, Parsippany, NJ, USA). The scale test tube of the above extract was then boiled for 20 min and allowed to cool to measure the final electrolyte leakage (C2). The permeability of the fruit cell membranes was calculated according to the following formula:

$$\text{Cell membrane permeability (\%)} = (C1/C2) \times 100.$$

2.8. Effects of ε-PL, SA, and MT Treatments on the Mycelial Growth and Inhibition Rate of D. passiflorae In Vitro

In vitro inhibition experiments were carried out with slight modifications following the methods described by Fan et al. [40]. The ε-PL, SA, and MT were added to the PDA medium to give final concentrations of 0.125, 0.25, and 0.5 mg/mL of ε-PL, 0.4, 0.8, 1.2, and 1.6 mg/mL of SA, and 1, 2, 4, and 8 mg/mL of MT. Fungal plugs (Ø = 5 mm) were inoculated in the middle of the medium with different concentrations. The plates were incubated at a constant temperature (28 °C) and humidity (85% RH). Colony diameters were measured daily using the crossover method and then photographed and recorded.

2.9. Statistical Analysis

The described parameters were all tested in triplicate. The SPSS 22.0 software (IBM Corp, Armonk, NY, USA) was used to perform one-way ANOVA and *t*-tests on the experimental data. All data in the figures are expressed as average values ± standard error (n = 3). The differences between treatments are denoted by asterisks, indicating significant (* $p < 0.05$) or highly significant (** $p < 0.01$) differences. Microsoft Office Excel 2021 (Microsoft, Chicago, IL, USA) was used to produce all graphs.

3. Results

3.1. Symptoms of Brown Rot on Passion Fruit

The changes that occurred in the freshly picked healthy passion fruits that developed brown rot during postharvest storage are shown in Figure 1. The initial symptom of the disease is the appearance of water-stained spots (Figure 1C), and these develop in round pale-brown patches with expanding edges (Figure 1D). Ultimately, the color deepens and the fruit rots completely (Figure 1E).

Figure 1. Symptoms of brown rot during postharvest storage of passion fruit: (**A**) 0 d; (**B**) 6 d; (**C**) 8 d; (**D**) 14 d; (**E**) 20 d.

3.2. Morphological Identification of Pathogens

The pathogen B4 was isolated from passion fruit brown rot, using a tissue separation method (Figure 2). When strain B4 was grown on the PDA medium, the entire plate (90 mm) was covered within 5 days. The colonies were white, round, and concentric, the mycelium was plush, short, and flat, and the back of the medium was yellow. Microscopic analysis identified α-conidia and β-conidia in the B4 samples. The α-conidia were fusiform, single-celled, colorless, transparent, and without septa, but with an oil ball at each end. The β-conidia were filamentous, linear, colorless, transparent, and without septa.

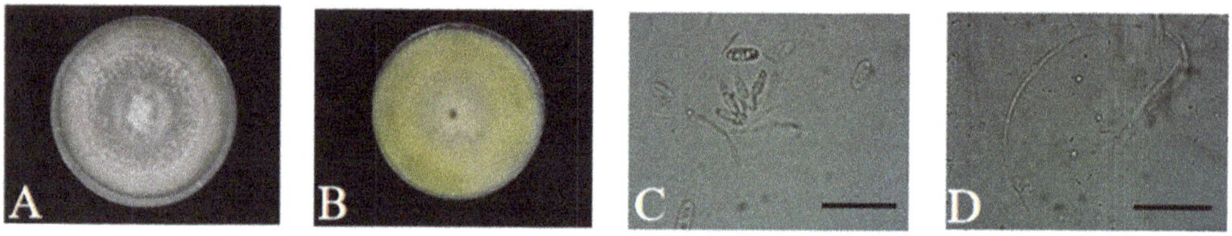

Figure 2. Colony morphology and spore morphology of the strain. (**A**) Front side of the pathogen after 5 days of growth. (**B**) Back side of the pathogen after 5 days of growth. (**C**) α-conidia (100×). (**D**) β-conidia (100×). Black bars represent 10 mm.

3.3. Pathogenicity Testing of Pathogens

Strain B4 was inoculated on healthy passion fruit samples and its pathogenicity was assessed (Figure 3). One day after inoculation, white mycelium began to grow at the inoculation mouth (Figure 3B). Three days after inoculation, the fruit had watery brown spots, which are characteristic symptoms of brown rot (Figure 3C). Five days after inoculation, the symptoms of brown rot on the passion fruit had increased (Figure 3D). The initial brown spots expanded in diameter and eventually covered the entire fruit. The brown spots in the middle of the fruit deepened in color, while the water-stained edges continued to expand. The disease development eventually caused the fruits to rot completely. These observations indicated that strain B4 was highly pathogenic to passion fruit and could cause extensive rot.

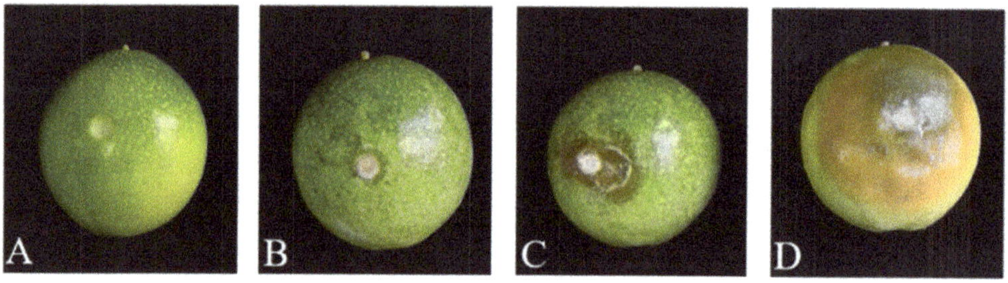

Figure 3. Pathogenicity test of isolated strains on passion fruit: (**A**) 0 d; (**B**) 1 d; (**C**) 3 d; (**D**) 5 d.

3.4. Molecular Biology of the Pathogen

According to the BLASTn analysis, strain B4 was 100% homologous to *D. passiflorae* (OQ087003.1) based on the ITS sequence and 100% homologous to *D. passiflorae* (MT409297.1, MT409300.1) based on the TUB sequence. Based on the TEF-1α sequence, strain B4 showed 100% homology with *D. passiflorae* (MT409348.1, MT409346.1). Sequence alignment based on the ITS, TUB, and TEF-1α genes displayed high similarity between strain B4 and *D. passiflorae* (Figure 4A–C). The phylogenetic analysis of the ITS sequences (Figure 4D) revealed that strain B4 clustered with *D. passiflorae* (OQ087003.1), while the phylogenetic analysis of the TUB sequences (Figure 4E) indicated that B4 was also clustered with *D. passiflorae* (MT409300.1, MT409299.1, MT409297.1). Similarly, the phylogenetic analysis of the TEF-1α sequences (Figure 4F) confirmed that B4 was clustered with *D. passiflorae* (MT409348.1, MT409347.1, MT409346.1, OR105938.1). The phylogenetic tree was constructed using neighbor joining on the combined dataset of the ITS, TUB, and TEF-1α sequences. According to Figure 4G, strain B4 and *D. passiflorae* with different accession numbers (MT409297.1, MT409300.1, MT409299.1) were clustered together in one branch, which indicates that strain B4 was most closely related to *D. passiflorae*. Therefore, combined with the pathogenicity test and morphological analysis, the pathogen B4 that was isolated from postharvest brown rot on passion fruit was identified as *D. passiflorae*.

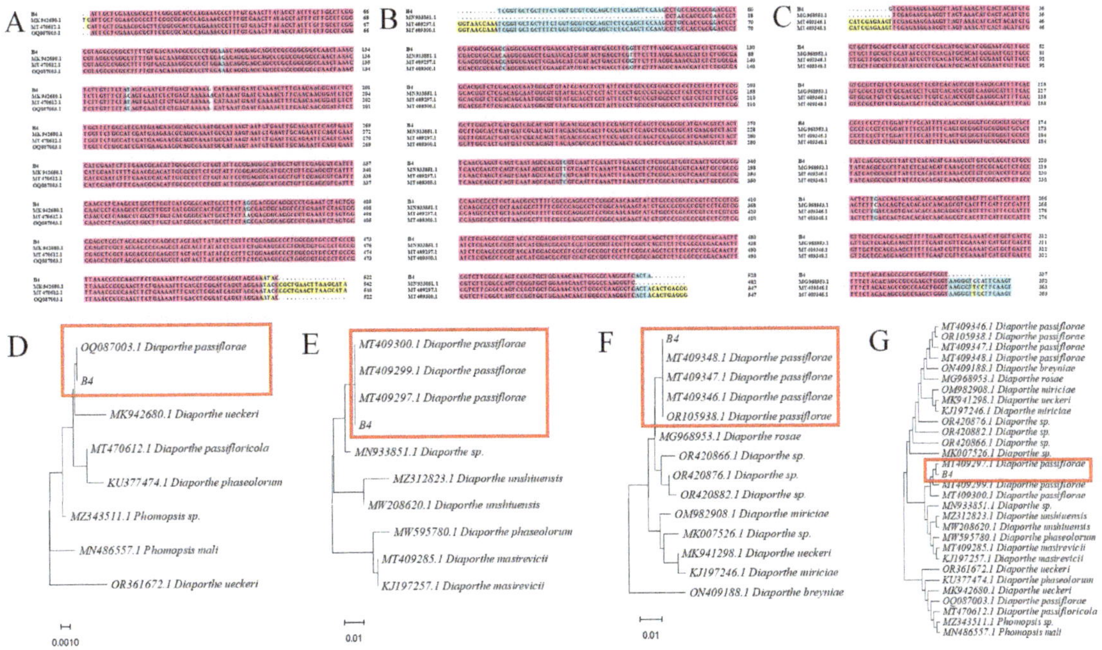

Figure 4. Molecular biological identification of isolated pathogen. (**A**) Sequence alignment based on rDNA-ITS sequence. (**B**) Sequence alignment based on TUB sequence. (**C**) Sequence alignment based on TEF-1α sequence. (**D**) Phylogenetic tree constructed for strain B4 based on rDNA-ITS sequence. (**E**) Phylogenetic tree constructed for strain B4 based on TUB sequences. (**F**) Phylogenetic tree constructed for strain B4 based on TEF-1α sequence. (**G**) Construction of a phylogenetic tree of strain B4 based on multiple sequences of rDNA-ITS, TUB, and TEF-1α. Branches including isolates obtained from the passion fruit are represented by red boxes.

3.5. Biological Characterization of D. passiflorae

D. passiflorae could grow in all four carbon sources tested and there were significant differences in the growth rates of the colonies (Figure 5A). The colonies grew relatively slowly on the control medium without sucrose and faster on the medium with the four carbon sources tested. Starch was used as the carbon source in the medium for the faster growth of the colonies. Mycelia developed, and the diameters of the colonies observed in the culture on day 4 were 39.56 ± 2.03 mm. Starch was determined to be the best carbon source.

D. passiflorae could grow on all three test nitrogen source media (Figure 5B). The colonies grew faster on the medium with peptone as the nitrogen source and the colony diameters were 78.83 ± 0.25 mm on day 4, which was highly significant ($p < 0.01$) when compared with the other experimental groups. The growth rate of the colonies was slowest on the base medium without sodium nitrate (control), as they were only 5 mm in diameter on day 4. These results show that peptone is the best nitrogen source for the growth of *D. passiflorae* colonies.

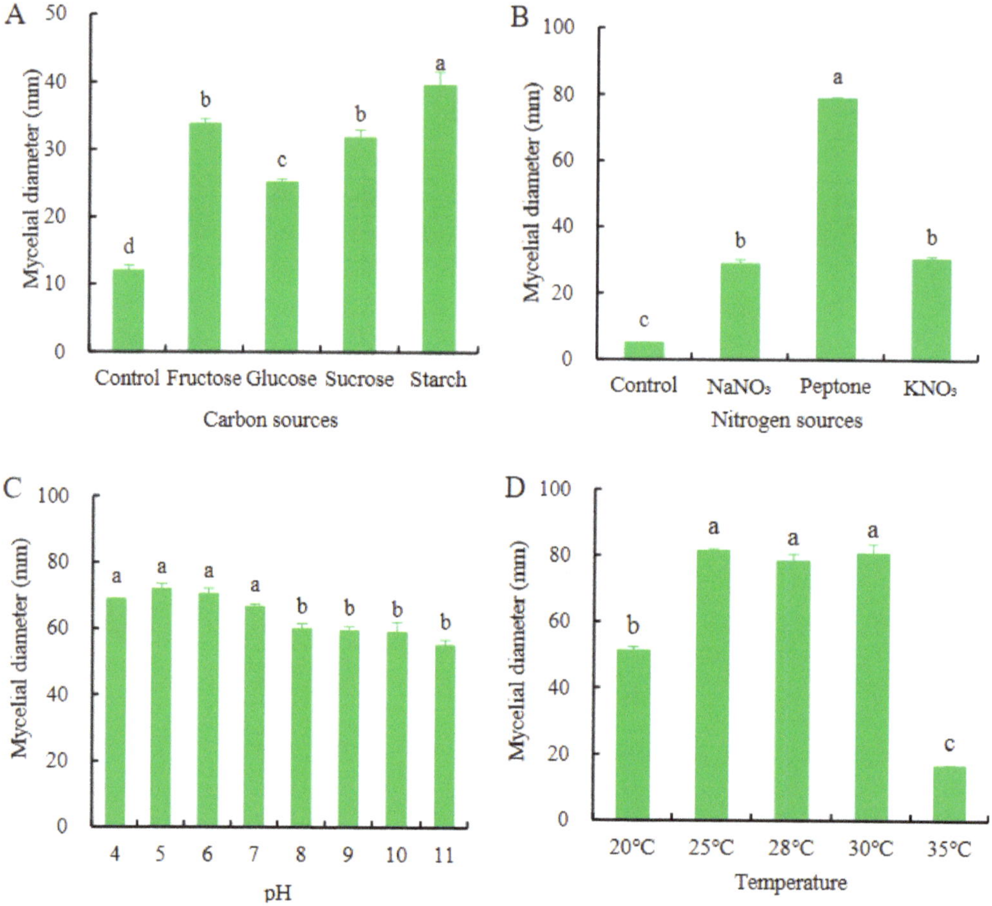

Figure 5. Biological characterization of *D. passiflorae*. (**A**) C-source. (**B**) N-source. (**C**) pH. (**D**) Temperature. The different letters above the columns indicate that the data were significantly different ($p < 0.05$) between groups.

D. passiflorae could grow in a pH range of 4–11. At a pH of >8, the growth rate of the mycelia was slow, and the colony diameters were small and sparse. At pH 5, the colony growth rate was the fastest, and the colony diameters were 71.83 ± 1.59 mm. Compared with the other experimental groups with pH values > 8, the difference was significant ($p < 0.05$). At pH values of 5 and 6, there were no significant differences between the experimental groups (Figure 5C). These results showed that the growth of the passionflower colonies was suitable under acidic conditions, whereas it was inhibited under alkaline environments.

D. passiflorae grew at all five temperature gradients (Figure 5D). The colonies grew fastest at 25–30 °C, and the colony diameters were 78.26–81.3 mm after incubation for 4 days. However, the colony growth was slowest at 35 °C, and the colony diameters were 16.50 ± 0.24 mm after incubation for 4 days. These results showed that 25–30 °C was the optimum temperature range for the growth of *D. passiflorae*.

3.6. In Vivo Assay

3.6.1. Effects of *D. passiflorae* on Spot Diameter

The diameters of the brown rot spots on the passion fruits inoculated with *D. passiflorae* increased with the storage time (Figure 6A,B). Compared to the control, the diameters of the brown rot spots were significantly ($p < 0.05$) larger in the passion fruits inoculated with *D. passiflorae* at 2–6 d after harvest and reached a highly significant ($p < 0.01$) level at 3–6 d. The diameters of the diseased spots on the fruits inoculated with *D. passiflorae* rapidly increased at 2 d, while the fruit lesions in the control group started only at 6 d. Additionally, on day 6, the diameters of the passion fruit disease spots in the control group were 93% smaller than those in the inoculated *D. passiflorae* group.

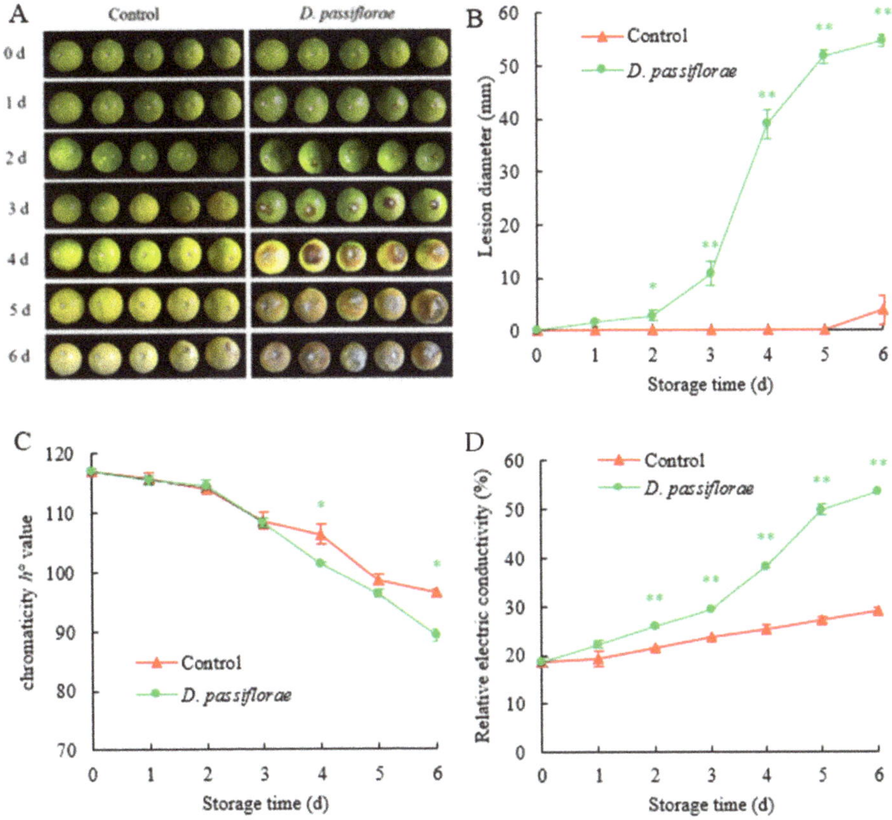

Figure 6. Effects of inoculation with *D. passiflorae* on fruit lesion diameter, $h°$ value, and cell membrane permeability. (**A**) Symptoms of passion fruit brown rot. (**B**) Disease spot diameter. (**C**) $h°$ value. (**D**) Cell membrane permeability. The marks * and ** represent the significant differences ($p < 0.05$ or $p < 0.01$) between the control group and the *D. passiflorae* inoculated passion fruits on each storage day. (▲) Control samples; (•) *D. passiflorae* samples.

3.6.2. Effects of *D. passiflorae* Inoculation on the Fruit Hue Angle of Passion Fruit

The color change of the passion fruit was measured using the value of the hue angle. The postharvest passion fruit surface $h°$ values continued to decline with the storage time, with the fruit gradually turning from green to yellow (Figure 6C). Compared to the control, the surface $h°$ values of the passion fruits inoculated with *D. passiflorae* were lower at days 4–6 and the differences were significant ($p < 0.05$) on days 4 and 6. Additionally, when compared with day 0, the h° values of the fruits inoculated with *D. passiflorae* on day 6 were reduced by 24%.

3.6.3. Effects of *D. passiflorae* Inoculation on Cell Membrane Permeability

The cell membrane permeability of the passion fruits increased with the storage time and showed a trend similar to that for the lesion diameter. Compared with the control group, the cell membrane permeability of the passion fruit inoculated with *D. passiflorae* was higher than that of the control group (Figure 6D). Additionally, it reached a highly significant level ($p < 0.01$) within 2–6 days. The cell membrane permeability of the control group was 29% on day 6, which was 46% lower than that with the *D. passiflorae* inoculation.

3.7. Effects of ε-PL, SA, and MT on the Mycelial Growth of D. passiflorae In Vitro

The ε-PL, SA, and MT treatments significantly ($p < 0.05$) inhibited the growth of the mycelia of *D. passiflorae*. As the concentrations of ε-PL, SA, and MT increased, the mycelial growth rate decreased continuously, and their inhibitory effects increased (Figure 7). The control group was completely overgrown on day 5, while the groups treated with the ε-PL, SA, and MT were not. The inhibition rates of 0.125, 0.25, and 0.50 mg/mL of ε-PL on the *D. passiflorae* were 54%, 79%, and 95%, respectively (Figure 7D). The inhibition rates of 0.4, 0.8, 1.2, and 1.6 mg/mL of SA on the *D. passiflorae* were 45%, 53%, 75%, and 88%, respectively (Figure 7E). The inhibition rates of 1, 2, 4, and 8 mg/mL of MT on the *D. passiflorae* were 29%, 53%, 65%, and 74%, respectively (Figure 7F). Therefore, ε-PL, SA, and MT all inhibited the growth of *D. passiflorae*, and ε-PL exhibited the best inhibitory effect (Table 1).

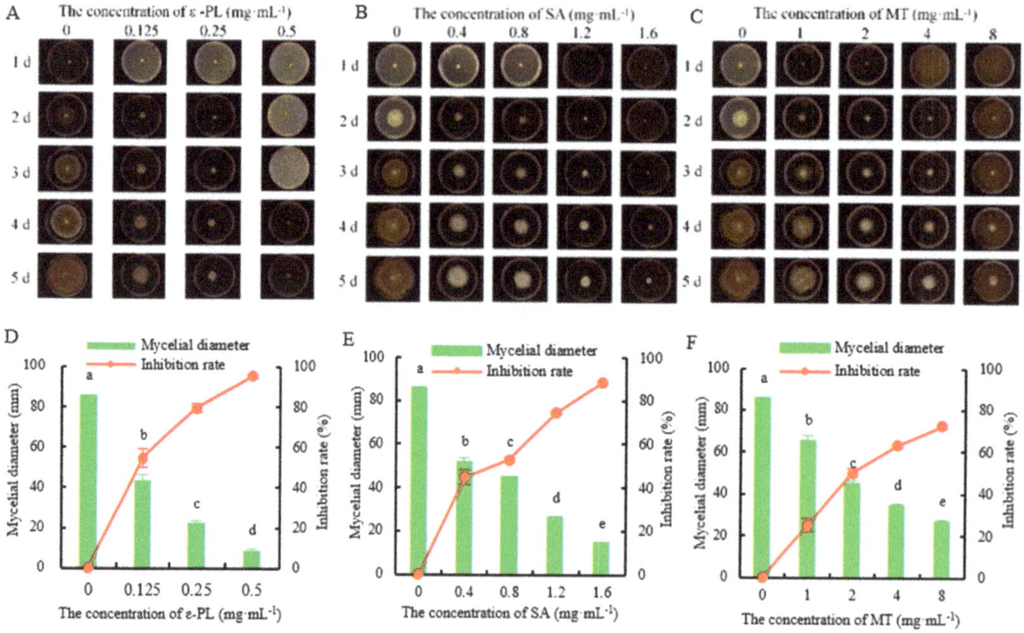

Figure 7. Effects of different concentrations of (**A**,**D**) ε-poly-l-lysine, (**B**,**E**) salicylic acid, and (**C**,**F**) melatonin on the mycelial growth and inhibition rate of *D. passiflorae* at 5 d in vitro. The different letters above the columns indicate that the data were significantly different ($p < 0.05$) between groups.

Table 1. Inhibition effects of ε-poly-l-lysine (ε-PL), salicylic acid (SA), and melatonin (MT) on *D. passiflorae*.

	The Concentration of ε-PL ($mg \cdot mL^{-1}$)				The Concentration of SA ($mg \cdot mL^{-1}$)				The Concentration of MT ($mg \cdot mL^{-1}$)			
	0	0.125	0.25	0.5	0.4	0.8	1.2	1.6	1	2	4	8
Diameter (mm)	86.00 ± 0.00^{a}	43.68 ± 2.71^{b}	22.54 ± 2.21^{c}	9.13 ± 0.70^{d}	52.08 ± 2.02^{b}	44.98 ± 0.40^{c}	26.66 ± 0.08^{d}	14.90 ± 0.46^{e}	65.59 ± 6.17^{b}	45.19 ± 1.22^{c}	34.77 ± 0.64^{d}	27.22 ± 0.56^{e}
Inhibition rate (%)	0 ± 0.00	52.25 ± 4.73	78.35 ± 2.10	94.90 ± 1.22	44.62 ± 3.36	52.96 ± 0.66	74.51 ± 0.14	88.35 ± 0.76	25.20 ± 3.28	50.26 ± 1.88	63.25 ± 0.96	72.57 ± 0.84

Different letters indicate significant differences ($p < 0.05$) between different concentrations in the same group.

4. Discussion

Passion fruit is a tropical fruit that is highly sought after by consumers owing to its high nutritional and medicinal values [7,9,41], and it is consequently widely cultivated in China. However, because of the susceptibility of passion fruit to various diseases, it has a short shelf life, which negatively impacts its commercial value. Based on the treatment method used and environmental conditions, microbial infections may occur after harvest, resulting in postharvest rot [42]. Pathogens can enter from various parts of the fruit, such as mechanically damaged wounds or the stomata. In the early stages of disease, spots appear only on the peel; however, with the passage of time, the pathogen spreads in the fruit and then gradually into the pulp, resulting in fruit rot. During the early stages of brown rot in passion fruit, the only symptom is small brown lesions that appear on the peel (Figure 1). However, the diameters of these lesions gradually increase over time and, finally, the whole fruit decays completely. *Diaporthe infercuda* [16], *Phytophthora drechsleri* [17], *Phytophthora nicotianae* [18], *Colletotrichum brevisporum* [19], *Lasiodiplodia theobromae* [20], and *Trichothecium roseum* [21] have previously been reported as the causal agents of postharvest rot in passion fruit. However, in this study, *D. passiflorae* was isolated as the dominant pathogen causing brown rot in passion fruit. Microscope analysis showed that the B4 colonies were white and fluffy with short mycelia, and further evaluations revealed that they had α-conidia and β-conidia (Figure 2). The morphological analysis showed that the conidia of strain B4 were similar to those previously reported for *Diaporthe* [43]. The fruit pathogenicity of B4 was further tested, and the results demonstrated that the fruits inoculated with the B4 strain all showed symptoms of disease consistent with those of brown rot under natural conditions in passion fruit (Figure 3). Additionally, the rDNA-ITS gene sequences obtained via PCR amplification were analyzed for homology and a phylogenetic tree was constructed. Finally, combining the results of the morphological analysis, pathogenicity tests, and molecular identification, the dominant strain B4 was identified as *D. passiflorae*. Notably, *D. passiflorae* has previously been isolated from kiwifruit [44], but its role in causing postharvest decay in passion fruit has not been previously reported.

The effects of environmental factors on the growth of the fungal pathogen have been studied through its biological characteristics. Fu et al. [45] found that the optimal temperature range for *Phomopsis mangiferae Ahmad* growth was 25–30 °C. Yan et al. [46] found that the optimum carbon source for *Diaporthe eres* growth was starch and the optimum nitrogen source was peptone. In a recent study, Ariyawansa et al. [47] identified six *Diaporthe* species associated with leaf spot disease of *C. sinensis* in Taiwan tea fields, and found that the optimal temperature for mycelium growth was 25–30 °C, with a preference for growth in a low-acid–alkaline medium. In this study, through experiments employing different carbon and nitrogen sources, acid and alkaline conditions, and temperatures, the preference and growth requirements of the *D. passiflorae* strain to specific environments were analyzed. The results showed that *D. passiflorae* could thrive in an acidic environment at room temperature. We found that starch is the most suitable carbon source for its growth, while peptone is the preferred nitrogen source. The observations in this study are consistent with those of Fu et al. [45], Yan et al. [46], and Ariyawansa et al. [47].

Infection with pathogens can cause fruit diseases and rot [48]. Gong et al. [49] and Sun et al. [50] found that the fruit lesion diameter and cell membrane permeability increased after fungal infection. In the study, the lesion diameter (Figure 6B) and the cell membrane permeability (Figure 6D) of fruit inoculated with *D. passiflorae* increased and the decay symptoms were more pronounced, while the chromaticity $h°$ values decreased (Figure 6C). Fruits uninoculated with *D. passiflorae* were not diseased on day 5 and maintained good-quality characteristics (Figure 6).

The development of the fruit industry has been seriously restricted by reductions in the commercial value of fruit due to microbial infection. Chemical fungicides have thus been widely used in fruit preservation to help maintain fruit quality and extend storage periods [24]. However, the long-term, irrational use of chemical fungicides can cause irreversible damage to human health and the environment [51]. Therefore, it is essential that we identify safe and effective treatment methods to inhibit the growth of pathogens and maintain fruit quality [52]. As a safe and nontoxic antifungal agent, ε-PL is widely used in food processing [53]. In recent years, it has been shown that ε-PL treatments can effectively inhibit the occurrence of postharvest disease in fruit. For example, the ε-PL treatment could effectively inhibit the occurrence of postharvest blue mold disease in apple, and it had an inhibitory effect on *Penicillium expansum* in vitro [26]. Additionally, ε-PL was found to have significant antifungal activity in vitro and could significantly inhibit the colony growth and spore germination of *Alternaria alternata* [54]. Similarly, SA and MT treatments could inhibit pathogen infection and thus maintain high fruit quality. SA treatments could reduce pathogen infections in mandarin fruit during cold storage and thus help maintain higher fruit quality [55]. It was found that the SA treatment could result in the protein leakage and lipid damage of the pathogen, thereby inhibiting the activity of *P. expansum* in vitro [56]. The occurrence of litchi downy blight was reduced and the resistance of litchi fruit was improved via the MT soaking treatment after harvest [57]. Meanwhile, it was found that the MT treatment of tomato fruits could suppress gray mold caused by *Botrytis cinerea* [58].

In the present work, the inhibitory effects of ε-PL, SA, and MT on *D. passiflorae* were studied. The results showed that 0.125 mg/mL ε-PL, 0.4 mg/mL SA, and 1 mg/mL MT could effectively inhibit the growth of *D. passiflorae* in vitro, and the inhibitory effect enhances with the increase in concentrations.

5. Conclusions

In summary, this study showed that the brown rot of passion fruit is caused by *D. passiflorae*. The infection with this pathogen causes significant postharvest disease. Furthermore, the research has clarified that the growth conditions of the fungal pathogen were determined by the biological characteristics of *D. passiflorae*. Moreover, the ε-PL, SA, and MT treatments inhibited the growth of *D. passiflorae* in vitro, among which the 0.5 mg/mL ε-PL treatment had the best inhibition effect. These findings not only improve our understanding of *D. passiflorae*, the passion fruit brown rot pathogen, but also provide valuable information for effective disease management.

Author Contributions: Methodology, H.W., M.L., Q.L. and Y.L. (Yu Lin); validation, M.L., Q.L. and X.J.; formal analysis, H.W., M.L. and Q.L.; investigation, H.W., Y.L. (Yuzhao Lin), and X.J.; data curation, H.W., Q.L., Y.L. (Yu Lin) and Y.L. (Yuzhao Lin); writing—original draft, H.W.; writing—review and editing, H.C. and Y.C.; supervision, Y.C.; project administration, Y.C.; funding acquisition, H.C. and Y.C. All authors have read and agreed to the published version of the manuscript.

Funding: This study was supported by the Natural Science Foundation of Fujian Province of China (Grant No. 2023J01902).

Institutional Review Board Statement: Not applicable.

Informed Consent Statement: Not applicable.

Data Availability Statement: The sequences have been deposited in GenBank (Figure 4). The data presented in this study are publicly available at NCBI. Publicly available datasets were analyzed in this study. These data can be found here: https://www.ncbi.nlm.nih.gov/.

Conflicts of Interest: The authors declare no conflict of interest.

References

1. Castillo, N.R.; Melgarejo, L.M.; Blair, M.W. Seed Structural Variability and Germination Capacity in *Passiflora edulis* Sims f. Edulis. *Front. Plant Sci.* **2020**, *11*, 498. [CrossRef] [PubMed]
2. De Jesus, O.N.; Lima, L.K.S.; dos Santos, I.S.; dos Santos, M.A.; Rosa, R.C.C. Bright Red Passion Fruit - Evaluation of Colorimetry and Physicochemical Quality for the Fresh Fruit Market. *Sci. Hortic.* **2023**, *317*, 112016. [CrossRef]
3. Huo, D.; Dai, J.; Yuan, S.; Cheng, X.; Pan, Y.; Wang, L.; Wang, R. Eco-Friendly Simultaneous Extraction of Pectins and Phenolics from Passion Fruit (*Passiflora edulis* Sims) Peel: Process Optimization, Physicochemical Properties, and Antioxidant Activity. *Int. J. Biol. Macromol.* **2023**, *243*, 125229. [CrossRef] [PubMed]
4. Zhou, Y.; Zhong, Y.; Li, L.; Jiang, K.; Gao, J.; Zhong, K.; Pan, M.; Yan, B. A Multifunctional Chitosan-Derived Conformal Coating for the Preservation of Passion Fruit. *LWT* **2022**, *163*, 113584. [CrossRef]
5. Li, C.; Xin, M.; Li, L.; He, X.; Yi, P.; Tang, Y.; Li, J.; Zheng, F.; Liu, G.; Sheng, J.; et al. Characterization of the Aromatic Profile of Purple Passion Fruit (*Passiflora edulis* Sims) during Ripening by HS-SPME-GC/MS and RNA Sequencing. *Food Chem.* **2021**, *355*, 129685. [CrossRef] [PubMed]
6. Niu, Y.-F.; Ni, S.-B.; Liu, S.-H.; Liu, J. The Complete Chloroplast Genome of Passiflora Caerulea, a Tropical Fruit with a Distinctive Aroma. *Mitochondrial DNA B Resour.* **2021**, *6*, 488–490. [CrossRef]
7. Liang, J.; Fang, Y.; An, C.; Yao, Y.; Wang, X.; Zhang, W.; Liu, R.; Wang, L.; Aslam, M.; Cheng, Y.; et al. Genome-Wide Identification and Expression Analysis of the BHLH Gene Family in Passion Fruit (*Passiflora edulis*) and Its Response to Abiotic Stress. *Int. J. Biol. Macromol.* **2023**, *225*, 389–403. [CrossRef]
8. do Carmo Santos, J.T.; Petry, F.C.; de Castro Tobaruela, C.; Mercadante, A.Z.; Gloria, M.B.A.; Costa, A.M.; Lajolo, F.M.; Hassimotto, N.M.A. Brazilian Native Passion Fruit (*Passiflora tenuifila* Killip) Is a Rich Source of Proanthocyanidins, Carotenoids, and Dietary Fiber. *Food Res. Int.* **2021**, *147*, 110521. [CrossRef]
9. Septembre-Malaterre, A.; Stanislas, G.; Douraguia, E.; Gonthier, M.-P. Evaluation of Nutritional and Antioxidant Properties of the Tropical Fruits Banana, Litchi, Mango, Papaya, Passion Fruit and Pineapple Cultivated in Réunion French Island. *Food Chem.* **2016**, *212*, 225–233. [CrossRef]
10. He, X.; Luan, F.; Yang, Y.; Wang, Z.; Zhao, Z.; Fang, J.; Wang, M.; Zuo, M.; Li, Y. *Passiflora edulis*: An Insight Into Current Researches on Phytochemistry and Pharmacology. *Front. Pharmacol.* **2020**, *11*, 617. [CrossRef]
11. Xia, Z.; Huang, D.; Zhang, S.; Wang, W.; Ma, F.; Wu, B.; Xu, Y.; Xu, B.; Chen, D.; Zou, M.; et al. Chromosome-Scale Genome Assembly Provides Insights into the Evolution and Flavor Synthesis of Passion Fruit (*Passiflora edulis* Sims). *Hortic. Res.* **2021**, *8*, 1–14. [CrossRef] [PubMed]
12. You, M.; Duan, X.Y.; Li, X.; Lou, L.J.; Zhao, Y.; Pan, H.H.; Gong, W.L.; Yang, L.R.; Xiang, Z.; Li, G.F. Effect of 1-Methylcyclopropene Combined with Chitosan-Coated Film on Storage Quality of Passion Fruit. *Sustainable. Chem. Pharm.* **2022**, *27*, 100679. [CrossRef]
13. Chen, F.P.; Xu, X.Y.; Luo, Z.; Chen, Y.; Xu, Y.; Xiao, G. Effect of High O_2 Atmosphere Packaging on Postharvest Quality of Purple Passion Fruit (*Passiflora edulis* Sims). *J. Food Process. Preserv.* **2018**, *42*, e13749. [CrossRef]
14. Cerqueira-Silva, C.B.M.; Jesus, O.N.; Oliveira, E.J.; Santos, E.S.L.; Souza, A.P. Characterization and Selection of Passion Fruit (Yellow and Purple) Accessions Based on Molecular Markers and Disease Reactions for Use in Breeding Programs. *Euphytica* **2015**, *202*, 345–359. [CrossRef]
15. Rizwan, H.M.; Zhimin, L.; Harsonowati, W.; Waheed, A.; Qiang, Y.; Yousef, A.F.; Munir, N.; Wei, X.; Scholz, S.S.; Reichelt, M.; et al. Identification of Fungal Pathogens to Control Postharvest Passion Fruit (*Passiflora edulis*) Decays and Multi-Omics Comparative Pathway Analysis Reveals Purple Is More Resistant to Pathogens than a Yellow Cultivar. *J. Fungi* **2021**, *7*, 879. [CrossRef]
16. Moreira, R.R.; Caus, G.; Gomes Figueiredo, J.A.; May De Mio, L.L. Phomopsis Rot Caused by Diaporthe Infecunda on Fruit and Flowers of *Passiflora edulis* in Brazil. *Australas. Plant Pathol.* **2020**, *49*, 141–145. [CrossRef]
17. Gil, J.G.R.; Tamayo, P.J.; Morales, J.G. Identification and Pathogenicity of Microorganisms Affecting Purple Passion Fruit in Colombia. *Rev. Ceres* **2017**, *64*, 250–257. [CrossRef]
18. Liu, Z.-L.; Zhou, S.; Huang, Y.; Yang, L.; Yan, Y.; Chen, G.; Sun, J.M.; Wu, S.; Chen, X. First Report of Fruit Rot Caused by Phytophthora Nicotianae on Passion Fruit in Guangxi Province, China. *Plant Dis.* **2021**, *106*, 336. [CrossRef]
19. Qiu, F.; Li, X.; Xie, C.P.; Li, J.; Zheng, F.Q. Identification of Colletotrichum Brevisporum Causing Fruit Rot in Yellow Passion Fruit (*Passiflora edulis* f. Flavicarpa) in China. *Australas. Plant Pathol.* **2021**, *50*, 229–232. [CrossRef]
20. Zhang, W.; Niu, X.L.; Yang, J.Y. First Report of Postharvest Fruit Rot on Passion Fruit (*Passiflora edulis*) Caused by Lasiodiplodia Theobromae in Mainland China. *Plant Dis.* **2021**, *105*, 1198. [CrossRef]
21. Li, C.; Zhao, J.; Wang, J.; Wang, X.; Xiang, W.; Zhao, J. First Report of Trichothecium Roseum Causing Postharvest Fruit Rot on Purple Passion Fruit in China. *Plant Dis.* **2022**, *106*, 3212. [CrossRef] [PubMed]
22. Xu, H.; Qiao, P.; Pan, J.; Qin, Z.; Li, X.; Khoo, H.E.; Dong, X. $CaCl_2$ Treatment Effectively Delays Postharvest Senescence of Passion Fruit. *Food Chem.* **2023**, *417*, 135786. [CrossRef] [PubMed]
23. Zhang, X.; Li, B.; Zhang, Z.; Chen, Y.; Tian, S. Antagonistic Yeasts: A Promising Alternative to Chemical Fungicides for Controlling Postharvest Decay of Fruit. *J. Fungi* **2020**, *6*, 158. [CrossRef]
24. Huang, X.; Ren, J.; Li, P.; Feng, S.; Dong, P.; Ren, M. Potential of Microbial Endophytes to Enhance the Resistance to Postharvest Diseases of Fruit and Vegetables. *J. Sci. Food Agric.* **2021**, *101*, 1744–1757. [CrossRef] [PubMed]

25. Wu, J.; Hu, J.; Jiao, W.; Du, Y.; Han, C.; Chen, Q.; Chen, X.; Fu, M. Inhibitory Effect of ε-Poly-L-Lysine on Fruit Colletotrichum Gloeosporioides through Regulating Organic Acid Metabolism and Exerting Membrane-Targeted Antifungal Activity. *Postharvest Biol. Technol.* **2023**, *200*, 112339. [CrossRef]
26. Dou, Y.; Routledge, M.N.; Gong, Y.; Godana, E.A.; Dhanasekaran, S.; Yang, Q.; Zhang, X.; Zhang, H. Efficacy of Epsilon-Poly-L-Lysine Inhibition of Postharvest Blue Mold in Apples and Potential Mechanisms. *Postharvest Biol. Technol.* **2021**, *171*, 111346. [CrossRef]
27. Jiang, B.; Liu, R.; Fang, X.; Tong, C.; Chen, H.; Gao, H. Effects of Salicylic Acid Treatment on Fruit Quality and Wax Composition of Blueberry (*Vaccinium virgatum* Ait). *Food Chem.* **2022**, *368*, 130757. [CrossRef]
28. Zhang, Y.; Li, S.; Deng, M.; Gui, R.; Liu, Y.; Chen, X.; Lin, Y.; Li, M.; Wang, Y.; He, W.; et al. Blue Light Combined with Salicylic Acid Treatment Maintained the Postharvest Quality of Strawberry Fruit during Refrigerated Storage. *Food Chem. X* **2022**, *15*, 100384. [CrossRef]
29. Yan, R.; Xu, Q.; Dong, J.; Kebbeh, M.; Shen, S.; Huan, C.; Zheng, X. Effects of Exogenous Melatonin on Ripening and Decay Incidence in Plums (*Prunus salicina* L. Cv. Taoxingli) during Storage at Room Temperature. *Sci. Hortic.* **2022**, *292*, 110655. [CrossRef]
30. Wang, F.; Zhang, X.; Yang, Q.; Zhao, Q. Exogenous Melatonin Delays Postharvest Fruit Senescence and Maintains the Quality of Sweet Cherries. *Food Chem.* **2019**, *301*, 125311. [CrossRef]
31. Chen, G.; Yang, L.; Luo, H.L.; Huang, Y.C.; Ju, Y.; Wei, Y.W.; Sun, J.M. First Report of Leaf Spot on Passion Fruit in China Caused by Alternaria Alternata. *Plant Dis.* **2023**, *107*, 1229. [CrossRef] [PubMed]
32. Mehraban, A.; Nasr, R.; Eslami, M.; Amrollahi, H. Molecular and Morphological Identification of Fungi Isolated from Sour Cherry, Plum and Apple Fruit Leather in Iran. *Gene Rep.* **2019**, *17*, 100500. [CrossRef]
33. Prakash, P.Y.; Bhargava, K. A Modified Micro Chamber Agar Spot Slide Culture Technique for Microscopic Examination of Filamentous Fungi. *J. Microb. Methods* **2016**, *123*, 126–129. [CrossRef]
34. Lu, J.Y. *Plant Disease Diagnosis*, 2nd ed.; China Agriculture Press: Beijing, China, 1997; pp. 32–59.
35. Qi, P.K. *Diseases of Fruits and Vegetables Storage and Transportation*, 1st ed.; China Agriculture Press: Beijing, China, 1994; pp. 3–8.
36. Liu, Z.Y.; Zhang, P.; Zi, L.L.; Zhu, T.S. Identification of the Pathogen Causing Leaf Margin Brown Spot on Jujube in Southern Xinjiang. *Hunan Agric. Sci.* **2022**, *9*, 22–26. [CrossRef]
37. Cui, L.; Yang, C.; Jin, M.; Wei, L.; Yang, L.; Zhou, J. Identification and Biological Characterization of a New Pathogen That Causes Potato Scab in Gansu Province, China. *Microb. Pathog.* **2021**, *161*, 105276. [CrossRef] [PubMed]
38. Zhao, Y.; Zhu, X.; Hou, Y.; Pan, Y.; Shi, L.; Li, X. Effects of Harvest Maturity Stage on Postharvest Quality of Winter Jujube (*Zizyphus jujuba* Mill. Cv. Dongzao) Fruit during Cold Storage. *Sci. Hortic.* **2021**, *277*, 109778. [CrossRef]
39. Shi, F.; Li, X.; Meng, H.; Wei, W.; Wang, Y. Reduction in Chilling Injury Symptoms by Hot Electrolyzed Functional Water Treatment May Function by Regulating ROS Metabolism in Satsuma Orange Fruit. *LWT* **2020**, *125*, 109218. [CrossRef]
40. Fan, S.; Li, Q.; Feng, S.; Lei, Q.; Abbas, F.; Yao, Y.; Chen, W.; Li, X.; Zhu, X. Melatonin Maintains Fruit Quality and Reduces Anthracnose in Postharvest Papaya via Enhancement of Antioxidants and Inhibition of Pathogen Development. *Antioxidants* **2022**, *11*, 804. [CrossRef]
41. Corrêa, R.C.G.; Peralta, R.M.; Haminiuk, C.W.I.; Maciel, G.M.; Bracht, A.; Ferreira, I.C.F.R. The Past Decade Findings Related with Nutritional Composition, Bioactive Molecules and Biotechnological Applications of *Passiflora* Spp. (Passion Fruit). *Trends Food Sci. Technol.* **2016**, *58*, 79–95. [CrossRef]
42. Hernández, A.; Ruiz-Moyano, S.; Galván, A.I.; Merchán, A.V.; Pérez Nevado, F.; Aranda, E.; Serradilla, M.J.; de Guía Córdoba, M.; Martín, A. Anti-Fungal Activity of Phenolic Sweet Orange Peel Extract for Controlling Fungi Responsible for Post-Harvest Fruit Decay. *Fungal Biol.* **2021**, *125*, 143–152. [CrossRef]
43. Gi, S.; Kim, W.; Yang, K.-Y. Emergence of Multiple Diaporthe Species Causing Kiwifruit Rot and Occurrence of Resistance to a Methyl Benzimidazole Carbamate Fungicide in South Korea. *Crop Prot.* **2022**, *158*, 106016. [CrossRef]
44. Li, H.; Yu, S.; Tang, W.; Miao, M.; Liu, Y. First Report of Diaporthe Passiflorae and Diaporthe Nobilis Causing a Postharvest Kiwifruit Rot in Sichuan Province, China. *Plant Dis.* **2019**, *103*, 771. [CrossRef]
45. Fu, Z.L.; Yang, S.Y.; Zhang, R.; Zhang, H. Study on the biological characteristics of Pointillomyces mangiferae. *South. China Fruits* **2017**, *46*, 74–76+80.
46. Yan, X.R.; Wang, X.; Hu, M.Q.; Dai, H.P.; Fu, J.F.; Li, T.L. Identification and biological characteristic of blueberry Diaporthe bud blight pathogen. *Acta Phytopathol. Sin.* **2015**, *45*, 556–560. [CrossRef]
47. Ariyawansa, H.A.; Tsai, I.; Wang, J.-Y.; Withee, P.; Tanjira, M.; Lin, S.-R.; Suwannarach, N.; Kumla, J.; Elgorban, A.M.; Cheewangkoon, R. Molecular Phylogenetic Diversity and Biological Characterization of Diaporthe Species Associated with Leaf Spots of Camellia Sinensis in Taiwan. *Plants* **2021**, *10*, 1434. [CrossRef]
48. Li, W.; Jiang, Y.; Hu, C.; Liu, G.; Li, Y.; Wang, S. Identification, Pathogenic Mechanism and Control of Rhizopus Oryzae Causing Postharvest Fruit Rot in Pumpkin. *Postharvest Biol. Technol.* **2023**, *204*, 112460. [CrossRef]
49. Zhang, J.; Liu, Q.; Chen, X.; Li, M.; Lin, M.; Chen, Y.; Lin, H. Slightly Acidic Electrolyzed Water Treatment Improves the Quality and Storage Properties of Carambola Fruit. *Food Chem. X* **2023**, *17*, 100555. [CrossRef]
50. Sun, Y.; Lu, R.; Pan, L.; Wang, X.; Tu, K. Assessment of the Optical Properties of Peaches with Fungal Infection Using Spatially-Resolved Diffuse Reflectance Technique and Their Relationships with Tissue Structural and Biochemical Properties. *Food Chem.* **2020**, *321*, 126704. [CrossRef]

51. Li, X.; Jing, T.; Zhou, D.; Zhang, M.; Qi, D.; Zang, X.; Zhao, Y.; Li, K.; Tang, W.; Chen, Y.; et al. Biocontrol Efficacy and Possible Mechanism of Streptomyces Sp. H4 against Postharvest Anthracnose Caused by Colletotrichum Fragariae on Strawberry Fruit. *Postharvest Biol. Technol.* **2021**, *175*, 111401. [CrossRef]
52. Strano, M.C.; Restuccia, C.; De Leo, R.; Mangiameli, S.; Bedin, E.; Allegra, M.; Quartieri, A.; Cirvilleri, G.; Pulvirenti, A. Efficacy of an Antifungal Edible Coating for the Quality Maintenance of Tarocco Orange Fruit during Cold Storage. *Crop Prot.* **2021**, *148*, 105719. [CrossRef]
53. Liu, H.; Chen, J.; Xia, Z.; An, M.; Wu, Y. Effects of ε-Poly-l-Lysine on Vegetative Growth, Pathogenicity and Gene Expression of Alternaria Alternata Infecting Nicotiana Tabacum. *Pestic. Biochem. Physiol.* **2020**, *163*, 147–153. [CrossRef] [PubMed]
54. Shu, C.; Cui, K.; Li, Q.; Cao, J.; Jiang, W. Epsilon-Poly-l-Lysine (ε-PL) Exhibits Multifaceted Antifungal Mechanisms of Action That Control Postharvest Alternaria Rot. *Int. J. Food Microbiol.* **2021**, *348*, 109224. [CrossRef] [PubMed]
55. Haider, S.-A.; Ahmad, S.; Sattar Khan, A.; Anjum, M.A.; Nasir, M.; Naz, S. Effects of Salicylic Acid on Postharvest Fruit Quality of "Kinnow" Mandarin under Cold Storage. *Sci. Hortic.* **2020**, *259*, 108843. [CrossRef]
56. da Rocha Neto, A.C.; Maraschin, M.; Di Piero, R.M. Antifungal Activity of Salicylic Acid against Penicillium Expansum and Its Possible Mechanisms of Action. *Int. J. Food Microbiol.* **2015**, *215*, 64–70. [CrossRef] [PubMed]
57. Zhang, Z.; Wang, T.; Liu, G.; Hu, M.; Yun, Z.; Duan, X.; Cai, K.; Jiang, G. Inhibition of Downy Blight and Enhancement of Resistance in Litchi Fruit by Postharvest Application of Melatonin. *Food Chem.* **2021**, *347*, 129009. [CrossRef]
58. Li, S.E.; Xu, Y.H.; Bi, Y.; Zhang, B.; Shen, S.L.; Jiang, T.J.; Zheng, X.L. Melatonin Treatment Inhibits Gray Mold and Induces Disease Resistance in Cherry Tomato Fruit during Postharvest. *Postharvest Biol. Technol.* **2019**, *157*, 110962. [CrossRef]

MDPI AG
Grosspeteranlage 5
4052 Basel
Switzerland
Tel.: +41 61 683 77 34

Journal of Fungi Editorial Office
E-mail: jof@mdpi.com
www.mdpi.com/journal/jof

www.ingramcontent.com/pod-product-compliance
Lightning Source LLC
Chambersburg PA
CBHW061457210926
51850CB00008B/1741
9783725867080